# Copernicus Books

Sparking Curiosity and Explaining the World

Drawing inspiration from their Renaissance namesake, Copernicus books revolve around scientific curiosity and discovery. Authored by experts from around the world, our books strive to break down barriers and make scientific knowledge more accessible to the public, tackling modern concepts and technologies in a nontechnical and engaging way. Copernicus books are always written with the lay reader in mind, offering introductory forays into different fields to show how the world of science is transforming our daily lives. From astronomy to medicine, business to biology, you will find herein an enriching collection of literature that answers your questions and inspires you to ask even more.

Virginia Shepherd • Charles Brau

# Beyond the Genius

## Nobel Stories of Passion and Controversy in Science

Springer

Virginia Shepherd
Vanderbilt University
Nashville, TN, USA

Charles Brau
Vanderbilt University
Nashville, TN, USA

ISSN 2731-8982        ISSN 2731-8990   (electronic)
Copernicus Books
ISBN 978-3-032-02734-4        ISBN 978-3-032-02735-1   (eBook)
https://doi.org/10.1007/978-3-032-02735-1



This Springer imprint is published by the registered company Springer Nature Switzerland AG
The registered company address is: Gewerbestrasse 11, 6330 Cham, Switzerland

If disposing of this product, please recycle the paper.

# Preface

This book has its origin in an Honors Seminar that we taught for several years at Vanderbilt University on the Nobel Prizes in Science. Intended for non-scientists, the course was popular in part because it satisfied a science requirement needed for graduation with a minimum of serious science. But it turned out to be more than that. We set ourselves two objectives: First, we had to teach some science; after all, the students were getting science credit. Second, we wanted the students to learn that science is done by human beings with all their flaws, and that progress in science is not a continuous process, but occurs erratically as scientists "grope in the dark" toward the truth. As Fred Hoyle put it, "I cannot understand what makes scientists tick. They are always wrong and they always go on." For each Nobel Prize that we and the students selected, we introduced the science behind the Prize ourselves, but we challenged the students to examine the scientists themselves and dig deep into their perseverance and drive; their personalities and foibles; the politics that each encountered along the path to their discoveries; and the impact and ethical questions surrounding their work. We probably learned as much from them as they did from us.

The material in this book derives from the research we did to prepare for the course and in the time since then. Much of the research was done in libraries, but we were able to travel to Stockholm to conduct research into archives that had been opened for some of our Nobel Laureates, and we traveled to some of the original research sites, including Berlin, Stockholm, and Copenhagen. We didn't have a chance to interview any of the Nobel Laureates ourselves since only one is still alive, but we did have the opportunity to talk to scientists who had worked with them.

In this book, we discuss eight Nobel Laureates and a ninth scientist who arguably deserved a Prize, plus a chapter on Alfred Nobel himself. The chapters are arranged chronologically based not on the award of the Nobel Prizes but on the time when the most important or controversial research was done. Each chapter is intended to be complete by itself so you can skip chapters or just read the chapters you are interested in, for example, Chaps. 2 (involving poison gas) and 4 (involving the atomic bomb), both focus on issues of science and war. The laureates themselves were chosen because each is controversial in some way, the controversies bringing the total picture into sharper relief. We hope to convey the idea that science, as an enterprise, is undertaken by real people, with all their complexity, flaws, and conflicts. We don't in this limited space address in great depth the larger issues, such as women in science or the Jewish migration in response to the Nazis except as they impact the individual scientists. But we do discuss to a limited extent the science behind the prizes, since without the science the stories would just be about ordinary people, interesting in their way, but no different from the rest of us. Just a note here to the reader: we want you to enjoy the stories; they are all exciting, surprising, tragic, and hopeful. So please don't stop because you might not understand the science; keep reading to get the complete picture of the laureate. We would also point out that this is not a book about the Nobel Prizes; there are many excellent books about Alfred Nobel and his Prizes. But we introduce Nobel and the Prizes because the awards are useful for identifying acknowledged geniuses, and because the awards themselves are interesting for what they reveal about science at the time, the scientists who get the awards, and even the motivations of the scientists who decide on the awards.

Finally, we wish to acknowledge the help that we have received that has made this work possible. Above all, we want to thank our students, especially the amazing, bright, challenging students we had in our honors seminars. We wish to thank the museums and laboratories that we visited in Berlin, Stockholm, and Copenhagen; we are sincerely grateful for the respect we were given and the access that we enjoyed there. Above all, we are grateful for the access and assistance that we enjoyed at the Swedish Academy, and the assistance that we have gotten since we visited there. Lastly, we gratefully acknowledge the help and encouragement we have received from the friends and colleagues who helped in the editing, and the editors who read and criticized the manuscript; we are forever grateful.

Nashville, TN, USA                                                    Virginia Shepherd
Nashville, TN, USA                                                    Charles Brau

# Contents

# 1

# The Merchant of Death

On 27 November 1895, Alfred Nobel signed his last will and testament (Fig. 1.1) at the Swedish-Norwegian Club in Paris. In less than one thousand handwritten words, Nobel outlined his plan to donate the majority of his vast fortune to a series of prizes for "those who, during the preceding year, shall have conferred the greatest benefit on mankind."[1] After listing small gifts to several relatives and friends, Nobel outlined his proposed plan to give the interest on his estate in five equal awards as follows:

> one part to the person who made the most important discovery or invention in the field of physics; one part to the person who made the most important chemical discovery or improvement; one part to the person who made the most important discovery within the domain of physiology or medicine; one part to the person who, in the field of literature, produced the most outstanding work in an idealistic direction; and one part to the person who has done the most or best to advance fellowship among nations, the abolition or reduction of standing armies, and the establishment and promotion of peace congresses.

After his death on December 10, 1896, Nobel's distant relatives and few friends eagerly awaited the reading of his will, anxious to hear what they would receive. To their horror, Nobel was leaving most of his estate—a huge fortune—to fund annual prizes to be awarded to people working in five areas: three in the sciences, one in literature, one in peace. What had he done? Why was he ignoring familial ties and donating to such an outlandish scheme? The

---

[1] Nobel's will is found on the Nobel Foundation website at https://www.nobelprize.org/alfred-nobel/full-text-of-alfred-nobels-will-2/.

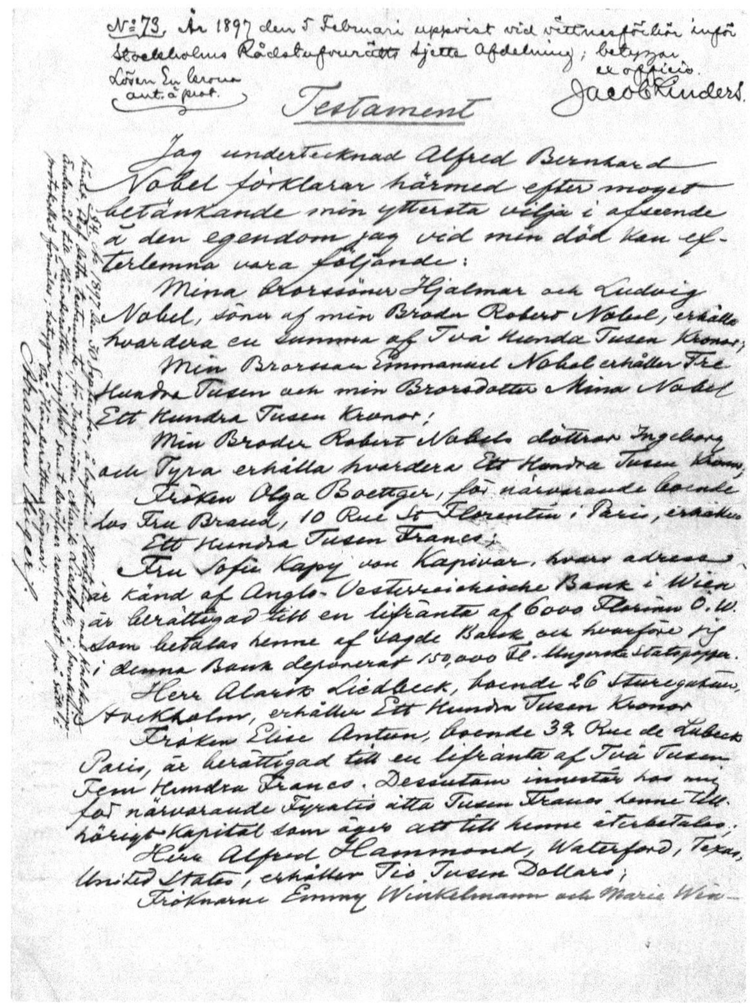

**Fig. 1.1** The handwritten will of Alfred Nobel. (Courtesy of the Nobel Foundation. Original converted to black and white)

distribution of his assets would be contested, but in the end the will stood mostly as written. The Nobel Prizes were established…at least on paper.

There were a number of complications to carrying out Nobel's wishes. His fortune was locked up in a variety of cities including real estate in Paris, San Remo, Italy, and Björkborn, Sweden; factories in several countries; negotiable securities held in various banks throughout Europe: Paris, London, Berlin, Saint Petersburg, Stockholm, and Gothenburg; and finally, a strongbox in Paris. After a magnificent funeral in Sweden, one of Nobel's executors, Ragnar Sohlman, hurried to Paris to gather the bank securities and the strongbox. Everything was eventually sent to Sweden where Nobel's will was probated and his plan for the Prizes carried out.

What kind of man was Alfred Nobel, and what possessed him to donate his vast fortune to this vision? Nobel certainly can be called an enigma: he manufactured dynamite and other explosives, yet he wrote poetry in five languages; he was the inventor of smokeless gunpowder, yet a close confidant of Bertha von Suttner, a leader in the peace movement; his explosives were enormously important for mining and construction, but they were dangerous to manufacture and use, accidents were common, with many fatalities, and people called him the "merchant of death." Between his last will written in 1895 and a previous will written a few years earlier, something had changed. In his earlier will he left most of his fortune to his distant relatives. In his last will, he left it to the world in the form of prizes to be awarded to those who had "conferred the greatest benefit on mankind." Amazingly, the legacy of these prizes has endured for more than 125 years. Within intellectual circles they are the most coveted of all prizes, and they are handed out each December by the King of Sweden himself. Even those like Hjalmar Branting, who criticized the prizes when Nobel's will became public, accepted an award when it was offered to him. But in a letter to Alerik Liedbeck, Nobel wrote of himself

> I am a misanthrope, and yet utterly benevolent, have more than one screw loose yet am a super-idealist who digests philosophy more efficiently than food.[2]

Nobel was a lonely hermit who gave all he had to the world to establish prizes for the things he loved most during his life. His legacy lives on.

## Alfred Nobel's Sweden

Although Nobel called several countries his home at various times in his life, including Russia, France, and Italy, he was most closely tied to Sweden, his place of birth and the place where his mother lived. Deprived by its northern latitude of sunshine and warmth, Sweden before the industrial revolution was a poor country. A peaceful country today, warfare seemed to be one of Sweden's major industries until a couple of decades before Nobel was born. In 1714, a new "Age of Freedom" opened, the economy was rebuilt, and Sweden joined in the Enlightenment. The Academy of Science was founded in 1739 and the Academy of Letters, History, and Antiquities in 1753. Nobel would later use both these Academies to award his prizes.

In the period from about 1840–1870, Sweden experienced a peaceful liberal revolution that included expanded rights for women and freedom of the press.

---

[2] Fant, K. (1991). *Alfred Nobel*, translated by M. Ruuth, Arcade. p. 2.

The world's first compulsory, free education system was instituted in 1842 (one year too late for Alfred). Industrialization grew rapidly, especially in mining and railroads. Both these industries would need dynamite, lots of dynamite. As the agricultural economy became more efficient, there was less need for manual labor on the farm so the cities grew. Poverty and sanitation became serious problems. The famine of 1867–1869, when Alfred was about 35 years old, was the last famine in Sweden. The year 1867, when Nobel was 34, was known in the south as *Storsvagåret* ("Year of Great Weakness") and, in the north as *Lavåret* ("Lichen Year") because of the bark bread made with lichen. Hastened by the famine, more than a million Swedes (about 20 percent of the population!) emigrated to the USA between 1850 and 1890. This was Sweden when Alfred Nobel was alive.

## Who Was Alfred Nobel?

Details are scarce about the life of Alfred Nobel. He never wrote an autobiography[3] and when his brother pressed him for some information he wrote:

> Alfred Nobel—his miserable existence should have been extinguished at birth by a humane doctor. Principal virtues: keeping his nails clean and never being a burden to anyone. Principal faults: that he has no family, is bad tempered, and has a poor digestion. One and only wish: not to be buried alive. Greatest sin: that he does not worship Mammon. Important events in life: none. Is that not enough and more than enough?[4]

Pressed further by his brother, he responded:

> Why do you tire me with biographical essays? No one reads essays except those about actors and murderers, preferably the latter, whether they have carried out their deeds on the field of battle or at home in a manner which makes people gape.

What is known about Nobel is largely extracted from his letters, which have been preserved, and information from his later years recorded by his

---

[3] Actually, he wrote a short autobiography in 1893 when an honorary Doctor of Philosophy was conferred on him by Uppsala University. It reads simply: "The undersigned was born on 21st October 1833; he acquired his knowledge in private studies and did not attend any secondary school. He devoted himself particularly to applied chemistry and discovered explosives under the names of dynamite, blasting rubber and smokeless powder, known under the name of Ballistite and C. 89. Has been a member of the Royal Swedish Academy of Science since 1884, of the Royal Society (London) and of the *Société des Ingénieurs Civil* in Paris. Since 1880 has been a knight of the Order of the Polar Star. He is an officer of the *Legion of Honour*. Sole publication: a paper in the English language, which was awarded a silver medal."

[4] Henriksson, F. (1938). *The Nobel Prizes and their founder, Alfred Nobel.* Bonniers Boktryckeri. p. 35.

friend and confidant Sohlman. This lack of personal information is at least partially explained by the very nature of Nobel himself: he was a lonely hermit, self-deprecating, and too consumed by his business responsibilities to bother writing about himself. What follows is derived from a variety of sources that tell us what little we know about Nobel the person.

Nobel's family appears to be descended from peasant stock in Östra Nöbbelöv, Sweden. The first in the family of whom anything specific is known is Petrus Olofsson, who was born around 1655. Olofsson studied law at Uppsala, using his musical talent to help pay his way. As was customary at the time, he Latinized his name to Petrus Olai Nobelius, after his place of birth (Nöbbelöv). Through his musical talent, this first Nobelius met Olaus Rudbeck (Nobel's third great grandfather), who was an influential figure during Sweden's "golden age" in the seventeenth century. An intellectual giant, Rudbeck was a scientist, writer, and professor of medicine at Uppsala who made contributions in many fields but especially anatomy and linguistics.[5] With Rudbeck's help, Petrus Nobelius was called to the Bench at an early age, and he soon married Rudbeck's daughter Wendela and joined in the social scene of Uppsala. One of Petrus' sons, Olof Nobelius (Noble's great grandfather), taught art at the University, and was highly regarded for his scientific drawings. Too poor to afford the education to become a physician, Olof's youngest son, Immanuel Nobelius (Alfred Nobel's grandfather), apprenticed to become a barber-surgeon, a profession that at that time belonged to the order of guilds. Owing to his poverty, Immanuel was obliged to become a non-commissioned officer in the Upland regiment for three years, where he was listed as Immanuel Nobell, which he later changed to Nobel. After leaving the army, he practiced medicine at various hospitals and was instrumental in introducing vaccination. Immanuel's son Emmanuel, Alfred Nobel's father, was born in 1801. Throughout these generations, the inventive and enterprising genes from the Rudbeck ancestry were apparent. Alfred's father had no formal education, but in his early 20s he taught himself engineering. He was responsible for several inventions, including plywood, which was made by shaving sheets of veneer from a log with a rotary lathe and gluing them together. Dismayed by the mass emigration of unemployed Swedes to America and elsewhere at that time, he described his plywood in a booklet with the title *An Attempt to Create Employment to Check the Emigration Fever Now Caused by the Lack Thereof.*[6] At times, his inventiveness led to great success, but at other times his entrepreneurial spirit led to disaster.

---

[5] Rudbeck also wrote the lengthy treatise *Atlantica*, which sets out in four volumes to prove the fantastic idea that Plato's fictional island of Atlantis, the cradle of civilization, was actually Sweden.

[6] Fant, K. (1991), p. 66.

Alfred Nobel was born in Stockholm, Sweden, in 1833, the third of four sons of Emmanuel Nobel. In 1832, the year before Alfred was born, Emmanuel had gone bankrupt and he struggled to support the family. When Alfred was four years old, Emmanuel moved first to Finland, then to Saint Petersburg, Russia, leaving the family in Sweden while he looked for opportunities to start a business. With Emmanuel away, the family lived in poverty: the house, as shown in Fig. 1.2, was cold and drafty, and the food was meager. Stockholm at that time suffered from poor sanitation, and thousands of infants died from cholera, diphtheria, and whooping cough. All this likely contributed to Nobel's frail condition and the poor health that accompanied him all his life. With help from her father, Alfred's mother kept the family alive. Looking back at the age of 18, Alfred wrote (in English):

> My Cradle looked a deathbed, and for years
> a mother watched with ever anxious care,
> so little chance, to save the flickering light,
> I scarce could muster the strength to drain the breast,
> and the convulsions followed, till I gasped

**Fig. 1.2**  Alfred Nobel's birthplace

upon the brink of nothingness—my frame
a school for agony with death for a goal.[7]

When he was eight years old, Alfred Nobel got his only formal education, consisting of two terms at Jacob's Apologist School. Free, compulsory education in Sweden was introduced only the next year. The circumstances in the school were rough and beatings were frequent, but Alfred got good grades. Because of his frail health, he couldn't join in games and sports with the other children, and he grew up a loner. Describing these years, his poem continues:

We find him now a boy. His weakness still
makes him a stranger in the little world,
wherein he moves. When fellow-boys are playing
he joins them not, a pensive looker-on;
And thus debarred the pleasures of his age
his mind keeps brooding over those to come.

Meanwhile, in Saint Petersburg, Emmanuel developed naval mines for the Russian government. With the profits from these inventions, he built a workshop and a foundry and became one of the leading engineers in Russia. Alfred and the rest of the family joined Emmanuel in Russia in 1842, when Alfred was nine years old. By this time Emmanuel was wealthy enough to have Alfred privately educated. His first tutor spoke only Swedish and his education at this point focused only on the humanities. After he learned Russian, Alfred got a second tutor who taught him science and the humanities.[8] He became fluent in five languages, and to improve his French he once translated a Voltaire text into Swedish, and then back into French to check his translation. He was particularly fond of English literature, physics, and chemistry, but he may have loved literature the most, especially the English poet Shelley. Later he destroyed almost all the poetry that he had written, but his eloquence, his use of words, is apparent in his voluminous letters, written in many languages, which have been saved.

But poetry wasn't practical, so during the period 1850–52 his father sent Alfred abroad to study in Sweden, France, Germany, and the United States to learn engineering. While he was in the U.S. he met John Ericsson, the inventor who later, during the American Civil War, built the first Union ironclad warship, called the *Monitor*. In Paris, Nobel met two people who would have

---

[7] Bergengren, E. (1960), translated by Blair, A. *Alfred Nobel, the man and his work*, Thomas Nelson and Sons. p. 9.
[8] Hendriksson (1938), p. 36.

a significant impact on his life. The first was a man named Ascanio Sobrero, who had invented nitroglycerine just few years earlier, in 1847. This liquid was a powerful explosive, but it was so unstable and so dangerous that Sobrero felt that it would have no practical value. It was extremely sensitive to shock and heat. A hammer blow would cause it to explode, and it would explode spontaneously at temperatures above 218 °C (424 °F). Since the chemical reactions used to produce nitroglycerine give off heat, the temperature of the mixture rises as the nitroglycerine is made. Thus, the preparation of nitroglycerine is fraught with danger. Having, perhaps, greater vision than Sobrero, Nobel recognized the potential in this powerful explosive if nitroglycerine could be "tamed."

It was also in Paris that Nobel met the only truly romantic interest of which we are aware, a Swedish girl working in Paris whom he met while he was visiting the city. He wrote:

I came to Paris—an ocean where Passion creates stormy weather and causes more wrecks than ever the salty waves did. Whoever is exploring this vast, boundless collection of sin and madness has to pay tribute to its false god, Sensuousness. My life, before then a desolate desert, came alive in felicity and hope. I had a goal, a heavenly goal: to win this lovely girl and to be worthy of her. I felt infinite happiness and we met again, and again and again, until we had become each other's heaven; and I became acquainted with the sweet compassion in her love and sealed it with a kiss, a chaste and holy kiss, of purest affection, even if no eye was there, except that of the Almighty, to watch over us. This could have ended in the usual way and provided joy and worry, but it was not meant to be; another bridegroom had a stronger claim—she is married unto death.[9]

She died of tuberculosis while he was still in Paris. Alfred Nobel was 18 at the time.

In 1852, when he was 19 years old, Nobel returned to Russia to join the family business. Soon the Crimean War (1853–1856) heated up and Russia was purchasing large quantities of naval mines to keep the British and French fleets at bay, so the business was booming. The pace of work was frantic, and Alfred's frail health couldn't keep up with his enthusiasm and his drive. Reassured by promises from the Tsar, Emmanuel took out large loans to meet the wartime demands. Unfortunately, when the war ended, the humiliating terms of the Treaty of Paris made it impossible for the Tsar to fulfill his

---

[9] Hendriksson (1938), p. 38.

promises, and Emmanuel went bankrupt again. He returned to Stockholm with Alfred's mother and his younger brother, Emil, while the three older brothers, Robert, Alfred, and Ludvig, remained in Russia to liquidate the business. In spite of the danger, Alfred set up a small laboratory in the family kitchen and began experiments with the nitroglycerine that he had learned about from Sobrero. Not only was it unstable and likely to explode at the slightest provocation, when it detonated it didn't explode reliably. Nevertheless, in 1861 Alfred got a loan from Société de Crédit Mobilier in Paris to develop nitroglycerine for use in construction. Alfred's career was off and running.

After experimenting with explosives on the Neva River, outside St. Petersburg, Alfred was awarded his first patent in October, 1863, for "blasting oil." This was a mixture of nitroglycerine and black powder (gunpowder) that while less powerful than pure nitroglycerine was more reliable than nitroglycerine alone. Alfred returned to Sweden in early 1863 to set up production of nitroglycerine and blasting oil, but he still needed a way to set it off reliably and safely. To do this, he developed a "blasting cap," just a wood capsule filled with black powder that could be ignited by a fuse. This proved to be a reliable way to detonate the blasting oil and, later, dynamite. In the end, the blasting cap may have been a more important invention than dynamite itself. In July 1864, he applied for a patent on the blasting cap, which was finally granted three years later.

While Alfred returned to Sweden to make blasting oil, his brothers Robert and Ludvig remained in Russia where they became immensely wealthy founding the Russian oil industry. As the story goes, Robert had a lamp shop in St. Petersburg and when he got a bad batch of lamp oil from Brussels, he saved himself from bankruptcy by learning how to refine the bad oil. Meanwhile, the younger brother, Ludvig, had set up a workshop for making, among other things, rifles for the Russian government. In 1873, when Ludvig got a large order, he sent Robert to Baku, in the Caucasus, to get walnut that they needed to make the rifle stocks. But while he was there, Robert noticed the naphtha, a crude lamp oil, that the local people were extracting from the ground using primitive techniques. Back in St. Petersburg, Robert convinced Ludvig to invest with him in the naphtha business, and in 1875 he returned to Baku, introduced new technology that included a seven-mile-long pipeline to transport the oil to the nearest port, and founded "The Nobel Brothers Naphtha Company." Called "Branobel" for short, it became one of the largest oil companies in the world, rivaling the Rockefellers and the Rothschilds. From time to time, when the brothers ran into financial difficulties, Alfred invested in the company himself, and when he died, part of his fortune came from those investments.[10]

Back in Stockholm, Emmanuel, Alfred, and Emil continued the development and production of nitroglycerine for blasting. On September 3, 1864, a massive explosion killed five people, including Alfred's younger brother Emil. The Stockholm newspaper reported that "In the capital people heard the violent sound of the explosion and saw a huge, yellow flame rise straight up in the air."[11] The walls of a nearby stone house were split open, windows were shattered throughout the neighborhood, and a woman living nearby lost her arm. Emil's body was barely recognizable. Alfred and several other workers were among the injured. Distraught by his son's death, Emmanuel Nobel had a stroke from which he never completely recovered. Alfred took over the family business in 1864 at the age of 31.

## The Dynamite Years

Alfred forged ahead. Following the explosion, the business was forced out of Stockholm, but blasting oil was powerful stuff, and demand for it grew in the mining industry and for building tunnels for the State Railways. With an investment from the millionaire J. W. Smitt, who recognized the potential importance of the new explosive, Nobel founded the Nitroglycerine Company. Because of the danger, initial production had to be carried out on a barge on nearby Lake Mälaren, but a year later they got permission to build a factory on a remote shore of the lake at Vinterviken. Before long, in order to meet increasing demand for his blasting oil, Nobel established factories in other countries, including Germany and Scotland.

Safety remained a problem. Although blasting oil exploded more reliably than nitroglycerine, its manufacture still began with nitroglycerine, which was then mixed with black powder (gunpowder). Often the blasting oil was delivered by drunk farm boys, and to keep factory workers from falling asleep at their job they were required to sit on one-legged stools (Fig. 1.3). More accidents were inevitable.

Nobel's first factory outside Sweden was at Krümmel, a small town in Germany about 30 km outside Hamburg. To minimize the hazard, or at least localize the destruction, the town officials demanded that the manufacturing

---

[10] These same oil fields were the objective of one of Hitler's vicious drives to get the oil he needed for his armies in World War II. He never reached Baku.

[11] Hendriksson (1938), p. 61.

**Fig. 1.3**  Nitroglycerine production

be done in wooden sheds separated by earthen walls to protect nearby sheds, and everything should be surrounded by an earthen wall to protect the town itself. Only one or two employees worked in each shed so that no more than one or two might be killed in an accident. Production there started on April 1, 1866, with about 50 employees.

In April, 1866, several crates of nitroglycerine were shipped to California for use by the Central Pacific Railroad. They planned to use it in the construction of tunnels through the Sierra Nevada Mountains for the transcontinental railroad. One of the crates exploded and destroyed a Wells Fargo office in San Francisco, killing 15 people. In response, the transportation of liquid nitroglycerine was banned in California. But nitroglycerine was too important to ban outright, so to eliminate the hazards of transportation it had to be produced on site. Seeing an opportunity to expand the business in post-Civil-War America, Nobel went to New York to promote nitroglycerine, but the newspapers were already spreading fear of the explosive. About the same time, a steamship exploded near Panama's Atlantic shore, and a ship exploded in Bremerhaven, Germany, killing 84. Showman and entrepreneur that he was, Nobel set out to try to suppress apprehension about the dangers of nitroglycerine. On May 4, 1866, he organized a demonstration in a quarry in upper Manhattan, New York. As reported in the *New York Times*:

> The professor was equipped with an abundance of the substance in crates and bottles as well as with matches, fuses, and quite a bit of courage. He opened a bottle and poured a small amount of its contents on a rock. He brought a match to it, and it burned like pitch. Unpunished, he threw a canister and a bottle, both filled with the explosive substance, from a high cliff to the ground, which demonstrated that impact alone would not make it explode.…The substance will explode only if the temperature is increased to 360 degrees Fahrenheit, a heat that cannot be generated during normal conditions…
>
> In a variety of entertaining and graphic ways, he kept experimenting with the explosive and convinced those present that, during the conditions there presented, nitroglycerine is safer to transport and handle than both gunpowder and cottonpowder, but its explosive force is considerably greater.[12]

In spite of Nobel's demonstration, the press and the public at large continued to fear nitroglycerine and began to call Nobel the "merchant of death." Two months later, on July 26, 1866, the United States Congress passed an anti-nitroglycerine law. Nevertheless, on August 14, 1866, Nobel was granted a U.S. patent for his blasting oil. Within limitations it could be used in the United States, and Nobel started production at the Giant Powder Company in California.

With its enormous power, blasting oil became an increasingly widespread and important invention for construction projects around the world. Inevitably, accidents continued: On Christmas Day 1867, blasting oil that had been illegally stored in the center of Newcastle upon Tyne exploded and killed eight people, and in June 1869, two wagons loaded with nitroglycerin exploded in the small Welsh village of Cwm-y-glo, killing the carters and three others. The debris, allegedly including a cart wheel, was thrown as far as the next village. The British government responded by passing the Nitro-Glycerine Act of 1869. As important as nitroglycerine was industrially, something safer was needed.

On July 12, 1866, while Nobel was still traveling in America, his factory in Krümmel blew up. When he returned to Europe, he went to survey the ruins. For a long time, Nobel had been looking for a safer form of nitroglycerine, but nothing worked better than blasting oil. But in the course of cleaning up after the accident, Nobel discovered that by mixing the nitroglycerine with a local variety of very fine, porous sand called *Kieselguhr*, it formed a putty-like

---

[12] Hendriksson (1938), p. 86.

substance that didn't detonate when dropped or heated. But his blasting caps would detonate it. He patented this invention in Sweden and elsewhere in 1867, and called it dynamite.[13] The coincidental development of diamond bits and pneumatic drills to make holes in rock made dynamite an instant commercial success in the construction and mining industries. To meet the demand, Nobel built factories in several countries, as shown, for example, in Fig. 1.4.

Unfortunately, the *Kieselguhr* that made the nitroglycerine safe reduced its power. There had to be something better, and Nobel was determined to find it. Nine years later, in 1875, he developed "blasting gelatin," a colloidal solution of nitrocellulose (gun cotton) and nitroglycerine. The idea came to Nobel one night when he cut his finger and tried to stop the bleeding with collodion. Collodion is a solution of nitrocellulose in ether and alcohol. It dries to form a flexible coating that can be used as a liquid bandage for cuts. Maybe a solution of nitroglycerine and collodion would work better than *Kieselguhr*. Nobel stayed up all night and had a sample ready by morning. It worked. Blasting gelatin had the power of nitroglycerine but the safety of dynamite. Since it was too powerful to be put in guns, the military applications were

**Fig. 1.4**  Women mixing dynamite in a factory in Scotland

---

[13] Bergengren (1960), p. 43.

limited, but in the construction of the St. Gotthard tunnel, blasting gelatin nearly doubled the rate of progress compared with dynamite.[14]

Dynamite had a further problem: it didn't store well and it became dangerous after a year or so. For this and other reasons, dynamite wasn't satisfactory for military use. Up to this time, black powder was universally used as gunpowder, but when guns were fired with the black powder, the battlefield became completely obscured by the smoke. In addition, the residue from black powder fouled the gun barrels, making them useless until they could be cleaned. Some sort of smokeless powder was desperately needed. Nobel and others set to work.

From 1873 to 1891, Nobel lived in Paris and had a laboratory in Sévran, about 10 miles north of the city. Paris was Nobel's favorite city, the place of his first romantic love, and the intellectual center of his world. It was in Paris in 1887 that he developed what he called ballistite, sometimes called the "crowning glory" of his research career. A form of smokeless gunpowder, it quickly replaced black powder in the armories of the world.[15] By that time the French were already using a smokeless powder developed by a French chemist, so when Nobel offered his product to the French gunpowder monopoly, pride and politics intervened and the French rejected it. Thereupon, Nobel offered ballistite to Italy, who immediately accepted. The French were understandably infuriated that he would sell arms technology to a foreign power, and a vicious press campaign ensued. The French government accused him of espionage, confiscated his small quantities of ballistite and much of his laboratory equipment, and threatened to throw him in jail. In response, Nobel moved himself and his laboratory (Fig. 1.5) to San Remo, Italy, just across the border in the Italian Riviera.

## War and Peace

By age 43, after 12 years at the helm of the family business, Alfred Nobel felt old. While visiting Vienna in 1876, he looked for an assistant with the advertisement: "A wealthy and highly educated old gentleman living in Paris seeks to engage a mature lady with language proficiency as secretary and

---

[14] Blasting gelatin also became a favorite of terrorists, such as the Irish Republican Army, who obtained it illegally.

[15] Curiously, ten years later, in the Spanish-American War at the battle of San Juan Hill (July 3, 1898), the Spanish artillery had smokeless powder while the Americans did not. As a result, the Spanish troops could locate Teddy Roosevelt's guns when they fired, but Roosevelt's troops couldn't locate the Spanish guns.

**Fig. 1.5**  Nobel's laboratory in San Remo

housekeeper."[16] Thirty-three years old, Bertha Krinsky was intelligent and well spoken. Coming from a family of minor nobility that had fallen on hard times, she had been employed as a tutor for Baron Karl von Suttner's four daughters, but she fell in love with their elder brother Arthur, who was a few years younger than she. His parents disapproved of the match and encouraged Bertha to answer Alfred Nobel's advertisement. After writing back and forth, she joined Alfred in Paris. They immediately became intellectual "soul mates." In her memoires she later wrote: "Alfred made a highly agreeable impression…his voice showing more melancholy than cynicism."[17] But she was with him only a week before Nobel left on urgent travel. By the time he returned three weeks later, she had gone back to Austria, selling her jewelry to pay for the ticket, and eloped with Arthur, to whom she had been secretly engaged. Nevertheless, the relationship with Baroness von Suttner lasted the rest of Nobel's life by correspondence; he saw her only twice. She became very influential in the peace movement and is probably responsible, at least in part, for the Nobel Prize for Peace. In fact, she actually won the Peace Prize in 1905.

Nobel's long correspondence with Bertha von Suttner, as shown in Fig. 1.6, represented a sincere but somewhat aloof participation in peace activities.

---

[16] Sohlman, R. (1983). *The legacy of Alfred Nobel*, The Bodley Head, originally published in Swedish as *Ett Testamente*. (1950). Translated by E. Harley Schubert. p. 54.

[17] Fant (1991), p. 112.

**Fig. 1.6**   Bertha von Suttner in 1906

Nobel contributed money to her peace congresses, but never became directly involved. In fact, he wrote her in 1891

> Perhaps my factories will put an end to war sooner than your congresses; on the day that two army corps can mutually annihilate in a second, all civilized nations will surely recoil with horror and disband their troops.[18]

There may even be some truth in this, for although his remark seems naïve in light of all that transpired in the first half of the twentieth century, there has been no world war since the advent of the atomic bomb. While he admired Bertha's dedication to peace and praised her novel *Lay Down Your Arms*,[19] he didn't believe that her approach would work: "Good wishes alone will not ensure peace."[20] Only the promise of a forceful response by the rest of the world would stop an aggressor, he felt, and in 1892 he wrote her that "a declaration should be made that any and every aggressor would have all of Europe against him."[21] Nevertheless, on January 7, 1893 he wrote her his first mention of a peace prize:

---

[18] Fant (1991), p. 265.

[19] von Suttner, B. (1889). Lay down your arms: The autobiography of Martha von Tilling. Edgar Pierson.

[20] Nobelprize.org. https://www.nobelprize.org/alfred-nobel/alfred-nobels-thoughts-about-war-and-peace/.

[21] Fant (1991), p. 269.

Dear Friend,

I wish Happy New Year to you personally and to the noble campaign you are pursuing so courageously against ignorance and stupidity!

I would like to bequeath part of my fortune for the establishment of peace prizes to be awarded every fifth year (let us say six times, for if at the end of thirty years we have not succeeded in reforming society, we shall inevitably revert to barbarism) to the man or woman who has contributed most effectively to the realization of peace in Europe. I do not refer to disarmament, which is an ideal we can reach only slowly and with caution, nor to compulsory arbitration. But we can and ought soon be able to go at least so far as to have all states, among themselves, pledge to intervene against anyone who breaks the peace.

This would be one means to make war impossible, and to force even the least reasonable and most brutal power either to make use of arbitration or keep still. If the triple alliance enveloped all states instead of just three, peace would be ensured for centuries.

I send the warmest greetings to you and your husband.

Your most affectionate

A. Nobel[22]

The Peace Prize is the only prize mentioned in the will that preceded the last will, which established all the prizes, and despite his apparent skepticism, peace congresses are specifically mentioned in the final will establishing the peace prize.

Bertha opposed the idea of prizes: those working for peace didn't need prizes, they needed support for their work. But 12 years later, von Suttner was herself awarded the 1905 Nobel Peace Prize. Nobel opposed the dismantling of armies "since there has to be an armed force to maintain order." He believed that the only workable way to achieve peace was through collective defense. In a speech at the Fifth Nobel Anniversary Dinner in New York, Einstein said

Alfred Nobel invented an explosive more powerful than any then known—an exceedingly effective means of destruction. To atone for this 'accomplishment' and to relieve his conscience, he instituted his award for the promotion of peace.[23]

It doesn't seem that Alfred Nobel actually felt that way at all.

---

[22] Fant (1991), p. 270.

[23] Einstein, A. (1956) *Out of my later years. Philosophical Library,* p. 213.

## The Later Years

Truly, Nobel was a citizen of the world. With factories and laboratories in many countries, his travel schedule was hectic and consuming. As he said "my homeland is where I am working, and I work everywhere."[24] As an inventor, he was brilliant. As a businessman he accumulated an enormous fortune. As a human being he was introverted, lonely, and suffered poor health, with chronic migraines, indigestion, and heart trouble. To his sister-in-law he wrote

> What a contrast between us. You, surrounded by love, joy, noise, pulsating life, caring and being cared for, caressing and being caressed, anchored by contentment; I, roving, without compass or rudder, like a useless and desolate wreck, without happy memories from the past, without the false but beautiful visions of the future that illusions create, without vanity, which is a blunt but ready and willing self-beautifier, without family, which is the only afterlife to expect, without friends for my heart or enemies for my bile, but with a self-criticism that shows every blemish in its unembellished ugliness and every inability in a merciless light. A self-portrait with such contours is not suitable reading for a home of joy and contentment and befits only the wastebasket to which it is headed.[25]

Living in Paris, Nobel still needed companionship and tenderness, which he got from his mother, but he also needed intellectual stimulation, which he could not get from his mother and found difficult to find elsewhere. As he once wrote,[26]

> I personally find the conversation of Parisians the dreariest thing I know, whereas it is delightful to meet cultured and not excessively emancipated Russian ladies. Unfortunately, they have an aversion to soap—but one must not expect too much.

Besides his mother, the only other woman in Nobel's life was Sophie Hess. When he met her in 1876, not long after Bertha had returned to Austria, Sophie was working as a salesgirl in a flower shop at a spa near Vienna to which Nobel had gone to "take the waters," which he did frequently for his health. Twenty years old, Sophie was quite pretty and coquettish, but irresponsible. Nobel was 43. As someone who knew them said, "It is difficult to understand how a man of such intellect and high ideals would become infatuated with this untalented and indolent little creature," but he was quite infatuated by her.

---

[24] Bergengren (1960), p. 71.

[25] Fant (1991), p. 5.

[26] Sohlman (1983), p. 54.

Although they never lived together, Nobel set her up in an apartment and paid for all her expenses at the fashionable spas she frequented; Sophie also complained of health issues. When he was traveling he wrote her nearly every day, addressing her in the diminutive as *Sophiechen*, or sometimes "My Dear Sweet Child," and signing his letters as "your grouchy-bear."[27] For a few years she filled a role in his stressful and lonely life, but she could never grow to provide the intellectual companionship he needed. The relationship was often painful and embarrassing for Nobel, and we see this in his many letters to her. She spent his money irresponsibly, and sometimes presented herself as Nobel's wife in order to buy things when they were not together. As early as 1878 he wrote:

> I therefore often force myself to be cold and reserved toward you, lest your affection takes deep roots. Perhaps you believe that you are fond of me, but it is only gratitude and perhaps respect, and that feeling is not enough to fill the need for attachment and love in your young soul. The time will come, and perhaps soon, when your heart will love another, … I feel and understand all that very well… perhaps I feel the weight of being abandoned in a desert of loneliness, and for years I searched for someone whose heart could join mine. But that cannot be the heart of a woman of twenty-one, whose outlook on life and whose spiritual life has little or nothing in common with mine.… Two such creatures are not suited to each other as lovers, although they can be and can remain good friends.…
>
> …You are still a child and don't think of the future at all, and so it is good that an old attentive uncle watches over you.[28]

Nearly two decades later, in 1886, he was still supporting her and he wrote

> Dear Sophie,
>
> I am sick, tired, and depressed. Under such circumstances you can easily comprehend that I long to be surrounded by people other than paid domestics. But I can't live with you, my dear child. The past is proof of that. You are too spoiled, or rather your upbringing has been spoiled.… That would be somewhat more bearable if you had wit, culture, or skills to balance your whims, but there is no great supply of those, it seems to me…
>
> My health is poor. I can't stand it much longer. But where am I to go? I don't know, perhaps under the earth!
>
> Sincere greetings,
> Alfred[29]

---

[27] Rummel, E. (2017). *A Nobel affair: Correspondence between Alfred Nobel and Sophie Hess.* University of Toronto Press. p. 33.

[28] Rummel (2017), p. 46.

[29] Rummel (2017), p. 128.

By this time he had, as he wrote, "adopted you, so to say, as an orphan and out of pure and noble considerations."[30] Increasingly, her letters begged for money; he certainly overindulged her in the beginning, and she never learned to live within any sort of budget. He finally set up a stipend for her in order to separate himself from her, and with his encouragement she ended up marrying the Hungarian cavalry officer by whom she had had a child. His last letter to Sophie was in 1895, but the demands from her and her Hungarian officer continued even after Nobel's death. She still held all the letters that Nobel had written her, and she threatened to sell them for publication until a settlement was reached.

But in spite of the troubled and often contradictory relationship between them, Nobel's letters to Sophie offer a window into his personal feelings, for he was quite open and candid with her. Besides the personal difficulties between them, a constant theme of his letters was the unrelenting stress of his business life and his declining health. With factories in many places, he was forced to travel a great deal, and from Glasgow he wrote.

> …you are better occupied than I who has to feed on indigestible English food, which has made me sick…. all the time I suffer from a headache, and of such intensity that I can't keep my thoughts together.[31]

As early as 1880 he wrote Sophie about retiring:

> …The clamour of the world suits my personality less than others, and I would be happy to retire to some corner and live there without great pretensions, but also without worry and suffering.
>
> Once these court proceedings are over, I am determined to retire from my business life. Of course this cannot happen all at once, but I will make a start as soon as it is at all possible.[32]

and in 1887 he wrote:

> You would not believe how empty I feel and how much my health suffers under that awful work. Sometimes I come back from Sévran and must go to bed at once, and day before yesterday I almost fainted in Sévran. Sometimes I feel like leaving everything behind and settling down in some small village to enjoy

---

[30] Rummel (2017), p. 137.
[31] Rummel (2017), p. 63.
[32] Rummel (2017), p. 59.

peace in my old days at last. I love work, but not the kind of work that destroys body and soul.[33]

His business ventures were very stressing and he worked as much as he could in his chemistry laboratories among toxic chemicals. Toward the end, as his health continued to decline, attacks of angina became more frequent and in October, 1887, he wrote Sophie.

I have been ill now for nine days and must stay in my room. I have no one around except a paid servant, no one who asks after me.… I feel I am much more seriously ill than Bonato [his doctor] believes. The problem is persistent and will not go away. If a man is fifty-four years old and all alone in the world, and a paid servant is the one who shows him the greatest benevolence, sad thoughts will occupy his mind, sadder than most people believe. I can read in the eyes of my servant how sorry he is for me, but of course I can't allow it to show.[34]

His health problems may have contributed to his interest in medicine and his founding of the Prize in Physiology or Medicine.

Throughout his life, as Nobel was inventing new explosives and developing factories to produce blasting oil and dynamite, patents, lawyers, and unscrupulous competitors were a constant drain on his energy. His exasperation is clear in a letter he wrote when he was having difficulties with patents in England:

Even with the luxury of a patent, protection in most cases is illusionary. I therefore suggest giving the patenting of chemical improvements the name 'Taxation of inventors to encourage parasites.'[35]

He was quite disillusioned by the people with whom he dealt in his businesses, and once wrote.

Lawyers have to make a living and can do so only by inducing people to believe that a straight line is crooked. This accounts for their penchant for politics, where they can usually find everything crooked enough to delight their hearts…[36]

---

[33] Rummel (2017), p. 131.

[34] Rummel (2017), p. 134.

[35] Fant (1991), p. 79.

[36] Sohlman (1983), p. 34.

England proved to be a particularly difficult place to work, and in 1887, the "cordite affair" hurt him both personally, since it involved people he previously considered colleagues, and financially. While working on ballistite, he had been communicating in a friendly way with Sir Frederick Abel and Professor Dewar who had been tasked by the British government to find a smokeless gunpowder. Using ideas that they got from Nobel, they developed a close relative of ballistite called cordite, and patented it. When he contested the patent, the verdict went against Nobel on a technicality, costing him not only the rights, but court costs amounting to more than $6 M today. His only consolation was the recognition by Lord Justice MacKay before the Court of Appeal that.

> Clearly a dwarf who succeeds in climbing onto the shoulders of a giant is able to get a better view than the giant himself…. In this case I can only sympathize with the original patentholder. Mr. Nobel made a great discovery, and produced what in principle was a very real novelty. Two skillful chemists then got hold of his specification, studied it carefully and, in the light of their own extensive knowledge, found that they could use practically the same materials—with one exception—to achieve precisely the same result. One wishes that it had been possible to reach a decision which would not have deprived Mr. Nobel of an exceedingly valuable patent.[37]

To his business problems at this time were added two very personal tragedies. In 1888 Nobel's brother Ludvig died, and in 1889, his mother died. Although he saw her only a couple of times each year, Nobel felt very close to his mother, and suffered her loss deeply.

Then, in 1890 and 1892, Nobel's French companies *La Société Général* and *La Société Central* nearly went bankrupt. The case of *La Société Central* was caused by mismanagement and speculation, and Nobel had to bail them out at great expense. But the case of *La Société Général* was caused by outright fraud by the managing director, who was a former French Senator. Of this, Nobel wrote

> A few days ago a demand was made on me for the tidy little sum of 4,600,000 francs on the pretext that I was responsible for the recent embezzlements. French law is peculiar, and directors who act in perfectly good faith can be held responsible, if it can be shown that there has been a lack of adequate supervision. The other members of the board and their lawyers are, it is true, of the opinion that there has been no such laxity, but when it comes to a lawsuit, Wisdom herself is

---

[37] Sohlman (1983), p. 52.

blind, and a judge's affliction with constipation, or the opposite, can often influence his views as to what is right or wrong....[38]

Nobel left Paris in 1891 following the ballistite affair and moved to San Remo. The warm climate was good for him, and he set up his laboratory there. By the early 1890s Nobel's depression seemed to be better, but his heart problems were more severe. He wrote Sohlman that:

My heart problems will keep me in Paris for another few days at least.... Isn't it the irony of fate that I have been prescribed nitroglycerine, to be taken internally! They call it Trinitrin, so as not to scare the chemist and the public.[39]

In July, 1893, he wrote his brother Robert:

Lately my eyesight has diminished at almost ten times the normal speed, which means that I have to go from one prescription to another in less than three months, instead of two years... The oculist has forbidden me all writing and reading for six months, but how is that possible ... and the tremble in my hand is making the writing less and less clear.... I have to look forward, with a certain melancholy, to a rapidly approaching time when everything will be dark in life. I am more philosophical about this than most, but that doesn't help.[40]

After years of struggle and disillusionment with competitors and bureaucracies, the stress, combined with his fragile health, took its toll on Nobel, shown in Fig. 1.7, and he wrote to his friend and colleague Alarik Liedbeck:

After the improvements with which I am involved have been completed, I intend to retire from all business life and live like an old spinster on my dividends. I will therefore eventually sell my various shares in my dynamite and other businesses. When I say that I want to live like an old spinster on my interest income, I do not mean I am going to be idle, but shall just choose a scientific rather than an industrial sphere for my work.[41]

Although he never quite disentangled himself from his businesses, late in life he did manage to devote more of himself to his scientific ideas and spent as much time as he could in his laboratories. In 1894, he bought the arms firm Bofors, in Karlskoga, Sweden in part because he hoped to use the facilities

---

[38] Schück, H., Sohlman, Osterling, A., Liljestrand, G., Westgren, A., Siegbahn, M., Schou, A., and Stahle, N.K. (1951). *Nobel, The man and his prizes.* The Nobel Foundation (Ed.) University of Oklahoma Press. p. 31.

[39] Sohlman (1983), p. 39.

[40] Fant (1991), p. 285.

[41] Fant (1991), p. 205.

**Fig. 1.7** Alfred Nobel, later in life

there to expand his research interests. Although he traveled less, he continued to invent, and spent time in his laboratories. Military applications, and even explosives, weren't the limits of Nobel's fertile mind. Right up to the end of his life, by which time he was living and working in San Remo, Nobel was also developing, among other things, synthetic leather and rubber, and had begun work on artificial silk. When he died, he had at least 355 patents in various countries.

His last letter was written to Ragnar Sohlman, to whom Nobel had by this time become so close:

San Remo
    December 7, 1896
    Mr. R. Sohlman
    Bofors
The samples you sent are exceedingly beautiful. The pure n/c powder seems splendid to me. Unfortunately my health is again so poor that I write a few lines with effort, but, as soon as I am able, I will come back to the subjects that interest us.
    Your affectionate friend
    A. Nobel[42]

---

[42] Fant (1991), p. 313.

An hour later he suffered a stroke and died three days later in San Remo on December 10, 1896. He died alone, with only paid servants, none of whom even spoke his native tongue; the stroke had left him able to speak only in Swedish. As he had written Sophie, "how unfortunate it is to have no one around whose loving hand will once close my eyes and whisper a true word of comfort."[43] No one dear to him was able to reach him in time.

## The Legacy of Alfred Nobel

As Stig Ramel writes in his forward to Ragnar Sohlman's book *The Legacy of Alfred Nobel,*

> On 27 November 1895, Alfred Nobel signed his last will and testament in Paris. Among its four closely-written pages, less than one referred to the donation which was destined to link his name with the supreme achievements of the modern world in science and literature and with the cause of peace. Nobel's conception of five prizes 'to those who during the past year have done humanity the greatest service' was ingenious and he bequeathed one of the largest fortunes of his century to its fulfilment.[44]

When he died, Nobel's fortune was worth 31 million Swedish Kroner, equivalent to more than $300 million today. It was an enormous sum at that time. But the will had many flaws and faced many obstacles. It consumed three and a half years of Ragnar Sohlman's life to probate the will and establish the prizes.

Sohlman (1870–1948, shown in Fig. 1.8) was a young chemical engineer when he became Alfred Nobel's assistant in 1893, just three years before Nobel died. Sohlman was brought up in Stockholm and graduated from the KTH Royal Institute of Technology in 1890. After graduation, he worked in the United States as an engineer at the dynamite plant of Hercules Powder Company in Delaware. In September 1893, much to his surprise, Sohlman was offered a job as assistant to Alfred Nobel. He found out later that he had been recommended by J. W. Smitt, who had earlier been the source of Nobel's first investment. The recommendation came from Nobel's nephew Ludvig Nobel, Robert's son, with whom Sohlman had gone to school. Sohlman's first job for Nobel was to arrange and catalog Nobel's reference library, but he was soon transferred to Nobel's laboratory at San Remo. There he worked with

---

[43] Rummel (2017), p. 146.
[44] Sohlman (1983), p. 7.

**Fig. 1.8** Ragnar Sohlman

Nobel on a variety of projects, including synthetic leather and rubber. The laboratory comprised a long building with electric generators, rooms for chemical experiments, and scientific instruments. With no more access to the French rifle range in Sévran, firing took place over the open sea.[45]

Although he worked for Nobel for only three years before Nobel's death, Sohlman became Nobel's closest friend and colleague. These were years of great stress for Nobel, with continuous negotiations and disputes that added to his chronic health problems. For relief he turned to his literary interests, and during this time he wrote the epic *Nemesis* and *The Patent Bacillus*, which was a parody of the cordite affair. His depression improved slightly when he and Sohlman would take carriage rides, since he and Sohlman could speak in Swedish, Nobel's native tongue. As he once said to him, "You know, Ragnar, I almost think of you as a young relation."[46] Sohlman relates one incident with Nobel that left him totally embarrassed.

> I had been given the job of trying to produce an organic compound, phthalic acid…. When I started the apparatus for this experiment, Nobel stood beside me to watch the result. Through my clumsiness in assembling the apparatus, warm caustic solution…soaked one of Nobel's trouser legs. We recognized the mishap at once…and Nobel, obviously shaken, left…in a hurry, while I stood

---

[45] Sohlman (1983), p. 19.
[46] Sohlman (1983), p. 39.

there both ashamed and horrified, fully expecting the sack. [Nobel went] to
Monte Carlo and Nice for a week, presumably to relax…[47]

Sohlman wasn't sacked, and Nobel not only forgot about the incident, he
was repeatedly almost embarrassingly generous to Sohlman. Despite the difference in their stations, Nobel asked Sohlman to call him by his Christian
name, and when Sohlman was married, not long before Nobel's death, he
doubled Sohlman's salary and bought a house for them in San Remo. Nobel
never had the opportunity to give him the house before he died.

Besides his work in the laboratory in San Remo, Sohlman also assisted
Nobel in Sweden, and he was in Sweden on December 8, when he received
word that Nobel had had a stroke. He immediately cabled Nobel's nephews
Emanuel and Hjalmar, and met them on the way to San Remo. They arrived
the night of December 10, too late to see Nobel before he died. A simple
ceremony was held at the villa before the coffin was sent to Stockholm.
Emanuel, the eldest nephew, took care of the arrangements, and Nathan
Söderblom, the Swedish pastor in Paris and a friend of Nobel, officiated. In
his eulogy, he said of Alfred:

…our dead brother may well seem to us, despite all his possessions and the
affection of his associates, to have been poor indeed. It was his choice, or his
fate, to live alone, and he died alone, without a hearth to cheer him or the hand
of a son or a wife to smooth his brow. And yet his was not a nature to be hardened by money, or to be embittered by loneliness; to the end of his life, he was
warmhearted and kind. In the life beyond, all that matters is to have lived nobly.[48]

A funeral with much pomp was held in the Great Church (*Storkyrkan*) in
Stockholm on the afternoon of December 29, 1896. The funeral procession
wound through the city, and the church was decorated with palm trees and
laurel garlands. Famous guests came from far away, although no representative
of the royal house attended. A funeral procession of 40 carriages carried
Nobel's casket to the North Cemetery, where he was cremated and buried.

Then came the highly anticipated reading of the will. Nobel had discussed
it with no one before his death, and no one knew of its location. The news
reached Sohlman in San Remo on December 15 that it had been found
deposited in a Swedish bank in Stockholm. Sohlman and Rudolf Lilljequist
were named as executors and it included the directives that upon his death his
veins were to be opened and his body cremated. From his father Nobel had

---

[47] Sohlman (1983), p. 22.

[48] Sohlman (1983), p. 42.

inherited a fear of being buried alive. Embalming, fortunately, had already accomplished the first directive, and cremation in Stockholm accomplished the second.

The will (Fig. 1.1) was hand-written with notes scribbled in the margins of the paper. It had never been reviewed by a lawyer. Nobel had written the will in Paris in the autumn of 1895, and signed it in Paris in early December in the presence of four Swedish witnesses. The will included the note:

> As executors of my testamentary dispositions, I appoint Mr. Ragnar Sohlman, resident in Bofors, Värmland, and Mr. Rudolf Lilljeq Lilljequist, of 31 Malmskillnadsgatan, Stockholm, and Bengtsfors, close to Uddevalla. As compensation for their attention and efforts, I grant to Mr. Ragnar Sohlman, who will probably devote most time to this matter, one hundred thousand crowns, and to Mr. Rudolf Liljequist, fifty thousand crowns[49];

Rudolf Lilljequist was an engineer with whom Nobel had worked and invested his money, and whom he respected. One hundred thousand crowns and fifty thousand crowns were sizeable sums in those days, equivalent to about $1 million and $500,000 today. Nobel apparently recognized the difficulties that Sohlman and Lilljequist would face. Fortunately, the two men got along, and Lilljequist, 14 years older than Sohlman, offered age and experience. They in turn engaged Carl Lindhagen, an influential and very capable attorney and politician in Stockholm. Probating the will and setting up the Nobel Prizes consumed the life of Ragnar Sohlman for more than three years, but in the end he was successful, and the Nobel Foundation and the Nobel Prizes owe their existence to the persistence and vision of this resourceful young man. The account of probating the will and establishing the Nobel Foundation and the Nobel Prizes is a story of anguish and adventure.

The will itself established five prizes as discussed above, with equal parts to the three areas of science (chemistry, physics, and physiology or medicine), one part to literature, and the fifth to peace.[50] The five prizes reflect the things that Nobel loved most in his life. Since Nobel was a prolific inventor and loved his time in the laboratory, it's easy to understand the prizes in physics and chemistry. The poor health that he suffered all his life would explain the prize in physiology or medicine. Nobel's love of literature, especially British

---

[49] https://nobelprize.org/alfred-nobel/full-text-of-alfred-nobels-will-2/.

[50] A sixth prize, called the Nobel Memorial Prize in Economic Sciences (officially called The *Sveriges Riksbank* Prize in Economic Sciences in Memory of Alfred Nobel) was subsequently established by the Bank of Sweden in 1968. It is administered by the Nobel Foundation, the Laureates are selected by the Swedish Academy of Sciences, and the awards are given at the same time as the Nobel Prizes.

literature, exceeded even his love of science, and explains his prize in literature. The prize for peace was undoubtedly inspired by Bertha von Suttner.

In the will Nobel includes the following stipulations for the awarding of the prizes:

> The prizes for physics and chemistry are to be awarded by the Swedish Academy of Sciences; that for physiological or medical achievements by the *Karolinska* Institute in Stockholm; that for literature by the Academy in Stockholm; and that for champions of peace by a committee of five persons to be selected by the Norwegian *Storting*. It is my express wish that when awarding the prizes, no consideration be given to nationality, but that the prize be awarded to the worthiest person, whether or not they are Scandinavian.[51]

Together, Sohlman, Lilljequist, and Lindhagen faced several problems from the outset. First, the will was imprecise, and it was not determined even in what country the will should be probated. The assets of the estate were not liquid, and in fact were in risky ventures. There was no "Nobel Foundation," so the money had been left to no one, to nothing. Some of the family, who were left only a small fraction of the estate, contested the will. The press and others criticized the will, calling it unpatriotic since it didn't single out Swedes for the prizes. The press also argued that it should not be used to make a few individuals wealthy. Finally, at the beginning of the negotiations, the institutions chosen to award the prizes weren't willing to accept the responsibility.

In addition to the problems facing the will itself, Sohlman and Lilljequist were concerned for the welfare and future of the employees of Bofors and Nobel's other firms. Some anxious employees at Bofors were already concerned about what would become of Bofors and Nobel's other firms, including the brothers' Naphtha Company in Russia, when Nobel's assets were liquidated. It was also not clear what would happen to Nobel's laboratories and the visions he had inspired in them.

The first job, then, was to establish the competent country and court in which to probate the will. In Sweden, at the time, there were no laws empowering the executors, relying instead on legal practice, whereas France had strict rules established by the Code Napolean that might present difficulties for the poorly written will. Since France also had broader taxes than Sweden, it was critical to get all liquid securities out of France quickly. After getting legal certification from Sweden, Sohlman and Lilljequist moved all the documents in Nobel's French accounts to three large safes placed in the major French

---

[51] Nobelprize.org.

bank *Comptoirs d'Escompte*, pending removal to a safer place. When Nobel's Russian relatives arrived in Paris and threatened to contest the will, the situation became urgent, and to avoid the French courts, the documents had to be removed from France without alerting the French authorities. First, the documents had to be moved to the Swedish Consulate. This was a risky undertaking and Sohlman describes the adventure thus:

> After packing the papers in a suitcase at the bank, [Consul General Gustav] Nordling and I took a horse-drawn cab to the consulate. I sat with a revolver at the ready in case of a direct attack or a prearranged collision with another vehicle, a trick not unusual among thieves in Paris at the time.[52]

For shipment to banks in England and Sweden, only the banking house of Rothschild would insure the contents, and even then, to a maximum of two and a half million francs in each parcel.

Only after the packages were out of France, at dinner one evening, could Sohlman inform the Russian relatives that it would be futile to contest the will in France. "This announcement acted like a bombshell."[53] Sohlman rejected the relatives' offer to settle the matter through arbitration.

The next step was to probate the will in Sweden. Although Nobel had been living in France, the Paris court declared itself incompetent in this case because Nobel had moved his domicile to Sweden, where he had a mansion with a permanent staff in Björkborn, near Bofors. The contest therefore moved to the court in Karlskoga, near Björkborn. To complicate matters further, it was soon discovered that Nobel's "last will" was not his only will; in the process of going through his documents, a previous will was found. In this will, which actually contained more explicit instructions, Nobel had included specific directives only for a peace prize, and the will gave more money to the relatives and to certain institutions, including hospitals and universities. Some in the family were understandably unhappy, and wanted to contest the will, going so far as to file for sequestration of Nobel's properties in Paris, Germany, Scotland, and Austria. But Alfred's nephew Emanuel, who was even wealthier than Alfred, immediately supported Alfred's desire to set up the prizes.

By January 1897, portions of the will regarding the awarding of the prizes had been made public, and the press and prominent public figures, both conservative and liberal, began to weigh in. The initial reaction was favorable. One paper described it as

---

[52] Sohlman (1983), p. 96.
[53] Sohlman (1983), p. 97.

a gift intended to further the progress of humanity and serve lofty purposes, indeed perhaps the greatest donation that any man hitherto has possessed the will or the means to provide.[54]

But critics soon emerged. The will was attacked as unacceptable on formal grounds since the executors represented no one. Nobel was deceased and the heirs—the Nobel Foundation and the prize winners—did not actually exist. It was feared that the institutions awarding the prizes would be compromised by the responsibilities involved and could be subject to bribery. Hjalmar Branting, the leader of the Social Democratic Party, attacked the "unfortunate choice of the Swedish Academy as prize distributor" for the prize for literature in an "idealistic direction," as specified in the will, and he criticized the Peace Prize, saying.

the only way goes through an international union of working classes in every country... and it is obvious that the masses should share in the proceeds of the Nobel fund in order to be able to continue and intensify the work for peace.[55]

He concluded that "A millionaire who makes a gift of this kind may personally be worthy of respect—but we are better off without either the millions or the donations." But in 1921, Branting had no qualms in accepting a share of the Nobel Peace Prize for his work in the League of Nations.

Finally, in May 1897, the Swedish government issued a decree to the Attorney General to validate the will. This negated all the claims by the plaintiffs. Now it became necessary to establish the Nobel Foundation and to secure the agreement of the institutions stipulated by Nobel to award the Prizes. Their cooperation was not immediately forthcoming. The Norwegian *Storting* accepted their assignment to administer the Peace Prize, although this required a trip by Sohlman to negotiate the terms. The *Karolinska* Institute accepted as well, provided that additional directions and clarifications were provided. The Swedish Academy, assigned to award the Prize for Literature, objected on the grounds that the members were not capable of appraising the literary output of the entire world; they were not prepared for the unpleasantness that might result from the selections; and they were concerned about the time and effort it would require and the effect it might have on the Swedish character of the Academy. Finally, the Academy accepted the responsibilities by a vote of 12–4 when they were assured that the rules of the award would

---

[54] Sohlman (1983), p. 83.
[55] Sohlman (1983), p. 85.

be established to protect them. For example, according to the statutes, no posthumous awards are allowed, no self-nominations are permitted, no protests can be lodged, and the records remain sealed for 50 years. The Swedish Academy of Sciences followed soon after with the same terms. A "Catch 22" nearly developed when the Swedish Academy refused to accept the assignment until the will was probated, and the will could not be probated until the Academies accepted the assignments. But on May 11, 1898, after it was assured that the will would be probated and the Academies were promised a role in working out the statutes of the Nobel Foundation, the Academies agreed to cooperate.

There still remained the claims of the other heirs, who sought to have the will annulled. Both the heirs and other influential groups in Stockholm exerted pressure on Emanuel Nobel, and in February, 1898, he was summoned to an audience with King Oscar. The King tried to persuade him to get the will changed. He was particularly troubled by the Peace Prize, given the troubles with Norway at that time. According to the King, "Your uncle was talked into this by fanatics, womenfolk mostly," to which Emanuel Nobel replied "Your Majesty may be thinking of Moltke's words 'eternal peace is a dream, but a beautiful dream."[56] The King went on to point out the difficulties with the will and the reluctance of the Academies, but Emanuel assured him that the Academies had agreed to accept their assignments. The King continued that it was Emanuel's responsibility to his family "to make sure that their interests are not jeopardized by your uncle's nonsensical ideas," to which Emanuel replied "Your majesty—I will not expose my family to the risk of reproaches in the future for having appropriated funds which rightfully belonged to deserving scientists." In the end, an out-of-court settlement was reached in which the heirs received lump-sum payments, and relinquished their claims. For his part, King Oscar later appreciated the aims of the Foundation and always personally handed out the Prizes himself at the annual ceremony. For their part, the Academies basked in the prestige of the Nobel Prizes.

During this time, some personal issues arose that interrupted Sohlman. The first was a carriage accident which landed him in bed for several weeks. The second interruption was caused by his required military service. To avoid holding up the work completely, "Recruit 114" was given a small building in the camp as his office, which he could use in the evenings. He had to install a telephone himself, and got permission to do this by agreeing to leave the phone for the non-commissioned officers' mess after he left. Sohlman also got

---

[56] Sohlman (1983), p. 132.

three days leave to travel to Oslo to meet with the Norwegian *Storting* to seek their acceptance of responsibility for awarding the Nobel Peace Prize. This happened during a time of tension between Norway and Sweden, which were at that time united under one king with two parliaments. This union actually dissolved only a few years later, in 1905, and was one of the reasons why King Oscar was opposed to the Peace Prize. Nothing is ever easy.

The statutes of the Nobel Foundation were finally established by a royal ordinance signed by King Oscar on June 29, 1900.

## The Awarding of the Prizes

The Nobel Foundation, which administers the Prizes selected by the various institutions, was established with two purposes in mind: first, to reflect the wishes of Alfred Nobel as expressed in his will; and second to protect the awarding institutions from criticism, or at least from litigation. The Foundation is governed by a board consisting of a chair and vice-chair appointed by the Swedish government, and representatives of the four institutions that award the Prizes: the Swedish Academy of Science (for chemistry and physics); the *Karolinska* Institute (for physiology or medicine); the Swedish Academy (for literature); and the Nobel Committee of the Norwegian *Storting* (for peace). Each of these bodies has a Nobel Committee to investigate the nominations and a Nobel Institute for research. The Swedish Academy and the *Storting* also have Nobel libraries.

Two critical questions now confronted the Nobel Foundation: who should be entitled to propose individuals to be considered for prizes and who should evaluate them? Regarding "who should propose works," it was decided that nominations would be solicited from the members of the Academies, previous Nobel Laureates, the faculties of the major academic institutions in Sweden, and selected academics from institutions around the world.[57] Some letters of nomination have been quite lengthy and detailed, others as short and simple as Einstein's nomination of Pauli for the 1945 Nobel Prize in Physics, which was submitted barely in time and comprised just two sentences in a telegram. Pauli got the Prize.

Nominations are invited in September for the following year and must be submitted by the end of January. In the early spring, each Nobel Committee selects those nominations that deserve further examination during the spring and summer, with the help of outside experts, and a member of the

---

[57] For example, one of the authors has been invited to submit a nomination.

committee is selected to write a report for each of the nominees. In September, the Nobel Committee members meet to discuss the reports on the various nominees. The outcome of this preliminary round is forwarded to the full Academy which makes the final decision at the beginning of October. The awards are then announced, one each day for a week, beginning on Monday.[58] Amidst great pomp and tradition, the awards are presented at a solemn and elaborate ceremony on December 10, the anniversary of Nobel's death. After the appropriate speeches, a gold medal bearing Nobel's image is presented to each awardee by the King. The awardees at this point are called Nobel Laureates.

Some scientists have no idea that their name will be announced. When the Nobel Prize for the invention of the transistor was announced, John Bardeen was allegedly so surprised that he dropped the pan in which he was cooking breakfast, and Walter Brattain called his colleagues at Bell Laboratories to see if it was a hoax. And in at least two cases the Swedes called the wrong person with the same name. But for the most part, scientists who have made out-standing and impactful findings have some idea that, at some point, they may be rewarded with a Nobel Prize. Some have previously received other prizes, such as the Wolf Prize and the Lasker Award, which are frequently precursors to the Nobel Prize.

Besides the gold medals, the Nobel Prizes include substantial financial rewards. If the Prize is awarded to more than one individual, the Laureates share the money in a manner determined by the individual committee. In the first year, 1901, the awards amounted to 150,782 SEK, equal to about $850,000 in 2022. In 2022 the Prizes were worth 10,000,000 SEK, equal to $970,000. Between these times, owing to wars and international financial conditions, the value of the awards has fluctuated, decreasing at times to little more than a quarter of the 1901 awards. The recipients can use the money in any way they see fit. In many cases, the Laureates have used some of the money for personal things and the rest to support research, philanthropy, etc. Several Nobelists have shared the money with a collaborator who they felt should have been recognized. For example, Frederick Banting (1923 Prize in Physiology or Medicine for the discovery of insulin) shared his half of the Prize with his colleague Charles Best. Not wanting to appear less generous, John McLeod, who shared the Prize with Banting, split his half with his colleague James Collip. Lisa Meitner, to whom Otto Hahn gave part of his Prize, gave her share to the Emergency Committee of Atomic Scientists, a charitable

---

[58] The Nobel Memorial Prize in Economics is handled a bit differently, and the winner is announced the following Monday.

association chaired by Albert Einstein after World War II. Günter Blobel (1999 Prize in Physiology or Medicine) gave his entire award (roughly $1 million) to the city of Dresden, Germany, for the restoration of the cathedral and the building of a new synagogue. Christine Nüsslein-Vollhard (1995 Prize in Physiology or Medicine) gave her entire award to establish a charity to provide support for women working in science in Germany. Albert Einstein (1921 Nobel Prize in Physics), who anticipated a future Nobel Prize, legally obligated any winnings to his first wife and two children two years before his Prize was awarded. Paul Greengard (2000 Nobel Prize in Physiology or Medicine) established an annual award for women in biomedical research with the "intent of countering the bias against women in science."

## The Nobel Prize Has Endured

Despite delayed prizes, the omission of worthy recipients, wrong prizes, and other issues surrounding the prizes each year, the Nobel Prize has endured as the single most prestigious prize in science. As might be expected, many of the prize selections have been controversial, and some seem, in retrospect, to have been wrong, or at least misguided. For example, António Egas Moniz, shared the Nobel Prize for Physiology or Medicine in 1949 for the "discovery of the therapeutic value of leucotomy in certain psychoses," commonly called a frontal lobotomy. This procedure has now been largely abandoned. One discovery that received a Nobel Prize was later found to be wrong. Johannes Fibiger was awarded the 1926 Nobel Prize in Physiology or Medicine "for his discovery of the *Spiroptera carcinoma*" [cancer caused by a parasite in the gut] in rats and mice." His experimental results were later proven to be mistaken; the cancer was caused by a vitamin deficiency, though it occurred at a site irritated by the worm. Erling Norrby called Fibiger's Nobel Prize "one of the biggest blunders made by the *Karolinska* Institute."[59]

Other Prizes have been criticized as trivial. In 1912, Gustaf Dalén was awarded the Nobel Prize in Physics "for his invention of automatic regulators for use in conjunction with gas accumulators for illuminating lighthouses and buoys." He won on the basis of a single nomination, and still stands as the least impressive award in any science category. It's been suggested that Dalén was chosen because the Committee deadlocked over more impressive

---

[59] Norrby, E. (2010). *Nobel prizes and life sciences.* World Scientific.

candidates such as Max Planck,[60] or because they wanted a Swede to get a Prize.

There are many examples of prizes not awarded in time before the scientist has died. Mendeleev, who was the first to organize the elements in the periodic table, died before he was honored, and Henry Moseley, whose x-ray spectra finally ordered the elements correctly, was killed in World War I before he could receive an award. Rosalind Franklin would most certainly have been included in the discovery of DNA together with Watson and Crick had she not died of ovarian cancer in 1958. A curious situation arose in 2005 when the Nobel Committee announced the award in Physiology or Medicine to Ralph Steinman of the Rockefeller Institute. Unfortunately, Steinman had died a few days before. The Committee ultimately decided to honor Steinman anyway since they had no knowledge of his death when they announced the award.[61]

Although it is stated in Nobel's will that the interest on the fund he established "shall be annually distributed in the form of prizes to those who, during the preceding year, shall have conferred the greatest benefit on mankind," many Prizes have been awarded years—or even decades—after the discovery. In fact, the lag between discovery and Prize has nearly doubled over the past 60 years, and scientists are now waiting more than 20 years.[62] In the first half of the twentieth century, it was common for Prize winners to be in their 30s, but that is unusual now. One possible reason for the lag is that "big splash" discoveries are less common, and the impact of these discoveries takes longer to assess. Even for a monumental discovery such as CRISPR, the lag between discovery and prize was still eight years. The argument is made that the value, or even the correctness, of the discovery must be determined before a Prize can be awarded. For example, to explain the spectrum of ultraviolet light from the sun Max Planck introduced the idea of quanta in 1900 "as an act of despair."[63] But the members of the Nobel Committee (and most other scientists) didn't believe in quanta at that time, and Planck wasn't awarded the Nobel Prize for his discovery until 18 years later. Einstein actually used Planck's quanta to explain the photoelectric effect, thus proving the reality of quanta just a few years after Plank's discovery, but Einstein himself wasn't awarded the Nobel Prize until 16 years later. Ernst Ruska experienced an even

---

[60]Feldman, B. (2000). *The Nobel Prize: A history of genius, controversy, and prestige*. Arcade Publishing, p. 134.

[61]https://en.wikipedia.org/wiki/Ralph_M._Steinman.

[62]Tozer, L. (2023). Scientists are waiting longer than ever to receive a Nobel. *Nature*. https://doi.org/10.1038/d41586-023-03086-3.

[63]Kragh, H. (2000). Max Planck: The reluctant revolutionary. *Physics World*, December 1. https://physicsworld.com/a/max-planck-the-reluctant-revolutionary/.

longer delay between the discovery and the Prize. Ruska conceived and tested the idea of an electron microscope (now called a transmission electron microscope) in 1929 when he was a graduate student in electrical engineering. But, despite the enormous importance of this invention, he wasn't recognized for the Prize until in 1986 the Nobel Committee wanted to award the Physics Prize to Gerd Binnig and Heinrich Rohrer "for their design of the scanning tunneling microscope." At this point, they could hardly ignore Ruska's much more valuable invention, so Ruska got half the 1986 award "for his fundamental work in electron optics, and for the design of the first electron microscope" some 56 years after the discovery. By this time, Ruska was 80 years old; he died two years later.

On the other hand, some awards seem to have been hasty, or even politically motivated. For example, Stanley Prusiner's 1997 award "for his discovery of prions—a new biological principle of infection" was made at a time when the discovery was still controversial, as discussed in a later chapter of this book. But just at this time, "mad cow disease" was terrifying the world, and the Nobel Committee of the *Karolinska* Institute was sufficiently convinced of the correctness of the discovery that they awarded him the Prize. They may have done this, at least in part, to stimulate more research into this threatening disease and to support political measures to prevent its spread. Likewise, Fritz Haber at the end of World War I, and Otto Hahn at the end of World War II were awarded Nobel Prizes for discoveries that had enormous wartime implications. Haber's Prize for synthesizing ammonia provided the only source of nitrates for the production of German munitions in World War I, and when he received his Prize his poison gases still lingered in the fields of Flanders. Hahn's discovery of nuclear fission led directly to the atomic bomb in World War II, and at the time of the award visions of mushroom clouds were still fresh in people's minds. Both discoveries and the two Laureates who made them were legitimate candidates for the awards, but it seems that the Nobel Committee also had a political motivation, hoping that these awards would contribute to the resurrection of science in Germany after the wars. The award to Haber was particularly controversial, causing Charles Barkla, the Scottish Physics Laureate who shared the stage with Max Planck and Fritz Haber, to suggest that "the Swedish Academy of Sciences may be qualifying for the Nobel Peace Prize."[64]

Under the terms of Nobel's will, a Prize cannot be awarded to more than three people. At first this wasn't a problem: the 1901 Nobel Prize in Physics was awarded to Wilhelm Röntgen for "the discovery of the remarkable rays

---

[64]Widmalm, S. (1995). Science and neutrality: The Nobel Prizes of 1919 and scientific internationalism in Sweden. *Minerva*, 33, 339–360.

[x-rays] subsequently named after him." But the following year, the Prize was shared by two people, and the year after that, by three. And while single individuals have received awards from time to time since then, the growth of the scientific enterprise and improvements in communication and collaboration have made it increasingly difficult to determine the three or fewer individuals most deserving of an award. Inevitably, worthy individuals are left out, and occasionally, less-deserving people are included. Lise Meitner, a physicist, collaborated for decades with Otto Hahn, a chemist, and actually initiated the research effort that led to the discovery of nuclear fission. But as a Jew she was forced to leave Germany just a few months before the discovery could be confirmed. Owing in part to limited communications during the war, and possible prejudices in the Nobel Committee, and in spite of many nominations, she was not recognized for the 1944 Prize in Chemistry, which went to Hahn alone.

Sometimes, the lower-level people who do the work and actually make the discoveries are left out and credit goes to their "bosses." For example, when Jocelyn Bell discovered "pulsars" (astronomical bodies called "neutron stars" that emit pulses of radio emission) in 1967 as a graduate student at the University of Cambridge, UK, her results were initially dismissed as radio interference by her supervisor, Antony Hewish. When her discovery was confirmed by the discovery of other radio sources, Hewish got the credit, and he shared the 1974 Nobel Prize in Physics "for his decisive role in the discovery of pulsars." Bell was forgotten until 50 years later when she earned a $3 million Special Breakthrough Prize in Fundamental Physics for her discovery of pulsars. In a similar fashion, painstaking work by Albert Schatz, a Ph.D. student at Rutgers University, led to the discovery of streptomycin. His advisor, Selman Waksman, received the 1952 Nobel Prize in Physiology or Medicine "for his discovery of streptomycin, the first antibiotic effective against tuberculosis" while Schatz was ignored.

Two nominees for Nobel Prizes in Literature (Jean-Paul Sartre) and Peace (Lê Đức Thọ) voluntarily declined their Prizes, and Boris Pasternak and Liu Xiaobo were forced by their governments to decline their awards. During the Hitler years, three scientists were forced to refuse their Nobel Prizes: Gerhard Domagk (1939 Nobel Prize in Physiology or Medicine), Richard Kuhn (1938 Nobel Prize in Chemistry), and Adolf Butenandt (1939 Nobel Prize in Chemistry). They later received their certificates and medals, but only a portion of the original Prize money.

While it's understandable for the Nobel Committees to miss or forget individuals who deserved the Prize, it's more difficult to understand why certain ideas or discoveries have been overlooked even after being nominated. Surely, if a gas regulator for lighthouses deserved a Nobel Prize for Gustaf Dalén in 1912, the invention of the airplane in 1903 should deserve an award to Wilbur and Orville Wright for having "conferred the greatest benefit on mankind," as stated in Nobel's will. They were nominated in 1909 and 1913, but never awarded the Prize. Arguably, the birth-control pill has "conferred the greatest benefit on mankind," but has never been recognized by a Nobel Prize. When it became available in 1960 it changed the lives of women by freeing them from the constraints of "*Kinder, Küche und Kirche*" (children, kitchen, and church), and by lowering the birth rate it may have saved the world from the starvation predicted by Malthus. But the pill was opposed by many, including the Vatican, and the Nobel Committee may have bowed to pressure. And what about digital computers? William Shockley, John Bardeen and Walter Brattain were awarded the 1956 Nobel Prize in Physics for "their researches on semiconductors and their discovery of the transistor effect," and Jack S. Kilby was awarded half the 2000 Prize "for his part in the invention of the integrated circuit." These discoveries are basic to the hardware of digital computers. But what about the software? The concept of a stored program dates back even before Alan Turing and his team at Bletchley Park were decoding German messages in World War II, but Turing showed that a so-called universal Turing machine could solve any computer problem. His work is the basis of all modern computers. Is this worth a Nobel Prize? And in 1950 Turing discussed artificial intelligence. He even proposed a test (the "Turing test") for intelligence in a computer: if a person conversing with the computer can't tell if the respondent is human or a computer, it's intelligent. With recent developments in AI, are we there yet? It's too late for Turing, now deceased, to get a Nobel Prize, but what about new discoveries in artificial intelligence? Or does this fall under the heading of mathematics—Turing was a mathematician—for which there is no Nobel Prize? Two recent Nobel Prizes seem to answer this question. In the first, the 2024 Nobel Prize in Physics was awarded to John Hopfield and Geoffrey Hinton "for foundational discoveries and inventions that enable machine learning with artificial neural networks." In the second, the 2024 Nobel Prize in Chemistry was awarded to David Baker "for computational protein design," and to Demis Hassabis and John Jumper "for protein structure prediction." These discoveries were made using AI programs they developed.

After all is said and done, despite its foibles, its mistakes, and its errors of omission, the Nobel Prizes remain among the most widely known and prestigious prizes in the world. There are other prizes, including the Abel Prize and the Fields Medal in mathematics, the Wolf Prizes in physics and chemistry, and the Lasker Award in medical science. All these and other prizes are well known and prestigious within their fields, but no other prize carries the prestige of the Nobel Prize.

Finally, it would be wrong to conclude this chapter without at least a brief mention of the Ig Nobel Prize. Its aim is to "honor achievements that first make people laugh, and then make them think." Awarded annually since 1991, the Ig Nobel Prize is clearly a spoof of the Nobel Prizes, but it's meant for fun, not to denigrate the "real" Nobel Prizes. In its way, it's taken quite seriously: the Ig Nobel Prizes are presented by Nobel laureates in a ceremony at Harvard University, followed by the winners' public lectures down the street at the Massachusetts Institute of Technology. In 2000, the Ig Nobel Prize in physics was awarded to Andre Geim and Michael Berry for the magnetic levitation of a live frog. Subsequently, Geim was awarded an actual Nobel Prize in Physics in 2010 for his discovery of graphene, and Berry has received other prestigious awards, including knighthood. Sometimes the amusing, if trivial, results recognized by an Ig Nobel Prize can even prove useful. For example, a study showing that the malaria mosquito *Anopheles gambiae* is attracted equally to the smell of Limburger cheese and human feet earned the Ig Nobel Prize in 2006, and it led to the use of Limburger cheese in mosquito traps.

All this said, more important than the Prizes are the discoveries and the scientists who made them. In the remainder of this book, we discuss eight Nobel Laureates and one who many say should have been a Laureate. The individuals and their discoveries have been chosen not only because they are important to science and to society, but also because they are controversial in ways that illustrate important issues for both science and society. They are ordered chronologically, in part because they reflect the progress of science in the more than a century since the Prizes were first established. They represent interesting stories in themselves and give us lessons that we can learn from.

## Bibliography


Fant, K. (1991). *Alfred Nobel.* Translated by Marianne Ruuth. Arcade.

Henriksson, F. (1938). *The Nobel Prizes and Their Founder, Alfred Nobel.* Bonniers Boktryckeri. (1993).

Sohlman, R. (1993). *The Legacy of Alfred Nobel.* The Bodley Head. Originally published in Swedish as *Ett Testamente.* (1950). Translated by Elspeth Harley Schubert.

# 2

## Bread from Air

## Fritz Haber—1918 Nobel Prize in Chemistry
### *for his method of synthesizing ammonia from its elements, nitrogen and hydrogen*

*At some point one needs to divorce the man from his deeds and ask the question: is the world better with Haber or without him?*[1]

It was 1915, early in the War to end all Wars. Fritz Haber and his German scientist corps opened the valves on the chlorine canisters, and watched the poisonous green fog slowly blanket the French and Algerian soldiers in their trenches. This was the horrific weapon that Haber had envisioned to save the Fatherland. Was it any worse to die of poisonous vapors than from bullets? But Fritz Haber was also the chemist who discovered how to capture nitrogen from the air and turn it into fertilizer, "bread from the air" as some called it, saving millions of people. This chemist of the war who unleashed poison gases was also the scientist of peace who fed millions. Haber was also a Jew, and despite his extreme patriotism, the Nazis rejected him and he died in exile. And as a final, tragic irony, the pesticide he developed to protect German agriculture was used in the ovens to exterminate millions of Jews in the holocaust. Do we tally the lives taken, and the lives saved? Is it the math that we count to determine if he is a monster or a savior? In the months following the end of the World War I, while the sick and wounded were still dying in hospitals and the poisons were still contaminating the earth in Flanders, the Swedish Nobel Committee did their sums and awarded Fritz Haber the 1918 Nobel Prize in Chemistry.

---

[1] How do you solve a problem like Fritz Haber? (n.d.). https://radiolab.org/podcast/180132-how-do-you-solve-problem-fritz-haber.

# The New Germany

Prior to 1871, there was no country named Germany. The land was a collection of dozens of small states, each with its own dialect. The people who lived along the major rivers (the Rhine, Spree, and Oder) considered themselves Silesians, Prussians, or Bavarians. In 1871, following a time of bloody wars, the French armies under Napoleon III were defeated, and France surrendered. Leaders from several German states met in the Palace at Versailles and declared Wilhelm I of Prussia as the King of a new German Reich. Germany had finally united, and Berlin became the capital of the new Germany. This new German empire, or *Kaiserreich,* was not an empire at all. It controlled no territories outside its borders, unlike France and Britain. But Germany was marching into modernity. Its railroads were the most efficient, its steel mills the most modern, and its military moving quickly beyond other European countries. As one American visitor reported at the end of the nineteenth century: "[H]ere was a Germany new to mankind. Hamburg was almost as American as St. Louis. …the green rusticity of Dusseldorf had taken on the sooty grime of Birmingham."[2] Importantly, industrialists, military officers, and academicians all preached the same gospel: Germans must unite in the service of German power.

An important transition that was occurring alongside the modernization of Germany was the rise of chemical industries. In the early twentieth century, major discoveries and development of chemicals led Germany to become the preeminent global industrial leader in soaps, detergents, paints, printing inks, staining dyes, glazes, iron and steel products, and photographic materials. The "Big Six"—Bayer, BASF, Agfa, Hoechst, Cassella, and Kalle—formed an *Interessengemeinschaft* or community of interests that would reduce competition among the members and establish a means of profit sharing. The community they formed—"IG" Farben—allowed each to still develop their own products outside the production of dyestuffs: Agfa (photographic materials); Bayer and Hoechst (pharmaceuticals).[3] One company in particular—BASF—would play a critical role in Haber's career.

---

[2] Charles, D. (2005). *Master mind: The rise and fall of Fritz Haber, A Nobel laureate who launched the age of chemical warfare.* Harper Collins, p. 10.

[3] Cornwell, J. (2003). *Hitler's scientists: Science, war, and the devil's pact.* Penguin Books, pp. 52–53.

## The Wandering Years

In December 1868, just before Germany's unification, a Jewish family in Breslau, Prussia, welcomed a new baby boy and named him Fritz, recalling Frederick the Great, the leader in the previous century. The first known picture of Haber at the age of three or four shows a young boy, perhaps presaging his later military demeanor (Fig. 2.1). The young Haber is wearing what almost looks "military" and he holds the barrel of a gun in his right hand. His eyes are expressive: he is looking forlornly at the camera, looking perhaps anxious. By the time of this picture, Fritz's mother had died in childbirth, and his father had a breakdown. Fritz lived with relatives, and his frowning expression in the picture perhaps hinted at the anxiety that would follow him throughout his life. Haber's father remarried, and Haber grew up in this new

**Fig. 2.1**  Haber at three years old

family, which eventually would include three additional stepsisters. Haber grew into an energetic teenager who was an enthusiastic but not particularly gifted student. He took the advantage of everything in Breslau, from theater to drinking in the beer cellars. At home, he and his father fought constantly, which emphasized the personality of the son. Haber was exuberant and found promise in everything. He constantly brought up new ideas, while his father seemed to have no imagination.

Haber attended Elisabeth High School in Breslau, which was oriented more toward the humanities. Mathematics and science were taught, but only as they complimented or related to the humanities. The chemistry lessons were modest at best, with no chemical experiments. Haber completed his *Abitur* (high school diploma in the USA) at the age of 17, with no firm plans as to what he wanted to do. His final high school leaving report described him as "sufficiently gifted, hard-working, advanced regularly…[and] has a sufficient knowledge."[4] This was an average report at best: his highest marks were in mathematics and history with scores of very good.

Haber stayed in Breslau for a short time but was desperate to leave. As he wrote a friend after graduating: "Nothing, absolutely…nothing satisfying to do, no stimulation, only irritation and tedium…I hope to achieve what I want when I first break free of the chains that wear me down…First, I'll go to university." He wrote another friend, "I am definitely not going to study at the Polytechnic School [in Breslau]…but at the University of Berlin, and afterward, in the summer, I will go to Freiburg or Heidelberg."[5] Thus began his wandering years.

Haber's father was against his decision to attend the university. His goal for his son was to complete a commercial apprenticeship and return to join the father's dyestuffs company. He felt that attending university was not in the tradition of the Haber family and was too expensive. After much prodding from his son and other family members, Haber's father relented, and Haber (Fig. 2.2) was off to study in Berlin. In another somewhat random decision, Haber stated to his father that he would study chemistry (not humanities, which he had excelled in at high school), and he was determined that this would be his life's work.

Over the next several years, Haber tried to find his niche. His friend Richard Willstätter described this time as "seven and a half years of detours and

---

[4] Stoltzenberg, D. (2004). *Fritz Haber: Chemist, Nobel laureate, German, Jew*. Chemical Heritage Press, p. 15.

[5] Stoltzenberg (2004), p. 18.

**Fig. 2.2**   Haber as a young gentleman

wanderings."[6] At the University of Berlin Haber hoped that his time would be filled with intellectual adventure. Instead, he found only disappointment: his physics lectures by Hermann von Helmholtz were confusing and unstimulating, and his chemistry lectures were trivial. He next studied for one year in Heidelberg with Robert Bunsen (of Bunsen burner fame). It was most likely during this time that he developed the habits of patient study and precise execution of experiments. Other than those experiences, his work with Bunsen was disappointing and did not further his chemistry knowledge.[7]

In 1889, he completed his one-year voluntary military service with the sixth Field Artillery. Haber liked the military discipline and customs that would fill his personal life but found military maneuvers tiresome and odious. One failure that he would remember for the rest of his life was his rejection for appointment as an officer. Although recommended by his superiors, they did not select him. This was possibly his first major rejection because he was a Jew.

In 1891, Haber returned to Berlin to continue his studies and complete his doctorate in the general field of organic chemistry, with an emphasis on chemical reactions. The work was good enough to be published in a respectable journal, and in May 1891 he took his final exams. He made it through these exams but declared that his performance had been miserable. In fact, he

---

[6] Charles (2005), p. 17.

[7] Goran, M. (1967). *The story of Fritz Haber*. University of Oklahoma Press, p. 8.

did well in philosophy, adequately in chemistry, and poorly in physics, but still managed to obtain his doctorate. During his doctoral studies, he became close friends with a fellow student, Roger Abegg, who was studying the new field of physical chemistry. Hoping to join this field with his friend, they both applied for assistantships in the laboratory of Wilhelm Ostwald, a leading figure in German chemistry. Abegg was accepted, Haber was not. This was Haber's first rejection by Ostwald; there would be several future negative interactions. At this point, Haber realized he had no real mentor to turn to, and he was questioning whether he would ever become what he hoped: a brilliant chemist who would have a significant impact on the field.

He was so demoralized by his poor showing in his doctorate work and the rejection by Ostwald that he felt that chemistry and academia were not in his future. He reluctantly returned to Breslau to prepare to join his father's company. His father convinced Haber that he should get some practical experience, and he arranged three apprenticeships in other companies for Haber that would provide him with practical knowledge to better prepare him to join his father's business. In rapid succession, Haber spent a few months in three different firms, then returned to Breslau, "resigned to life as a dye merchant."[8] His dreams unfulfilled, his wanderings over, he was ready to learn his father's trade. Haber's work with his father lasted for a little over one year and ended badly. Besides the constant arguing, the final straw was a poor deal by Haber that left his father with materials that he could not sell. Haber's father finally agreed that the best place for his son was not in business, but in academia. Through connections with a friend at the university in Jena, a third-tier university, Haber accepted a position as an unpaid and unhappy assistant in the organic chemistry laboratory. Although he was relatively unproductive during these years, it was here that Haber decided he would remain in academia and find a more prestigious position.

## The Karlsruhe Years

In 1894, Haber finally found the scientific environment that he needed to support his rise in chemistry. Haber was hired as an assistant in the technical-chemistry institute at the Technical University of Karlsruhe and was assigned two mentors who helped develop his career: Carl Engler, Director of the Institute for Chemistry, and Hans Bunte. The first few months at Karlsruhe were difficult, since his training had been so limited. He also had no idea what subfield of chemistry he wanted to work in, but instead branched out into several. For many scientists, this might have been the wrong approach: that is, not focusing efforts on a single problem. However, Haber had found the collegial

---

[8] Charles (2005), p. 23.

and scientific milieu that he needed to expand his entry into chemistry. In his first four years he established a reputation in a variety of areas, including technology of dyes and printing, autooxidation, and electrolysis of iron. This latter area had a very practical application: it was known that stray currents from streetcars led to corrosion of underground gas and water pipes. By switching to alternating current, the problem was solved. One of the most important areas of study during his years at Karlsruhe was the thermodynamics of gas reactions. This knowledge was at the heart of his eventual investigation of the combination of nitrogen with hydrogen to form ammonia using the nitrogen in the air.

Haber's first test in his field came at a major scientific conference in 1898, while he was still only a *privatdozent* (equivalent to an assistant professor in the USA) at Karlsruhe. Ostwald hosted the conference at the Institute of Physical Science in Leipzig, the very institute that Haber had applied to with Abegg but failed to get accepted. Haber believed he was ready for the challenge, and he eagerly jumped onto the stage to give his talk. The other attendees were mature scientists, well known in their field. His science was not particularly noteworthy, but he presented his data enthusiastically and rapidly. Many in the audience applauded, impressed by both the presentation and Haber's confidence at such a young age. There were some, however, who were not impressed by the chutzpah of this young man and loudly expressed their displeasure. This first exposure on a national stage revealed Haber as "an impetuous scientific outsider fighting for respect and acceptance" from his colleagues.[9]

Shortly after accepting the position at Karlsruhe, Haber jumped into the field of dyes and fibers, and soon became an expert in the field. He continued his study of electrochemistry and published a highly regarded textbook less than four years after joining the University. At the end of his fifth year, Haber was promoted to *Ausserordentlicher Professor*, one step below the level of full professor. His reaction to this accomplishment was to work even harder, exhibiting extreme ambition and drive, a characteristic that would define him for the rest of his life. Unfortunately, these times were also characterized by nervous conditions and stomach ailments that caused him discomfort. As a result, from this point forward until the end of his life Haber frequently retreated to a sanatorium to calm his anxieties.

In 1906, the position of chair of physical chemistry opened up and the senate of the Karlsruhe Institute recommended Haber to fill the position. Haber immediately accepted the offer. He was now the Director of the Institute for Physical Chemistry and Electrochemistry. He had finally achieved his goal at Karlsruhe. By 1910, Haber's group included 65 professors, *Dozents*, unpaid assistants, and doctoral students. They studied a wide variety of problems, including flame reactions and combustion; the connection between heat

---

[9] Charles (2005). p. 37.

changes and observed electron emissions; and formation of amalgams. They also ventured into practical applications, such as the quality of glass used for manufacturing wine bottles. His selection of Karlsruhe had another very important aspect that furthered his life in the field of chemistry: the university had established a strong relationship with one of Germany's largest chemical companies, *Badische Anilin & Soda-Fabrik* (BASF). Those long months that he had spent in apprenticeships preparing to work in his father's company would prove critical for his future career.

As a young professor in Karlsruhe, Haber had almost everything he wanted in his professional life, but his personal life still lacked any close female companionship. He had first met a beautiful young woman in Breslau—Clara Immerwahr—when she was 21 and he was 23 (Fig. 2.3). At that point in their lives, neither he nor Clara were yet ready for marriage. Then, in 1901, Haber pursued Clara with marriage in mind. There was an upcoming conference in Freiberg, and through his friend Abegg (now in Breslau) he invited both Abegg and the young woman chemist he had met ten years ago. Clara Immerwahr would change Haber's life in unexpected and tragic ways. Her life gets lost in the story of her husband, but it's a story that must be included to understand her role in the shadow of her formidable husband.[10]

**Fig. 2.3**   Clara Immerwahr

---

[10] Friedrich, B., and Hoffmann, D. (2017). *Clara Immerwahr: A life in the shadow of Fritz Haber*. In Springer eBooks (pp. 45–68). https://doi.org/10.1007/978-3-319-51664-6_4.

# Clara

Clara Immerwahr was born on June 21, 1870, at the estate of Polkendorf, just outside Breslau in Prussia. Clara's family was educated and considered part of the Jewish middle class in Breslau. The city itself was growing in both population and culture, and by the time Clara was in her teens the city had several colleges and universities. Breslau also had the third-largest Jewish population in Germany. This Jewish community was academically oriented and formed the aristocratic class. Although the Immerwahrs participated in activities within the Jewish community, they rarely attended the synagogue. The Immerwahr family's political leanings were liberal, although after the unification of Germany, they adopted a certain degree of German nationalism and patriotism. Up to this point in Clara's story, her family's identity was almost the same as that of Haber's family.

During the decade since she first met Haber at the age of 21, Clara had accomplished extraordinary things. At a time when women could not attend *Gymnasium* (the U.S. equivalent of high school), she—like Marie Curie and Lise Meitner—had to hire a tutor to provide all the education she should have received in *Gymnasium*. She then attended lectures at Breslau's university as a guest, having gotten permission from each professor. After attending these lectures for a year, she took and passed the exam that earned her the title of a *Gymnasium* graduate. She was then able to return to the university and pursue a degree in chemistry. In December 1900, a crowd, mostly of women, crammed into the university auditorium to hear Clara's dissertation defense. She passed easily and was honored in the evening newspaper with the headline "Our First Feminine Doctor." Sadly, while the dean of the school loudly announced that "Science welcomes each person, irrespective of sex, confession, race, and nationality," he went on to state that "women should continue to find their most beautiful and holy duty within the shelter of the family."[11] Sexism was alive and well and would continue to be so for many more decades. But Clara was not quite ready to live in the "shelter of the family." She stayed at the university and became a laboratory assistant.

It was at this point in Clara's life that Haber began writing to her, clearly with the intent of marriage. Initially Clara was hesitant, and refused his proposal of marriage, stating that she didn't feel that marriage was right for her. She finally accepted his marriage proposal, but with caution. She believed that marriage—like other life experiences—should afford the ability to continue to develop "one's abilities to their fullest...And so I decided to get

---

[11] Charles (2005), p. 50.

married…because I felt that otherwise a page of the book of my life, and a chord in my soul, would lie fallow and untouched."[12]

In 1901, the married couple moved to Karlsruhe. Haber was a young professor at the University, while Clara—the newly crowned Doctor of Chemistry—became a professor's wife, responsible for running the household and fading into the shelter of the family. During these first few years of marriage and Haber's increasing need to immerse himself in ever-expanding projects at work, Clara's identity was reduced to housewife, with no ability to embark on a career of her own. In 1902, their son Hermann was born, tying Clara even more to the domestic issues at home. Haber was spending more time at the university, leaving much of the child rearing to Clara. As she wrote in one letter: "Fritz is so scattered, if I didn't bring him to his son every once in a while, he wouldn't even know that he was a father."[13]

Haber's colleagues only knew Clara from the dinner parties that Haber gave. The guests described her as a quiet figure, always in the background, as Haber would entertain his guests with constant stories, jokes and rhymes. One guest commented that he also had heard the rumors that she constantly nagged Haber, which of course pushed her husband to immerse himself even more in his world that excluded Clara. In support of Clara, her second cousin wrote: "She sacrificed her profession for him, and she never really found the necessary substitute for it in family life. She had no interest in playing a prominent social role, nor was she particularly good at it."[14]

For Haber, marriage was one more step toward stability and social stature in his life. For Clara, it was an identity crisis. She became increasingly unhappy. She expressed her feelings in letters to her (and Haber's) friend Richard Abegg who had mentored her while a student at Breslau. In her letter dated April 23, 1909, she wrote:

Consider the other side! What Fritz has achieved in these 8 years, I have lost—and even more. And what's left fills me with the deepest dissatisfaction… Even if external circumstances and my own particular temperament are partly to blame for this loss, what's mainly responsible, without a doubt, is Fritz's overwhelming assertion of his own place in the household and marriage. It simply destroys any personality that's incapable of asserting itself against him even more ruthlessly.[15]

---

[12] Charles (2005), p. 51.
[13] Charles (2005), p. 69.
[14] Charles (2005), p. 71.
[15] Charles (2005), pp. 114–115.

# A Spy in America

Shortly after his son was born, Haber was selected by the German Bunsen Society to travel to America as a representative of German science to enhance German-American relations. Unofficially, he was sent to essentially spy on his Western rivals. For 16 weeks he traveled across America, covering thousands of miles from the east to west coast. He visited universities, aluminum and carbide factories, copper and lead refineries, and hydroelectric power stations. Ironically, the apprenticeships that his father had forced him to undertake in his youth had provided an understanding of the industrial world, and Haber turned out to be a remarkably good industrial spy. He thrived on conversations, digested new information rapidly, and took copious notes of everything. He enjoyed every aspect of his trip, amazed at the size and scope of every place he visited, from the Brooklyn Bridge in New York to the factories in Niagara Falls. Back in Berlin Haber gave a series of speeches to scientific audiences. Although most of his U.S. hosts assumed that the trade secrets they were sharing would remain secret, Haber provided details of the cities and their mass transportation, the vast western lands, and the industrial innovation occurring throughout the country to his German colleagues. At one point he stated:

> I have never witnessed such a bustling scene of economic activity. The impression grows as one enters the inner harbor and becomes overwhelming in the…city, where the means of transportation dwarf anything with which we are acquainted…[16]

In a speech a few months later, he told his audience:

> For a long time we underestimated the advances of the United States…it seems that Bismarck's phrase—that the Germans fear nothing but God—is being gradually amended in business circles with the additional phrase: And the United States, a little bit.[17]

In the end, Haber did not view the Americans as rivals; he felt that he had met soul mates, who were supporters of technical progress.

---

[16] Charles (2005), p. 58.
[17] Charles (2005), p. 57.

# The Guano Wars

It was 1898 as the attendees waited for the first annual address by the newly appointed President of the British Academy of Science. These inaugural speeches were generally long, espousing the outstanding achievements during the past year. Sir Willaim Crookes, the new president, was a newly knighted physicist and the discoverer of thallium, the most recent element in the periodic table. This year, instead of droning on about his science, Sir William had a different plan. He decided to shake up his audience. Clearing his throat, he welcomed everyone, then stated: "England and all civilized nations stand in deadly peril." The audience fell silent, craning to hear every soft-spoken word. He continued: "if nothing is done, great numbers of people are soon going to starve to death. As the number of mouths multiply, food sources dwindle"[18] (paraphrasing the well-known theory espoused a century earlier by Malthus). There was only one way to stop this impending tragedy: the world must create vast amounts of fertilizer—new fertilizer by the thousands of tons. "Finding new ways to make fertilizer…chemical manures—was the great challenge of [our] time."[19]

Crookes finished his talk by warning that "thousands of people, then hundreds of thousands, then millions, would begin to die." He continued: "It is through the laboratory that starvation may ultimately be turned into plenty. It is the chemist who must come to the rescue…" He went on to state that the stores of nitrates necessary for fertilizer were running out and would end perhaps as early as the 1920s. He was convinced that the only way to conquer this challenge was to find a way to fix nitrogen from the air, and this discovery "was a matter of life and death." Unless someone discovered a process, "the great Caucasian race will cease to be foremost in the world." This last racist statement was common in this era; most took it for granted that England was the pinnacle of civilization. And perhaps in response to Crooke's challenge, they would be the chosen ones to discover the "chemical manures."[20]

Every civilization from the earliest time has searched for the best approach to growing sufficient crops to feed their populations. Farms that did well had serendipitously discovered that crops grew better in areas that had the highest accumulation of manure. Healthy farms therefore doused their crops with

---

[18] Hager, T. (2008). *The alchemy of air: A Jewish genius, a doomed tycoon, and the scientific discovery that fed the world but fueled the rise of Hitler.* Crown Publishing Group, p. 4.

[19] Hager (2008), p. 7.

[20] Hager (2008), p. 7.

manure from their domesticated animals. However, as populations grew it simply wasn't enough. More manure was needed each year, spawning a new industry called manure gathering. Workers would scour the countryside for animal manure, clear city streets of horse manure, then sell what they found to farmers. But it simply was not enough as populations continued to grow. This is exactly what Robert Malthus predicted in 1798: population growth is exponential while growth of the food supply is linear, resulting in famine or war. Obviously, something had to be done quickly to keep pace with the growing number of people.

For most crops, the most important element is nitrogen. Low nitrogen means low crops, and high nitrogen means high crops. But unfortunately, it's not that simple. Oddly, we are swimming in nitrogen: the air around us is 78% nitrogen. So, we could just get what we need from the air, right? Not so fast. Atmospheric nitrogen is stuck in a triple-bonded conformation; every nitrogen atom is bound to another nitrogen by a triple bond that is almost impossible to break apart, freeing the nitrogen to bind to another element—that is, "fix" the nitrogen. The manure that early farmers used was a source of fixed nitrogen that stimulated the frantic manure gathering that occurred. Uncultivated land is also high in fixed nitrogen, but once those lands have been cultivated, the usable nitrogen is quickly gone.

Long before Crooke's talk, farmers had been searching for additional sources of fertilizer for their crops. One source that had been known as far back as the Incan populations was bird droppings, or guano, specifically the guano from seabirds that inhabited a few tiny islands known as the Chinchas, off the coast of Peru. Pelicans, cormorants, and boobies produced this guano that contained the highest content of nitrogen of any known animal droppings. What was so unique about the Chinchas Islands that made them sought after in the guano trade? It turns out that most other islands where seabirds left their excrement were in areas of high rainfall, and the fertilizer was quickly washed away. The Chinchas were unique in having very little rain, they were surrounded by high populations of sardines (seabird food), and there were few if any predators. The population of birds was so thick that the droppings would pile as high as 200 feet. Seabirds reigned supreme, and once the secret was out, the guano wars began.

The value of the white guano powder had been known by the Incas and tribes before them. They would paddle to the islands, scoop up the powdery substance, and take it back to the farmers. It was so valued that killing a seabird could be punishable by death. The value of guano was ignored for centuries by the Europeans who focused on stealing other valuables from these South American lands. Finally, in 1804, Alexander von Humboldt, a German scientist, traveled to Peru to investigate the seabird guano trade. He quickly realized the value of guano for European farmers who were struggling to maintain soil fertility and, at the same time, keep up with the demands of the ever-growing population. Humboldt brought back samples of guano to Germany, where the chemists confirmed that this was a huge discovery that could transform farming. Both England and America joined the battle to claim the guano trade in the early 1800s. By the 1840s, they were buying all the Peruvian guano that was available. The guano boom was on. Wars broke out as demand quickly outpaced supply. The U.S. even went so far as to pass an act that allowed confiscation of any unowned island that might have even the prospect of guano. As the guano demands grew, the guano supply began to disappear. By the late 1800s, the guano was gone, completely mined out. Today the Chinchas once again belong to the birds. It is a protected wildlife sanctuary with only limited access to the guano that accumulates each year.

Then word spread that there was another source of fertilizer known as saltpeter, located in South American deserts. Legend has it that a group of natives were heading for the Pacific coast to do some trading. But to reach the ocean, they had to cross a desert known as the Tarapaca. They set up camp and built a small fire on the desert floor. To their amazement, it seemed that the very earth caught fire, with sparks and flames surrounding them. They assumed that it was the work of the devil, but a local priest convinced them it was simply a rather poor form of a salt called saltpeter that was used in gunpowder. The priest also knew that this saltpeter could make crops grow better. But there was a much better supply of higher quality saltpeter in another Peruvian desert—the Atacama.

The floor of the Atacama Desert is a strange mixture of a variety of chemicals: nitrates, chromates, perchlorates and more. These had apparently built up over thousands of years, but the exact reason for this accumulation isn't known. The extraction of the saltpeter was difficult and complicated. First, the top protective rock layer had to be removed by blasting it apart. Then the next layer, caliche, was dug out and hauled to refineries. The caliche was pulverized and boiled in huge vats, and the pure saltpeter that crystallized during the process was collected. But, as with guano, the Chilean nitrate era was nearing its end. In 1907, 200 refineries were up and running, but by 1940, almost all the refineries were gone. The refineries were not disappearing because of a lack of saltpeter—there was plenty left in Chile. The caliche era ended because of the work of a German chemist: Fritz Haber.

# Bread from Air

Even with the Atacama saltpeter, chemists were under pressure to find the chemical manure that Crookes had called for, but many of the first attempts by scientists failed miserably. Various approaches were tried to fix atmospheric nitrogen. For example, scientists knew that lightning produced oxides of nitrogen (NOx) during thunderstorms. Perhaps they could burn the nitrogen out of the air like lightning bolts did. Several scientists in both Europe and the USA began to tackle this approach, using electricity to trap artificial lightning bolts inside a box. Unfortunately, electricity was too expensive to be useful, and the nitrogen came out in the form of corrosive nitric acid. Relying on lightning to fix nitrogen turned out to be neither practical nor controllable.

One of the greatest chemists of the first decade of the twentieth century, German physical chemist Wilhelm Ostwald attempted to fix nitrogen for use as a fertilizer and as an explosive. If war broke out, he knew it would be almost impossible for Germany to compete with Britain to gain access to the Atacama saltpeter in Peru. Ostwald, like Crookes, was convinced that Germany must make its own fertilizer soon. He began to focus on a chemical method to fix nitrogen from the atmosphere by combining it with hydrogen gas to create ammonia. Ostwald reasoned that fixing nitrogen to form ammonia was going to take a balance of heat, pressure, and a catalyst. He built a small pilot reaction chamber that used pressure produced by a bicycle pump. He tested a variety of catalysts and found that iron wire bits would work. Just two years after Crookes' speech and nine years before Haber, he put together this rudimentary unit, passed hot air and hydrogen into the unit, and got an appreciable amount of ammonia. Importantly, Ostwald had been one of the first German scientists to build close ties between German academic science and German industry, and through these ties he approached the industry giant BASF with a patent application. On March 12, 1900 Ostwald wrote to BASF that he had "discovered a method for combining free nitrogen…with hydrogen gas to make ammonia…The method has been tested in the laboratory…Furthermore, it is easy to get nitric acid from ammonia using atmospheric oxygen…"[21] Ostwald continued in his autobiography: "Thus I am justified in calling myself the intellectual father of this industry."[22] BASF was interested but wanted their chemist Carl Bosch to test Ostwald's machine. Unfortunately, Bosch's tests showed that the ammonia Ostwald thought he

---

[21] Stoltzenberg (2004). p. 80.

[22] Stoltzenberg (2004), p. 81.

was producing was the result of contaminants in the machine. Ostwald withdrew his patent application and dropped completely out of the ammonia race.

It's likely that Haber entered the ammonia race to compete with the very best chemists (and outdo them), to address the most important problem of the twentieth century, and use his chemistry knowledge to solve the problem. His first studies used a high voltage arc to fix nitrogen, following others who had attempted ammonia production by this method. These early investigations drew the attention of a number of chemical companies; Haber eventually agreed to share all of his results on ammonia production using the electric arc method with BASF. Although his results were superior to those of others who had tried it in prior years, the method was still too expensive and resulted in only small amounts of ammonia. In 1903, an Austrian company approached Haber and asked if it would be feasible to use a catalyst to combine large quantities of nitrogen and hydrogen to produce ammonia. Haber took on the challenge (with significant monetary support) and began with a simple setup to see if the method might work. He passed the gas mixture over iron at a high temperature but at ambient pressure. Ammonia was produced but was destroyed quickly by the high temperature. Haber shared his results with the Austrian company, informing them that his results were mostly negative and that the temperature was too high to allow large yields of ammonia. Haber dropped this attempt and began to consider fixing nitrogen through oxidation.

At the same time (1906) Walther Nernst, a professor at the University of Berlin, had also joined the ammonia hunters and had begun to analyze data involved in the formation of ammonia. Specifically, Nernst had calculated precisely how much energy was involved when a nitrogen molecule combined with three hydrogen molecules to form ammonia:

$$N_2 + 3H_2 \rightleftharpoons 2NH_3$$

When Nerst compared his data with the results in Haber's published report, Nernst believed that Haber should have found even less ammonia than he had reported. Nernst wrote to Haber that he (Nernst) would be presenting these data at the upcoming Bunsen Society meeting in the spring of 1907. Haber's anxiety and stomach issues started with a bang. This was terrible news. He was still a junior professor compared to Nernst, and to have his chemistry challenged in public by someone of Nernst's stature was unbearable. Haber began to recheck his data. One difference in their methods was that Nerst had used increased pressure, so Haber added higher pressure to his setup. He also brought in a well-known chemist—Robert Le Rossignol—from Ramsay's group in England. Together, they increased the pressure and made careful measurements. The results were closer to Nernst's, but Haber and Le Rossignol kept getting more ammonia than Nerst had reported. Haber was close—very close—to a process that would use high pressure and temperature with a catalyst to produce ammonia. He was sure of his results, and the potential for commercial uses.

Haber nervously presented his results at the Bunsen meeting, followed by Nernst's presentation of his conflicting data. The interaction that followed did not go well. Nernst called Haber's numbers highly erroneous. He added that it was too bad that Haber's data were wrong, because if right they could potentially lead to a commercial application. Nernst added: "I would like to suggest that Professor Haber now employ a method that is certain to produce truly precise numbers."[23] Haber was furious. Not even someone as prominent as Nernst had a right to publicly accuse another scientist of inaccurate data. As would be Haber's reaction to such situations then and in the future, he threw himself back into his ammonia research. As a practical matter, the differences hardly mattered; neither Haber's nor Nernst's calculations gave much hope for a process that could lead to commercial levels of the much-needed ammonia. The pendulum swung in Haber's favor when Robert La Rossignol was added to the group. La Rossignol was a master at engineering equipment, and he would build the equipment for ammonia synthesis necessary for solving Crooke's challenge. Haber was also thinking about what they might try experimentally. His mind shifted to a possible scenario: he knew that small amounts of ammonia could be produced with high pressure; what if he combined high pressure with high heat, adding a catalyst that would speed up the reaction to avoid the loss of ammonia caused by the high temperature?

At this point, and with some success, Haber felt comfortable enough to approach an industrial partner. Karlsruhe had a strong relationship with BASF, and when Haber approached them with his proposal, they were convinced, and signed a contract in March 1908. About the same time, to Haber's astonishment, Nernst announced he was still in the game, stating that his work would make "the entire nitrate trade obsolete."[24] Nernst would never accomplish this goal, but its effect was to speed up Haber's work and exacerbate his stomach and anxiety problems.

Haber and La Rossignol had to conquer several experimental challenges to begin producing ammonia from air. First, they knew they had to run the reaction under extreme pressures—up to 200 atmospheres, 100 times the pressure in an automobile tire. This pressure level would burst their valves and the steel

---

[23] Hager (2008), p. 69.
[24] Charles (2005), p. 91.

reaction chamber. La Rossignol ran the reaction through a thick-walled quartz tube encased in iron, with valves that would withstand these pressures. These valves also allowed the compressed mixture to be cooled quickly as the mixture left the chamber. They achieved this through heat exchange with the new incoming mixture. Then there was the catalyst. Haber tried a variety of hard-to-find chemicals provided by a local industrial partner. He found that osmium worked the best; he later replaced the osmium with uranium, which was cheaper. Finally, in March, Haber and La Rossignol saw success. Haber put the osmium sheet into the pressure chamber and pumped in the nitrogen/hydrogen mixture. He squeezed and heated to the limits of the steel apparatus. Success! About 6% of the total input was ammonia, and when it cooled, the ammonia liquified. Calling his laboratory group together, they saw the wonder of about 1 ml of liquid ammonia in the container. Fig. 2.4 shows a picture of Haber's original working model.

When Haber presented his report to BASF, the director of research was skeptical, to say the least. He couldn't believe that BASF would consider putting money into this small laboratory experiment that used incredibly high pressures that had little hope of being replicated on a large industrial scale. BASF once again called on Bosch, their high-pressure guy. He listened to

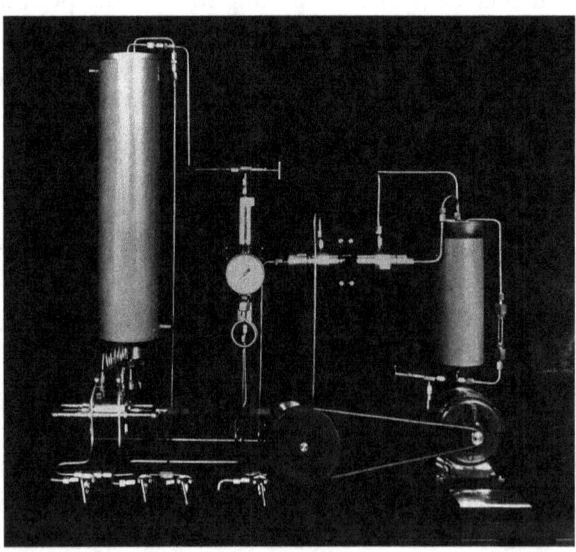

**Fig. 2.4** Haber's ammonia producing apparatus

Haber's report, then simply said to his bosses: "I think it can work. I know exactly what the steel industry can do. We should risk it."[25]

The next step was for Bosch to visit Haber's laboratory and view a run for himself. Of course, on the big day, someone in the Haber group had tightened a connection with a bit too much pressure and a seal started leaking. It would have to be fixed before the machine would work. It was a tight moment. Bosch might leave any minute. They were helpless to do anything until all the connections were tight. Haber and La Rossignol kept at it, and finally nodded that the repair was complete. The heaters and pumps were warmed up, the gases began to circulate, the critical pressures and temperatures were reached, and ammonia began to flow. Calculations revealed that between 6% and 8% of the incoming nitrogen went out as ammonia. A major success! BASF was now on board—with their expert Bosch—to adapt the process to an industrial scale. What they had witnessed was, of course, a small demonstration, but it marked a major turning point in history. One historian compared this production of ammonia to "the Wright Brother's flight at Kittyhawk, or Edison's discovery of a successful light bulb."[26]

In March 1910, Haber felt that, as a university professor, one of his duties was to publish his results. BASF, like any industry, wanted to keep the discovery very secret. Haber finally convinced BASF, and on March 18, 1910, Haber presented a lecture at Karlsruhe entitled *Making Nitrogen Usable*. After discussing previous attempts at fixing nitrogen, Haber got to his main point: "the preparation of ammonia through direct combination of the elements nitrogen and hydrogen." He then described the "Haber" method of combining the gases at high pressure and high temperature in the presence of a catalyst (osmium). He declared his results would form the basis for synthetic ammonia produced by BASF on an industrial scale. In fact, Germany would build two of the world's largest ammonia plants during the coming war years.

By this time, Haber had spent 17 years in Karlsruhe. He was happy there, with many friends. Most of his colleagues and members of his laboratory revered him. Richard Willstätter wrote in his memoirs many years later that Haber's time in Karlsruhe was his most brilliant, underscored by the exemplary atmosphere in the institute and by many important scientific discoveries in addition to the ammonia synthesis. While at Karlsruhe, Haber had moved from the bottom of the academic ladder to the top, transforming himself

---

[25] Hager (2008), p. 98.
[26] Charles (2005), p. 100.

from a "scientific castaway to conquering hero."[27] Quotes about Haber by his colleagues give us an excellent picture of the man as researcher and teacher[28]:

> Hermann Staudinger: "He could talk about everything interesting, about art, literature, and about whatever was happening in the world…Haber's lectures were always excellent. He was an unbelievably fast thinker. He was always interested and always hit the nail on the head…He analyzed my lectures nicely and drew my attention to several points. He always found the right point."
>
> Karl Holderman: "As a person, Haber was deeply devoted to his students and was always happy to help with advice or by doing something for them. He gained great respect and admiration."
>
> Franz Richardt: "A seminar with Haber…was wonderful and extremely enjoyable…He always listened attentively to his students…; he countered quickly and often and loved to spice the conversation with a joke that fitted the situation…"

But there was one person's opinion—the one closest to Haber—that should have made a difference. On April 23, 1909, Clara sat down at her writing desk and wrote a letter to their mutual friend Roger Abegg. It had been one month since the euphoric day, the day they saw the first drips of ammonia produced in the laboratory. Since that day, Haber had been working feverishly to prepare the reports for BASF. In her letter, Clara bared her soul at what she had lost while Haber had gained. She wrote[29]:

> If I wanted to sacrifice even more of the small life that remains to me here in Karlsruhe, I'd let Fritz desiccate into the most one-sided…researcher that one could ever imagine…I think that even a genius shouldn't be permitted such behavior.

It would be Clara's last letter to Abegg. Almost a year later, in April 1910, Abegg was piloting a hot-air balloon that crashed, killing him. It would remove an incredibly important person in Clara's life that might have prevented her eventual tragic ending.

---

[27] Charles (2005), p. 34.

[28] Stolzenberg (2004), p. 67.

[29] Charles (2005), p. 115.

## Tons per Hour

BASF subsequently patented Haber's discovery of the fixation of ammonia. The company knew that they had a process under their control with the potential for producing fertilizer that could save millions of lives—and, of course, make millions of *Reichsmarks* for the company. Carl Bosch, only 35 years old, accepted responsibility for scaling up the synthesis of ammonia. The challenges were huge: he had to find the right materials for the equipment that would stand up to the high pressure and temperature; find a cheap, efficient catalyst; and identify a way to control the pressures and temperatures. The work was exhausting; for example, by 1912, about 6500 experiments with 2500 different substances had been run to find the best catalyst. Bosch and his team built a prototype plant that by early 1911 was pumping out two tons of ammonia per day. It was an incredible achievement. And the project just kept getting bigger. The final step was to build bigger ovens and a full-scale factory. The site chosen was Oppau, a small village in southeast Germany, and on May 7, 1911, the Bosch team began construction. Oppau opened for business in September 1913, 28 months after the start of the plant. As was later noted, Bosch's team was building the "biggest scientific effort in history, comparable in scale to the Manhattan Project in WWII"[30] (Fig. 2.5). Within a year, the Oppau plant was producing tons of ammonia per hour. Oppau was not just the biggest ammonia plant in the world—it was also the beginning of Bosch's career in high-pressure chemistry. In 1914, Bosch knew that the Oppau production was still far below what German farmers needed, and

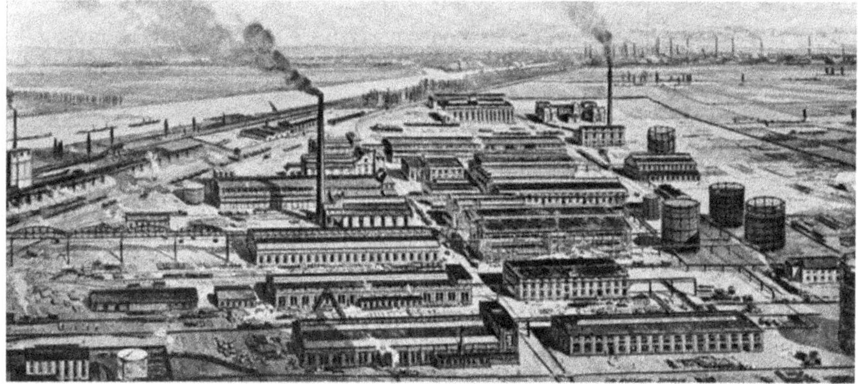

**Fig. 2.5**  Oppau plant in 1913

---

[30] Hager (2008), p. 122.

Bosch wanted to expand. But these plans were put on hold—the Great War had begun.

## The Road to Berlin

The concept of a research institute that would bring together Germany's greatest scientists had been in negotiations since the early 1900s. With internal and external donations, construction of the first two institutes began around 1910. Specific to Haber, Leonard Koppel, an acquaintance of Haber's, agreed to provide the bulk of the money if Haber was named the director of the first of two institutes: the Kaiser Wilhelm Institute (KWI) for Physical Chemistry and Electrochemistry. Haber accepted the offer, but it would be another year before he arrived in Berlin. His boss at Karlsruhe tried to keep him, but Haber was by now totally invested in building his new institute. Finally, in October 1911, Haber, Clara and Hermann moved to a small apartment in Dahlem, awaiting the completion of the primary residence. With much pomp and circumstance and with the Kaiser leading the way, the official grand opening of the Institute for Physical Chemistry and Electrochemistry took place on October 23, 1912, at the same time as the opening of the KWI for Chemistry, with Emil Fischer as director (Fig. 2.6). It would be well into 1913 before scientific work would begin in the new buildings. During this era, the period of classical physical chemistry seemed to be ending with the appearance of quantum ideas and atomic theory. The first publication was a paper by Haber "dealing with ultraviolet and infrared characteristic frequencies of crystals and the connection between heat of reaction and quanta of energy."[31]

On the personal side, Clara retreated further into her shell, having left friends behind after 17 years in Karlsruhe. One bright spot for both Clara and Haber was the recruitment of Albert Einstein to Berlin in 1913, bringing with him his wife Mileva Marić and their two sons. Haber had pushed for the recruitment of Einstein, understanding what a scientist such as Einstein could bring to the rapidly growing scientific community in Berlin. Haber and Einstein immediately struck up a strong friendship (Fig. 2.7), although the two men were completely different. As Hager notes:

Haber was…the perfect German: promilitary, pro-Kaiser, ever more the stiff-necked Prussian-style patriot. Einstein was free-thinking…cosmopolitan, and a

---

[31] Stolzenberg (2004), p. 117.

**Fig. 2.6** The opening of the Kaiser Wilhelm Institutes with the Kaiser leading the procession. (Courtesy of Archives of the Max Planck Society, Berlin, VI. Abt., Rep. 1, Haber, Fritz IX/9)

bohemian pacifist who enjoyed making fun of precisely the traits Haber…[embodied].[32]

As Einstein wrote in 1913, remarking about his friend's growing egotism:

Haber…unfortunately [needs] to be seen everywhere…Unfortunately I have to accept that this otherwise so splendid a man had succumbed to personal vanity and not even of the most tasteful kind.[33]

Oddly, the one thing that they both shared was their position as Jews and outsiders. They were both respected for their positions as Germany's most prominent Jewish scientists. Einstein understood that and hated it, but Haber still worked for acceptance.

Unfortunately, the Einsteins' marriage had been in trouble for several years and was finally ending. In July 1913, Mileva moved into the Haber home with her two sons. Mileva became Clara's only friend in Berlin, and Haber was in the unfortunate position of intermediary during the Einsteins' divorce.

---

[32] Hager (2008), p. 132.

[33] Hager (2008), pp. 132–133.

**Fig. 2.7** Haber and Einstein. (Courtesy of the Archives of the Max Planck Society, Berlin; VI_001_Haber_Fritz_VI_2)

Einstein, who pretended to be detached from personal sentiments, could not hide his distress at losing his sons. Mileva was taking them back to Zurich. Haber accompanied Einstein to the departing train and tried to console the sobbing physicist during that day and night.

## The Manifesto

I was one of the mightiest men in Germany. I was more than a great army commander, more than a captain of industry. I was the founder of industries; my work was essential for the economic and military expansion of Germany. All doors were open to me.[34]

Fritz Haber, as quoted by Chaim Weitzmann.

---

[34] Ushi (2024, January 7). *The light and darkness of Dr. Fritz and Mr. Haber.* Museum of the Jewish People. Retrieved January 7, 2020, from https://anumuseum.org.il/blog/fritz-haber/.

On June 28, 1914, a Bosnian Serb named Gavrilo Princip assassinated Archduke Franz Ferdinand, heir to the Austro-Hungarian throne. In retaliation, Austria-Hungary declared war on Serbia, beginning a cascade of nations entering the war. After weeks of tension following the assassination, on August 4, Germany declared war against Russia. France and Great Britain declared war on Germany after Germany invaded Belgium, The War to End all Wars had begun, and the German people united behind their Kaiser. Anticipating that there would be another war, Germany had developed the Schlieffen Plan. In this plan, Germany would wage war on two fronts—France and Russia—with the decision to attack France not on their shared border but by going through neutral Belgium to avoid major French defenses. The Schlieffen plan declared that Germany would accept no resistance from Belgium; any resistance would result in executions and massive destruction. If the army perceived that one sniper shot at them from one building, the Germans would destroy the entire village. As World War I began, Germany activated the Schlieffen Plan. As they pushed through Belgium, they killed over 6000 civilians, and set many buildings on fire. One of the worst instances of destruction was the sack of Louvain, where over five days Germany brutally murdered more than 250 people and destroyed over 1200 homes. As one British newspaper wrote:

> The Germans have committed an atrocious act which will turn the heads of every civilized nation in the world against them. They have utterly destroyed the peaceful and historic old city of Louvain, the Oxford of Belgium. The beautiful Hotel de Ville…the stately church of St. Pierre, the famous University, all are gone. Even the library of 70,000 volumes and priceless manuscripts was committed to flames by the ruthless barbarians who have set forth to spread "German culture" throughout the globe.[35]

In October 1914, 93 German professors released a public statement called the Manifesto to the Civilized World. Better known as the Manifesto of the 93, it focused on excusing the Germans for their aggression and supported the "honor of Germany in her hard struggle for existence…Have faith in us! Believe that we shall carry on this war to the end as a civilized nation."[36] The text rejected the accusation that Germany caused the war:

---

[35] *An atrocious deed.* (1914, Aug 29). Hull Daily Mail. https://www.newspapers.com/imageview/794160015/?match=1&terms=an%20atrocious%20deed.

[36] vom Brocke, B. (1914, Oct 4). *Scholarship and militarism: The appeal of 93 to the civilized world.* Retrieved July 8, 2021, from https://ghdi.ghi-dc.org/docpage.cfm?docpage_id=1721.

Neither the people, the Government, nor the Kaiser wanted war. Germany did her utmost to prevent it…Only after overpowering forces that long were lurking on our border fell upon the German people from three sides did they arise as one.

The Manifesto document spent considerable time denying claims that the Germans had killed Belgium citizens or burned any buildings.

It is not true that our troops treated Louvain brutally. Furious inhabitants having treacherously fallen upon them in their quarters, our troops with aching hearts were obliged to fire a part of the town as a punishment. The greatest part of Louvain has been preserved. The famous Town Hall stands quite intact.

The only reason they did not burn down the Town Hall was that the German officers used it as their headquarters.

Among the 93 intellectuals who signed the document were 12 current or future Nobel Prize winners including Haber, Ostwald, Paul Ehrlich, Max Planck, and Nernst. The manifesto completely backfired. It only enhanced the anti-German belief that these scientists had succumbed to a German nationalism. And it only worsened Haber's situation in the eyes of his scientific colleagues. Einstein avoided letting the war create a rift between Haber and him. However, when Haber signed the Manifesto, Einstein broke with Haber over their political differences. After the defeat of Germany, a report in 1921 in *The New York Times* found that of 76 surviving signatories, 60 expressed regret. Some even claimed not to have seen what they had signed. There is no record that Haber ever regretted signing the manifesto.

## Bread or Gunpowder

Haber was ready for war. He believed it was his destiny to "save" his beloved Germany. But he found that at the age of 45, he was too old and too Jewish to be considered for a military command. He therefore volunteered to help assess Germany's weapon stores. The army had marched into the war in a state of ignorance about where their weapons and ammunition would come from. The estimates were that they had stockpiled ammunition for only six months, expecting a short war with a quick victory. But by August, the *Wehrmacht* had stalled outside Paris; it would be a long war. Haber took a leading role in organizing the industrial, military, and scientific worlds. He had quickly risen in the ranks of the German war machine and was prepared to do anything in his power to save his beloved Fatherland.

In his speech to his Karlsruhe colleagues after the first production of ammonia, Haber's primary focus was on the use of ammonia for fertilizer. But he made one additional key comment suggesting that he was already thinking about other uses for ammonia: he described the "extraordinary need for bound nitrogen, mainly for agricultural purposes and to a much smaller extent for the explosives industry…"[37] Although the military was confident that they had sufficient explosives to win the war (perhaps six months' worth), both Bosch and Haber realized Germany would run out of explosives if the war lasted many more months (or years).

Haber approached BASF with the idea of turning Oppau into a plant producing the basic raw materials needed by the military: nitric acid that could make gunpowder and explosives. The need became critical as Germany began using far more explosives at the beginning of the war than had been anticipated. In November 1914, the Germans estimated that approximately 30 kilotons of fixed nitrogen would make enough munitions for an entire year of warfare. In fact, by 1915, Germany was burning through almost 30 kilotons of ammunition every ten weeks. The situation had become critical. As Haber had predicted, the side that won would be the side with the greatest access to fixed nitrogen.

Bosch presented BASF with his proposal: convert the Oppau plant from production of ammonia to sodium nitrate—saltpeter—which could then be easily transformed to nitric acid for gunpowder and explosives. BASF signed the deal; they were now a defense military industry, producing munitions for the war. Bosch did not like this new role. He and his group had been supplying ammonia for fertilizer that Germans desperately needed for their farms, and now there was no more fertilizer; it had become gunpowder. But BASF converted the Oppau plant and began producing nitrates at a rate of 7500 tons/month, more than enough to fulfill the needs of the military.

Then, something unexpected happened: flimsy, primitive airplanes from France flew over Oppau and began to drop small bombs on the plant. There was not a lot of significant damage, but the continued minor repairs due simply to wear and tear within the giant plant, together with the bombings, began to wear down Oppau. The solution was to build a second plant farther away from France that the planes couldn't reach. A small village near Leipzig called Leuna was chosen for the second plant. For Bosch, Oppau looked like a test run: Leuna would be the masterpiece. Leuna would become something like a secret weapon, switching easily from farming during peacetime to arming during war. The plant development began in May 1916, and incredibly

---

[37] Hager (2008), p. 108.

Leuna was up and running by April, 1917, producing 36,000 tons/year at its opening. Almost all the output was fed into the German war machine, supplying all of Germany's needs for war munitions, far out of the reach of the French bombs.

## The Prussian *Geheimrat*

Haber worked hard during the war to keep in the good graces of his country's leadership. He was relentless in strengthening his contacts with industry and governmental offices. He transformed his institute into a dedicated military research center, forcing as many laboratories as possible into research to benefit the military. The institute was rich with government funding, staffed with hundreds of workers, and surrounded by barbed wire. Soldiers were on guard around the clock. Haber had completely evolved into a one-man military complex; he wore his uniform to work to complete the vision of himself as the perfect German patriot. In fact, Haber was cold-blooded in his quest for a German victory. A fellow scientist commented about Haber during this period: "Haber's actions continued to contradict Montesquieu's belief that knowledge makes men gentle. His boundless ambition seems to have made him determined to win the war single-handed."[38] The British chemist and onetime Haber collaborator J.E. Coates probably had the best description of Haber during the war years. Coates wrote:

> The war years were for Haber the greatest period of his life. In them he lived and worked on a scale and for a purpose that satisfied his strong urge towards great dramatic vital things…To be a great soldier, to obey and be obeyed—that, as his closest friends knew, was a deep-seated ideal…It must not however be supposed that he exalted and enjoyed war as such, but the coming of the war brought out another side of his nature and transformed him into a Prussian officer, autocratic and ruthless in his will to victory.[39]

As the war began, Haber wasn't convinced that Germany had sufficient explosives and believed that the ammonia plants should be converted to producing nitrates for munitions. But Haber also had another personal goal: he did not believe that explosives alone would win the war. He wanted complete victory, and he believed that the development of a weapon that would be "so

---

[38] Hager (2008), p. 155.
[39] Charles (2005), p. 153.

devastating that just the thought of it would terrorize the enemy" was the answer for a German victory.[40] The weapon he chose was based in chemistry, a weapon that would break the morale of the enemy. That weapon was poison gas. He knew that the Hague convention in 1899 specifically banned the use of poison gas of any sort. Haber's plan was to design a delivery system that would not be banned by the Hague convention.

Haber promoted his ideas in the highest levels of the German ministry and became the head of a new chemicals section in the Prussian war ministry, with a title of captain that was more honorary than military. Captain Haber eagerly began the development of his super weapon. He decided he would need a heavy gas that "would advance like ground fog, surging into trenches as it advanced."[41] He envisioned rows of large canisters set along the front, with each canister containing the heavier-than-air gas (Fig. 2.8). The canisters would be opened all at the same time, with the gas pushed by the wind along the ground eventually ending up in the enemy's trenches, forcing the soldiers out. Haber wanted a killer gas, not the tear gas irritant. He also believed that this method of opening valves on the canisters would get around the Hague ban because no projectiles would be involved. Haber and his institute workers settled on chlorine, which was heavier than air and highly toxic. Haber truly

**Fig. 2.8** Haber and his gas canisters at the front. (Courtesy of the Archives of the Max Planck Society, Berlin, VI. Abt., Rep. 1, Fritz Haber IX/9 VI. Abt., Rep. 1, Haber, Fritz IX/9)

---

[40] Hager (2008), p. 147.
[41] Hager, (2008), p. 159.

believed that this approach would force the enemy to surrender, and Germany would march victorious into Paris. When challenged about the use of poison gases, Haber responded that "the war would end more quickly, and countless lives would be saved"[42].

## The Poisonous Cloud

Haber was known to his colleagues as *Geheimrat,* or privy councilor. He was not officially a military officer, but an academic, a professor, who also saw himself as a Prussian officer. Haber gathered his institute scientists into an elite company of engineers known as the *Pionierkommando 36;* several would go on to Nobel fame including Otto Hahn (1944); James Franck (1925); and Gustav Hertz (1925). Cornwell describes these scientists as the "new German warriors"[43]—the scientist/soldiers who would be the first scientists to actually use their weapon in war.

At 5 pm on April 22, 1915, Haber and his gas troop were dug in along the German front at Ypres, facing French and Algerian soldiers. They carried with them over 5700 canisters, each weighing roughly 200 pounds containing chlorine gas in liquid form. They had waited several days for this moment when the winds were perfect to lift the poison away from them and toward the enemy. Military meteorologists who had carefully monitored the wind direction gave the order to release the gas. Wearing gas masks for protection, they opened the valves on the canisters, releasing a greenish cloud that traveled toward the Allied lines (Fig. 2.9). Within minutes, 160 tons of chlorine gas drifted toward the French, engulfing all the soldiers who were downwind. The surprise use of chlorine gas ruptured the French line, causing terror and panic and the chaotic retreat of the French soldiers. Within minutes, the wall of gas killed over 1000 French and Algerian soldiers and wounded over 4000 more. An excellent account of what that day might have looked like is described by Latabut in his book "*When We Cease to Understand the World*"[44]:

When they awoke on the morning of Thursday, April 22, 1915, the soldiers saw an enormous greenish cloud creeping towards them across no-man's-land. Twice as high as a man and as dense as winter fog, it stretched from one end of the horizon to the other, as far as the eye could see. The leaves withered on the trees

---

[42] Hager, (2008), p. 160.

[43] Cornwell (2004), p. 61.

[44] Labatut, B. (2021). *When we cease to understand the world.* New York Review of Books.

**Fig. 2.9**  An aerial view of the chlorine cloud

as it passed, birds fell dead from the sky; it tinged the pastureland a sickly metallic colour. A scent like pineapple and bleach filled the throats of the soldiers when the gas reacted with the mucus in their lungs, forming hydrochloric acid. As the cloud pooled in the trenches, hundreds of men fell to the ground convulsing, choking on their own phlegm, yellow mucus bubbling in their mouths, their skin turning blue from lack of oxygen…As noted by one of the soldiers who opened the canisters: The wind kept moving the gas towards the French lines. We heard the cows bawling and the horses screaming…Then everything was quiet again. In a while it had cleared and we walked past the empty gas bottles. What we saw was total death. Nothing alive. All of the animals had come out of their holes to die…When we got to the French lines the trenches were empty but in a half mile the bodies of French soldiers were everywhere…You could see where men had clawed at their faces, and throats, trying to breathe.

A soldier who witnessed the attack wrote this[45]:

[I watched] figures running wildly in confusion over the fields. Greenish-gray clouds swept down upon them, turning yellow as they traveled over the country blasting everything they touched and shriveling up the vegetation…Then there staggered into our midst French soldiers, blinded, coughing, chests heaving, faces an ugly purple color, lips speechless with agony, and behind them in the gas soaked trenches, we learned that they had left hundreds of dead and dying comrades.

---

[45] Van Pelt, R. (1915, April 26). *Roar of Krupp guns heard for 30 miles*. Retrieved March 25, 2021, from https://www.willametteheritage.org/april-26-1915/.

A British army surgeon recorded an important report at the front[46]:

It has been my painful duty today to witness the effects of a strong dose of the poisonous gases being employed by the Germans, and so appalling are they that I wish I knew how to make known this infamy throughout the civilized universe…As an army surgeon one has had to face many duties that have shocked the roughest natures. But I declare on my honor that never in the course of experiences…have I met with cases of inhuman torture to equal this poisonous gas procedure of the Germans. It is, in fact, a slow, painful process of drowning brought about by the total destruction of the lung tissue. It is most easily realized by the lay mind by comparing it with the effects of injecting a burning acid fluid into the lungs and so killing your victim by long drawn-out painful suffocation…Surely my professional brethren throughout the world will rise to a man and put a stop to this blot upon science.

The Haber marriage had reached a boiling point as Haber started his work on poison gas just as World War I had begun. As he left for Ypres to test his new weapon, Clara was in despair. She was shocked that chemistry—the science she loved—was being used to create horrible weapons. A few days after Ypres, Haber returned to Berlin for a brief time with family, and a celebration for friends to applaud the success of the Ypres attack. It was during this party that Haber's wife Clara took his service revolver, went out into the garden, and shot herself in the breast. Two hours later, she died in her son's arms. The very next morning, Haber left Berlin to travel to the eastern front for another experimental attack, this time against the Russians. It's not entirely clear what Haber felt or what his reaction was to his wife's suicide. There is some speculation that he simply didn't care; that his duty to his country was more important. There is also some recorded information that Clara's suicide affected him deeply. Haber's godson and historian Fritz Stern recorded the most likely scenario:

The oft-repeated assertion that Haber reacted coldly or indifferently to his wife's death is erroneous. He left for the eastern front because he was ordered to and because allaying the emotional pain with work was the only thing he knew how to do. Haber did not understand his wife, which is not the same as saying that he did not love her.[47]

---

[46] *Daily Telegraph* (1915, Thursday Mar 6), p. 8.
[47] Hager (2008), p. 165.

Haber later wrote in a letter to a friend describing his grief[48]:

> For a month I doubted that I could keep going. But now the war, with its dreadful images and constant demands on all my powers, has made me calmer…I have no time to look left or right, to reflect or sink into my own feelings…I hear in my heart the words that the poor woman once said, and, in a vision born of weariness, I see her head emerging from between orders and telegrams, and I suffer.

There are several versions of why Clara might have taken her own life: a protest about gas warfare; depression because of a domestic life that she was forced to live and thus unable to follow her own career; or the suspicion that her husband was having an affair. There is one account that Clara and Haber had argued over gas warfare a few days before her death. Other recorded details suggest Clara found her husband and Charlotte Nathan, the business manager of Haber's club, in an embarrassing situation at the celebration that night at the Haber home. It should be noted that Charlotte and Haber had certainly noticed one another, and they later married in 1917 (Fig. 2.10). One member of Haber's gas brigade—James Franck (Nobel Prize, 1925)—wrote later that he felt that gas weapons played a role in Clara's death. He recalled: "[Clara] was a good human being who wanted to reform the world. The fact that her husband was involved in gas warfare certainly played a role in her suicide."[49] Some of Haber's friends only mentioned Clara's death in passing or not at all. Haber's closest friend Richard Willstätter's memory is typical: "Haber was completely a man of duty…the spring day in 1915 on which he returned home for a short visit…was the day on which his wife died. On that same evening Captain Haber traveled to the eastern front, where he was expected."[50]

The next poison gas test run used a mixture of phosgene and chlorine against the Russians in Galicia. Otto Hahn, who was present for this attack, later described that he felt ashamed on seeing the horrific effects on the Russians. He actually attempted to revive some soldiers, to no avail. In the final test in July 1917, the poison gas unit returned to Ypres and used mustard gas (Yellow Cross), a poison that didn't blow away with the wind. It stuck to everything, including soldier's skin and clothes. The Allied troops reported a shimmering cloud and a strange peppery smell. Within 24 hours men developed horrific blisters and sores, and some started coughing up blood. Mustard

---

[48] Charles (2005), p. 169.
[49] Charles (2005), p. 166.
[50] Charles (2005), p. 165.

**Fig. 2.10** Haber and Charlotte on their marriage in 1917. Haber's son Hermann is on the left

gas killed or injured on contact, causing painful blisters and blinding soldiers who clawed frantically to remove it from their eyes. Those who inhaled the gas died. And since the gas could be absorbed through the skin, gas masks were useless. It could take up to six weeks to die; it was a terrible way to die.

Despite the apparent success of the gas attacks, the result disillusioned Haber. He had fully believed that his new weapon would be so expansive and so successful that he would single-handedly win the war for his beloved Germany. Haber argued for a much larger attack than the test run at Ypres. He stated after the war that Germany would have won if they had listened to him and conducted a much larger scale attack. It's possible that the German high command realized that most of the top officers did not support gas weapons. In fact, Haber could find only one officer in the first attack who supported him and led the troops in their use of poison gas. After the war, Haber continued to justify the use of gas warfare, even though he was now carrying the label father of chemical warfare. He stated that France had first used gas warfare (artillery shells filled with tear gas—an irritating gas, but not debilitating). He also claimed that he had not gone against the Hague convention because he didn't use artillery shells to release the gas. And he countered with the argument that this type of warfare caused fewer tragic injuries, such as lost limbs. But even Haber was not prepared for the horrific injuries that the soldiers endured.

The poison gas brought out a tremendous outpouring of horror at the killing and maiming of the soldiers. A famous poem by Wilfred Owen entitled "*Dulce et Decorum Est*" (from—as Owen states—an old line from a Roman poem meaning it is sweet and fitting to die for one's country) presents a vignette from the front lines describing a group of British soldiers who have been attacked with chlorine gas. Two verses of the poem provide the imagery of this horror[51]:

> Bent double, like old beggars under sacks
> Knock-knee, coughing like hags, we cursed through sludge
> Till on the haunting flares we turned our backs,
> And towards our distant rest began to trudge.
> If in some smothering dreams, you too could pace
> Behind the wagon that we flung him in,
> And watch the white eyes writhing in his face,
> His hanging face, like a devil's sick of sin,
> If you could hear, at every jolt, the blood
> Come gargling from the froth-corrupted lungs,

In a moving tribute, the British Government commissioned a painting by John Singer Sargent to contribute a central painting for a Hall of Remembrance. Sargent's painting was based on the dressing room at the village of Bailleulval where they treated soldiers who had been exposed to mustard gas (Fig. 2.11). In the center of the painting is a line of soldiers being led

**Fig. 2.11** "Gassed" by John Singer Sargent (original colors: yellow and brown)

---

[51] Owen, W. (1921). *Dulce et decorum est*. Retrieved April 6, 2021, from https://www.poetryfoundation.org/poems/46560/dulce-et-decorum-est.

by a medical orderly. Their eyes are bandaged, and each man grasps the shoulder of the man in front. One soldier is leaning over to vomit on the ground. A second group of blinded soldiers is in the background, and more gas-affected soldiers lie in the foreground. The painting measures over nine feet tall and 21 feet wide. It currently hangs in the Imperial War Museum. Newspapers throughout the world described it as extraordinary, monumental, a masterpiece, and epic. British broadcaster/historian Jon Snow called the painting one of the "10 Best British Artworks About War,"[52] and *The Guardian* listed *Gassed* among "1,000 Artworks to See Before You Die."[53] Upon viewing the painting for the first time, Winston Churchill commented: "With all its brilliant genius and painful significance… how the field of national psychology must have been harrowed by events which had taken place during the war."[54]

## The Prize

The normal process for the awarding of Nobel Prizes involves the submission of nominations from the scientific community, followed by reports by committee members, then a recommendation to the awarding body for the person who should receive an award that year. The Swedish Academy of Science ultimately selects the prize winners. Alfred Nobel's will establishing the coveted Nobel Prize stated that "The Nobel Prize is awarded to people deemed to have conferred the greatest benefit to humankind." He ended his will with this statement: "It is my express wish that when awarding the prizes, no consideration be given to nationality, but that the prize be awarded to the worthiest person."[55]

The war presented several complications for the Royal Swedish Academy. Politics swirled around the awards: many viewed Germany as the leader in the sciences in the early 1900s. Germans had won over one-third of the physics and chemistry prizes, more than any other country. Scientists and students

[52] Sargent, J.S. (2018, Jun 3). *Gassed.* Retrieved July 18, 2021 from https://www.theworldwar.org/exhibitions/john-singer-sargent-gassed.

[53] *The Guardian.* (2022, October 19). 1000 artworks to see before you die. https://www.theguardian.com/culture/2008/oct/27/1000-artworks-to-see-before-you-die-art.

[54] Churchill, W. (2018, Jan 31). *First World War Centenary*, 1914–1918. Retrieved March 6, 2021, from https://1914centenary.com/2018/01/31/sargents-gassed-headlines-usa-exhibition/.

[55] The Nobel Prize: *Alfred Nobel's Will.* Retrieved June 21, 2020, from https://www.nobelprize.org/alfred-nobel/alfred-nobels-will/#:~:text=Alfred%20Nobel's%20will,could%20be%20awarded%20in%201901.

from other countries flocked to Germany to study with some of the world's finest minds. Germany's model of close collaborations with the government and industries was viewed as a model that many countries wished to emulate.

When Germany lost the war, many scientists in the Allied countries wanted Germans ousted from any postwar collaboration. Sweden, who awarded Nobel prizes and had remained neutral during the war, had reasons to support German reemergence, but there was also pressure to oust them in postwar years. When the Swedish Academy selected three Germans for the 1918 and 1919 chemistry and physics awards, there was outrage from the Allied countries. An excellent article by Sven Widmalm[56] in 1995 discusses the issues that Sweden faced, and how they dealt with the awarding of prizes during and immediately following the conflict.

An important member of the Royal Swedish Academy of Sciences who was intimately involved in the prizes during the war years was Svante Arrhenius. In 1900, he assisted in setting up the Nobel Institutes and was elected a member of the Royal Swedish Academy in 1901, a position he held until his death in 1927. He was a member of the Physics Committee, and a de facto member of the Chemistry Committee. Arrhenius was the first Swede to earn a prize in physics in 1905. He thus was an important voice in the Nobel wartime decisions.

Arrhenius was a forceful advocate for international cooperation in science and hoped that Sweden's position as a neutral nation would help maintain that cooperation. Indeed, it was often stated that the Nobel prize gave Sweden some responsibility to mediate collaboration. But the outbreak of the war dramatically affected the peaceful collaboration that Arrhenius hoped would continue in all of Europe. The question of Sweden as the arbiter of neutral conduct in science came directly into focus in 1919 as the Academy awarded Nobel prizes in science to three Germans, including Fritz Haber.

Arrhenius and his colleagues had conflicting views about Germany and what should happen postwar. Wilhelm Ostwald, a German physicist who had won the chemistry prize in 1909, stated prior to the beginning of the war that collaboration in science would pave the way toward eternal peace in Europe. He was one of the 93 signees of the 1914 "Manifesto" supporting and excusing Germany's wartime aggression. Ostwald stated that the war would be a good thing, implying that once Germany won, German science—that is, the "superior science"—would stimulate a new higher order throughout Europe.

---

[56] Widmalm, S. (2012). A superior type of universal civilization: Science as politics in Sweden, 1917–1926. In R. Lettevall, G. Somsen, & S. Widmalm (Eds), *Neutrality in twentieth-century Europe: Intersections of science, culture, and politics after the First World War.* Routledge eBooks (pp. 79–103).

A number of scientists from the Allied countries were on the opposing side, including another close colleague of Arrhenius, the British scientist William Ramsay. When Germany invaded Belgium, Ramsay became an extreme opponent of involving German scientists in any kind of collaboration ever again, adding that their defeat would finally spare the science community of the mediocre science produced by Germans.

With the outbreak of war, attitudes changed dramatically: most scientists on the Allied side believed that science conducted in German laboratories was used almost exclusively to prepare Germany for their military aggression. Indeed, Haber had turned his institute into a research center devoted entirely to military research, with his laboratory specifically devoted to poison gas development. After the war, the Allies organized meetings and conferences to determine how Germany should be treated regarding international scientific cooperation. At a meeting held in London on October 9, 1918, the Allies decided that Germany and the Central Powers should be excluded from international scientific collaboration based on their atrocities during the war. Arrhenius and others in the Allied countries agreed that the German model of science should be emulated, but with the removal of Germany, other countries could fill this model.

The award, therefore, to three physics and chemistry science prizes in 1919 to Germans was a shock to many outside Germany. If the Allied countries would be leading the world in rehabilitating science, why then did Sweden honor these three Germans? More specifically, why honor Haber, who was potentially on the war criminals list, and if not a war criminal, at least someone who had committed atrocities specifically outlawed by the Hague Convention rules? Several members of the Swedish science community and, more specifically, the Nobel Academy were strongly pro-German. Obviously, it was this latter faction that held sway—and the Nobel prizes in chemistry and physics in 1918 went to two Germans (Haber and Planck), and the physics prize in 1919 to Johannes Stark (German). Additionally, the 1920 chemistry prize went to Walther Nernst (German). There was no chemistry prize in 1919.

Haber had been nominated several times for the chemistry prize, first in 1912, then 1913, 1914, 1915, and finally in 1919. Each nomination was specifically for his synthesis of ammonia from atmospheric nitrogen, excluding in later years any mention of his work on poison gas and its practical application during the war, and the use of his ammonia synthesis process to produce nitrates for wartime munitions. By 1919, the Swedish Academy certainly knew of his potential war crimes. Why then did they decide to honor him with a prize?

Earlier, in 1916, the Nobel committee had finally decided that "Haber's method for synthesizing ammonia was now so refined that its practical value could no longer be disputed."[57] However, the committee still decided not to award the prize to Haber for at least two reasons. First, Haber's ammonia work was kept totally secret by BASF, allowing no one from the outside to evaluate the process. And because of this secrecy, the committee could not determine if Bosch or anyone else should be included in the prize. Second, the committee felt they could not in good conscience award a prize to someone whose discovery was being used as a weapon in the European war. The committee, therefore, decided not to award chemistry prizes in 1916 and 1917.

With several nominations in 1918, including two from members of the committee, Haber once again had to be considered for a prize. Peter Klason prepared the report for the committee. He stated that Haber's process for nitrogen fixation "would be of great benefit to all mankind," but its military use "had made it possible for the Central Powers to wage this long war."[58] The committee took a preliminary vote. The chairman, Professor Olaf Hammarsten of Uppsala University, opposed Haber's candidacy, stating that the situation had not really changed since 1916, and the reasons opposing an award were still valid. In fact, it had become clearer that Haber's process had contributed to a greater extent to the war than realized in 1916. It's interesting to note that Hammarsten was pro-German and had hoped for a German victory. However, he felt it was politically unwise for the committee to award the prize to Haber, and he made a formal reservation against a Haber prize in 1918. The committee report attempted to minimize the military impact of Haber's discovery and focus on Nobel's wish that the award would be for the greatest utility to humankind. In the end, the committee voted the nomination down.

In 1919, two pro-Haber members of the committee nominated Haber. For some reason, every member of the committee now voted to award Haber the reserved 1918 prize. The official statement was that new information had become available, putting Haber in a more favorable light. However, the chairman, Hammarsten, who was against an award for Haber said that there was still no new information about potential co-awardees, particularly Bosch. But, amazingly, at the last minute, Hammarsten changed his position, and voted for Haber, stating that this might be the last opportunity to honor him. In the end, the committee concluded that a decision concerning Haber had to be made in 1919; further postwar complications might make it impossible

---

[57] Widmalm (2012), p. 348.

[58] Friedman, R. M. (2001). *The politics of excellence: Behind the Nobel Prize in science*. W.H. Freeman and Company, p. 112.

to award Haber with a Nobel Prize. Pro-German sentiment on the Nobel committees most likely wanted to rehabilitate German science, and to reestablish the strong bonds between Germany and Sweden. Members of the committee stated to the press that their job was to award discoveries that would have the greatest benefit to mankind. "Haber's invention would greatly enhance food production; the fact that he had worked for the German Ministry of War was neither here nor there."[59]

Interestingly, the conversation returned to Arrhenius and his role in this controversy. He had made a formal reservation against any decision to award a German scientist with a Nobel prize. He still struggled to remain neutral and hoped that the Swedish scientists would likewise remain neutral. He feared that awarding Germans so close to the war would come back to hurt Sweden. A scientist and political leader on the left—Hjalmar Branting— agreed with Arrhenius and stated even more forcefully that the Swedish Academy "had shown itself unworthy to award international world prizes."[60] Arrhenius did achieve at least part of his goal to remain neutral and play a role in international science as a mediator between the Allied and Central powers. Sweden joined the International Research Council in March 1920, and Arrhenius could at last work for rehabilitation of Germany from inside this organization.

Haber was elated when he heard the announcement that he and two other Germans had won Nobel prizes. As he wrote in a letter to Willstätter:

> I think it was a deed of greatness on the part of the Swedish Academy to elect three Germans—and only Germans—as prizewinners. My heartfelt wish is that it may lead to renewed international understanding.[61]

It led instead to outrage from scientists from the Allied countries. Two French scientists refused to accept their prizes. Ernest Rutherford attended the 1920 Nobel ceremony but refused to shake Haber's hand. Theodore Richards, who won the chemistry prize in 1914, canceled his plans to attend the ceremony and stated that he wanted no contact with Haber until he and the other German winners publicly retracted their signing of the wartime Manifesto. The protests continued well into the 1920s, with many of the Allied laureates refusing to attend awards ceremonies. As one scientist stated: "It seems that

---

[59] Widmalm (2012), p. 252.
[60] Widmalm (2012), p. 355.
[61] Charles (2005), p. 196.

the Nobel Prize panel forgot that on April 22, 1915, Haber and his gas troop…were the first to unleash deadly chlorine gas."[62]

The outrage extended unfortunately to attacks on the Nobel committee itself. Hjalmar Branting, the soon to be prime minister—and a winner of a Nobel Peace Prize—stated in a newspaper article that was widely published that Haber's prize "served only to equate culture with barbarism." He continued: "Would a comparable French or British chemist, who had worked directly for the war effort, have been given a prize at this time while the sores are still dripping blood?"[63] Members of the committee spoke out, defending their position, claiming that Nobel's will only requires them to evaluate the scientific merit and benefit for humankind. The conservatives also believed that even though the imperial German government had fallen, "Germany's elite culture remained." Therefore, it was imperative that the Swedish Academy honor three Germans to honor that elite culture (which, of course, made no sense to the outraged opponents throughout Europe).

German scientists were elated that the committee had confirmed the superiority of German science. The dismay expressed by other countries extended far and wide. Many scientists were too shocked to comment, while editorials in scientific journals and newspapers expressed disbelief. Some wondered if it was possible that the committee didn't know about Haber's gas warfare, but the massive coverage of his wartime activities made this theory highly unlikely. As a final statement of outrage, one French newspaper reported: "Has then the Academy of Sciences in Sweden, shut behind Nordic mists, seen nothing, heard nothing, and understood nothing?"[64]

## The Postwar Years

At the end of the war, Haber returned to Berlin and his family, but by this time he was a psychological wreck. His marriage to Charlotte was failing, his country had been soundly defeated, and his efforts to win the war with his chemistry had failed. As one friend observed: "He was overwhelmed by the outcome of the war, and for several months, nervously exhausted."[65] Then, to make his life even more miserable, in 1919, he got word that his name was on war criminal lists because of his work on chemical weapons. He believed that

---

[62] Ronald, S. (2023). *Hitler's aristocrats: The secret power players in Britain and America who supported the Nazis.*, 1923–1941. St. Martin's Press.

[63] Friedman (2001), p. 113.

[64] Friedman (2001), p. 114.

[65] Hager (2008), p. 184.

if he stayed in Germany, he would be arrested and put on trial. He gathered up his family, bought a forged passport, and fled to Switzerland. Haber tried to disguise himself by growing a beard and basically stayed under the radar as much as possible. As talks of war criminals died down, Haber determined it was safe to return to Berlin and continue his career as the Director of the KWI for physical chemistry. Then the unthinkable happened—not only had he not lost his reputation as one of Germany's top chemists, but he was notified that he had won the 1918 Nobel Prize in chemistry for his synthesis of ammonia. There was no reference to his work on poison gas, gunpowder, or explosives. His belief that he would be held in disrepute for his poison gas warfare was overshadowed by his development of a process for producing ammonia.

Haber also hoped that Germany would one day reemerge as a powerful nation. And, against the stipulations of the Treaty of Versailles, he continued his work secretly on gas warfare and other banned projects. Unbelievably, he even tried to fund a KWI for Chemical Warfare to be directed by him but he was unsuccessful. He was also involved as an advisor in several gas weapons projects in Spain and Russia. Haber and the Kaiser, who was by then living in exile in Holland, also continued their plans in anticipation of a future revenge war against the Allies. In June 1927, the Kaiser wrote Haber that he was interested in a plan for gassing large cities. Many of Haber's former associates also continued to work on chemical weapons in the German military-funded Imperial Biological Institution for Agriculture and Forestry. In addition, during the war, Haber's group had worked on the development of insecticides, and had produced an effective chemical that they called Zyklon B. The parent chemical was Zyklon A; Zyklon B had been developed by adding an irritant to ensure that the farmers wouldn't be accidentally poisoned. Now, postwar, the scientists tested toxic chemicals including Zyklon B on mice, rats, rabbits, and guinea pigs. The mandate from the high officials was to test the potential new chemical weapons. After several failed ventures into selling these chemical weapons on the black market, the projects were shut down, and this was Haber's final work in the weapons underground.

One of the most quixotic projects that Haber undertook after the war was driven by his desire to help pay off the huge reparations imposed on Germany by the Allies. Haber had read a paper by Arrhenius that there were tiny amounts of gold in seawater, and he calculated that every ton contained about 6 mg of gold. Even if the calculations were inexact, Haber still believed that sufficient gold could be collected to pay Germany's debts. Therefore, in the spring of 1920, Haber gathered a few of his colleagues and announced his plan to mine gold from the ocean. He designed an ambitious project to take measurements of seawater to verify the concentration of gold, traveling first to the USA. He funded the installation of a secret laboratory in an ocean liner to take samples throughout the trip. The laboratory had its own power, water, and gas, with exquisite equipment to collect and store the samples. The group returned to Germany and found not 6 mg, but a disappointing level of one thousandth of that amount. Haber continued. He set sail for South America in 1923, again collecting and processing samples in his secret onboard laboratory. His team measured over 5000 samples. All measurements continued to show far less gold than Arrhenius had reported. After five years of trying, he gave up. He closed his notebooks and never talked about it again or published any of the data.

After the failure of his gold scheme, Haber focused on rebuilding his research group. The Institute became his life. He hired talented researchers and gave them the freedom to define their own physical chemistry projects. The institute became a mecca for physical chemists throughout the world. Important findings during these years covered such fields as colloids, chain reactions, charged particles, and gases. He started a regular colloquium featuring keynote speakers. One visitor called the institute "The very empyrean of science."[66] He became an elder statesman in the field of physical chemistry as his wartime infamy faded from memories.

The environment at home was a totally different situation. As Hager noted: "in his institute he was a genial father figure" while at home he was a demanding tyrant.[67] Once Haber left work, he was exhausted and appeared unhappy with his life. By 1927, his marriage to Charlotte was in serious trouble. They had a second child (Ludwig) in the early 1920s, but that seemed to make the situation worse. Charlotte tried hard to keep Haber connected to her and the children, but it was no use. He was drawn to his work, his friends, and long-established habits that he was simply too old to give up. The marriage and its acceptance by others were complicated. From comments made by friends, it appeared that Charlotte was unliked because of her boisterous manner, the age difference between her and Haber (20 years), and the way she continued to pull him away from what he truly enjoyed in life. Some of Haber's friends were still loyal to Clara and felt that Charlotte was trying to erase that memory. There was also conflict with Hermann, Clara's son. Charlotte tried one last time to save their marriage. She proposed a six-month cruise around the world, taking Haber even further away from the life he knew and loved. The attempt backfired, and finally in 1927, Haber wrote Charlotte a letter:

---

[66] Hager (2008), pp. 215–218.
[67] Hager (2008), p. 215.

> Your friends are not my friends; your inclinations are not mine. Even when we are together, we live for our own individual selves, and your attempts to change are as futile as mine. Let us call ten years enough. I can't do it anymore.[68]

Haber and Charlotte divorced in 1927, and Charlotte and their two children moved to England.

At the beginning of 1933, Haber's life, like that of many Jews, changed forever. Adolph Hitler became chancellor in January of that year, and on April 1, 1933, the Nazis called for a nationwide boycott of Jewish businesses. The elimination of the Jews was just beginning. The rules decreed that anyone who had a grandparent who was Jewish was themself Jewish and would be treated as such. Then Hitler's "Law for Restoration of the Professional Civil Service" demanded the removal of anyone in government service who was not Aryan. This applied to all Institutes within the Kaiser Wilhelm Society (KWS), including Haber's Institute. Haber was spared (at least for a while) since he had served in World War I. But he would have to fire all his Jewish employees, almost 25 percent of his institute staff. Haber was stunned, too outraged to do anything for several days. Several of his scientific colleagues who were Jewish had already decided to leave Germany (Einstein and Franck fled to the USA) while others stayed at least for a while, perhaps too long, as in the case of Lise Meitner. Finally, Haber resigned, with the request that he be allowed to stay long enough to clean up the Institute's business, identify a successor, and find positions for many of his Jewish staff.

It was at this point in Haber's life that he realized that he had created a false existence for himself and for his family. He was baptized a Christian in 1892, but no baptismal service would ever erase the fact that he was a Jew. Everything he had done for his beloved Germany was not enough: his Nobel Prize, his Iron Cross, all his scientific efforts to save his country. All were gone; he was a Jew. In an effort to console his friend, Einstein put his dilemma into scientific terms: "It is somewhat like having to abandon a theory on which you have worked your whole life. It's not the same for me because I never believed in it in the least."[69] In a moving statement to his friend and colleague, Haber wrote: "I was German to an extent that I feel fully only now, and I'm filled with incredible disgust."[70]

Haber's goal after resigning from the institute was to find another position outside Germany. He sent out letters to several European cities, as well as

---

[68] Charles (2005), p. 213.

[69] Hager (2008), p. 241.

[70] Hager (2008), p. 241.

Palestine. The responses were few and slow in coming. His colleagues knew that it would be difficult to find a suitable position since Haber had become a top administrator, unused to working in a laboratory. Even more significant perhaps was his lingering reputation as a war criminal. Finally, in the fall of 1933, he was offered a position at the University of Cambridge, with a laboratory and no teaching responsibilities. He was elated; he knew that by accepting this position, he could also qualify for British citizenship. After the way he and his colleagues had been treated in Germany, he stated that one of his "most important goals in life [is] that I do not die as a German citizen..."[71] He also continued conversations with Chaim Weitzmann who was establishing a new center in Palestine and was urging Haber to consider a position there. Haber first traveled to England to learn more about their offer. Faced with the cold and rainy climate, and the hatred still felt by many toward him because of his gas warfare, Haber decided he needed the warmer, more inviting climate of Palestine. He succeeded in getting only as far as Switzerland when he had a major heart attack. He was carried off the train and died later that evening.

There was a brief ceremony in memory of Haber on February 1 at the crematorium of the Hörnli Cemetery in Basel. Richard Willstätter came from Munich to deliver the eulogy. Haber had written his will in 1933 that he wanted to be cremated with his ashes buried in the cemetery in Dahlem where Clara's ashes were. He realized that the current situation in Germany might make it impossible for this to happen, but he hoped that once it was safe, some person in his family should try to make this happen. He also wanted his gravestone to have his name and the date of his birth (December 9, 1868) and his death. He also suggested, if possible, to add "He served his country in war and peace as long as was granted him."[72] On September 29, 1934, Haber's urn was buried in the free section of Hörnli Cemetery, and Clara's urn was subsequently brought from Germany and buried in the same grave.

Following the war, Max Planck was the grand old man of physics and the most revered scientist in Germany. Planck had won a Nobel Prize in 1918. He was head of the Kaiser Wilhelm Society (KWS), and he hated the Nazis. He had done what he could to direct science but avoided outward obedience to the party he hated. However, there was one duty he felt compelled to do, even if it meant going against the Nazis and facing potential punishment. It had been over a year since Haber had died, with his ashes left in Switzerland. Planck felt strongly that it was insulting to Haber's memory that so little had

---

[71] Hager (2008), p. 247.
[72] Stoltzenberg (2004), p. 300.

been done to memorialize his work. So, in January of 1935, Planck decided that the KWS would sponsor an anniversary memorial. Planck had invitations sent out for the event to be held in Dahlem on January 29, 1935, exactly one year after Haber's death. The Ministry of Education immediately sent out an edict that no German state employee could attend. Interestingly, the event would almost coincide with the anniversary of Hitler's appointment as chancellor on January 30, which had been recognized in the last several years as a national holiday and celebration. An event celebrating a Jew so close to this Nazi holiday would be unconscionable. Other Nazi Ministers followed the education minister, and it looked like no one would come. The ministry hoped that the event would have essentially no attendees so that the world would see how few Germans cared about a dead Jew. Planck forged ahead, arranging for several speakers in a hall that would seat 500 within the KWI. It was amazing—even with Nazis standing at the door ready to jot down names of those attending—the hall filled up. Bosch was there with many of his colleagues who knew Haber; Lise Meitner sat near the back; Otto Hahn gave a eulogy. There were also many who simply were not willing to take the risk of attending: Haber's son Hermann; Einstein and Franck who were in the USA; and Max von Laue who was only a short walk away but one of those banned civil servants. Most remarkable was the fact that the room seemed to be filled with mostly women: these were the wives of all the scientists who felt that they couldn't risk attending. Planck ended the service with these words: "We reward loyalty with loyalty, and dedicate this hour to the honor of Fritz Haber: a great scholar, an honorable man, a fighter for Germany."[73] No local newspapers covered the event, but the New York Times noted that leading German scientists had gathered to honor a great colleague in spite of the Nazi threats.

Finally, a tragic irony to the Haber story is his involvement in the synthesis of the insecticide Zyklon A. Workers outside Haber's group had continued to study Zyklon A, and refined the chemical so it was easier to handle, and called it Zyklon B. For 20 years, the chemical was used for eradicating insects, it's original purpose. Then the Nazi SS realized that this could be the method that they were looking for to eradicate not insects, but Jews. Zyklon B soon found a place in the showers at the concentration camps, and it became "a tool of death on a scale beyond all normal imagination."[74]

---

[73] Charles (2005), p. 245.
[74] Charles (2005), p. 246.

# Epilogue

## Carl Bosch

> Carl Bosch…was a man of contradictions: a business mogul who won a Nobel Prize and an ardent anti-Nazi who founded and led the most infamous Nazi [industrial war plant].[75]

At the end of World War I, Carl Bosch (Fig. 2.12) was ready to switch his factories at Leuna and Oppau from making nitrates for war munitions to producing ammonia for fertilizer. The farmers were hurting, and food for the German populace was in short supply. A more important issue was that BASF and its patents and factories were in danger of being taken over by the French. Leuna was safe for now, but Oppau was just inside the French zone. Bosch knew that even if the French captured the patents, this would not be enough for them to replicate Oppau, but if they seized the factory and reviewed its operation, then the ammonia secret would be out. BASF and Germany would lose its hold on the secret process, and could lose a huge amount of money, sorely needed as post war Germany attempted to rebuild. Bosch assumed that the fate of Oppau would be a part of the peace negotiations. What he didn't

**Fig. 2.12**   Carl Bosch

---

[75] Hager (2008), p. xv.

realize until he became part of the German delegation to the peace conference in Versailles was that Germany was not there to negotiate the peace: they were there to accept the peace. Bosch decided at that point that he had to act quickly or everything he had worked for would end up in French or British hands. So, during the Versailles meeting, he met secretly with a French group to negotiate a deal that would hand over all the knowledge needed to build a Haber-Bosch plant (so the world would now have his secret), but he and BASF would keep Oppau and Leuna. For now, the plants were secure, and both continued production of ammonia to feed the population.

By 1923, Germany had stopped paying reparations to the Allies. The Weimar government said it was a matter of survival and would begin its repayment once inflation was under control. Some of the nations agreed to give Germany some time, but France did not. They marched into Germany's heartland, taking everything they could. Bosch once again shut down all machines in Oppau and refused to comply with French demands. BASF (now Farben) executives including Bosch were tried by the French in absentia, and Bosch was found guilty, fined, and sentenced to eight years in prison. Bosch fled to Heidelberg, just out of French jurisdiction. When the French realized it was losing huge amounts of money, they gave up and returned home, and Bosch restarted his plants. By 1924, after the French had left, fertilizer was again selling, and Farben was making a profit.

In late 1923 Bosch traveled to the USA to examine the Standard Oil model and determine if Germany could replicate this model. He came back with two important insights. First, in contrast to Germany which had few cars, the USA had major highways jammed with cars. Bosch recognized that the real money for Germany was not in making automobiles but in producing what made them go: he wanted Germany to produce synthetic gasoline. The second insight was that BASF was a small business compared to Standard Oil; Bosch wanted to replicate this giant. Bringing all the businesses together in one large conglomerate could enhance the ability to finance Bosch's ideas. Hence, the German pharmaceutical companies, dye industries, and ammonia plants all coalesced into the overarching conglomerate called *Interessengemeinschaft Farbenindustrie Aktiengesellschaft*, literally the Interest Community of the Dye Industry, Inc. This was shortened to IG Farben or just Farben. Bosch was named its director in 1925. Now he was positioned to retool one of his plants to make synthetic gasoline.

Bosch and a collaborator had worked out a way to make gasoline from coal; this would be Farben's new product that would bring in huge profits. But, just as Farben was beginning to ramp up production of synthetic fuels, two disasters happened in close succession. First, in the late 1920s, a new oil field was found in Oklahoma. Then in 1929, the Great Depression began, hitting Germany especially hard. Farben income dropped precipitously. By 1930, income from ammonia fell by one third, and by 1933 by half again. Bosch was able to continue some production from his plants, at least keeping total collapse from happening. Then in 1931, Bosch was notified that he had won a Nobel prize for his high-pressure chemistry.[76] Even though he had been overlooked in 1918 for a prize with Haber, he was now a Nobel laureate. His good fortune was short-lived, as the era of Hitler began in 1933.

Bosch knew that he would have to make deals with the Nazis to keep his plants running, but he was against the Nazi regime and made no effort to hide his feelings. The secret police had gathered information about Bosch for Hitler. They reported that Bosch "was a typical southwest German liberal, a highly educated friend of the Jews, but a man who…put business and science above ideology."[77] In a one-on-one meeting with Hitler early in the Nazi reign, Bosch attempted to protect his Jewish employees. He talked about the civil service law and the damage it would do to German science if the Jewish chemists and physicists were forced to leave. Hitler merely shouted, "Then we'll just have to work one hundred years without physics and chemistry."[78]

Bosch made one more anti-Nazi move. When he heard about Haber's resignation in 1933, he wrote a note to his former collaborator:

I heard with great regret in Berlin how very oppressed you feel personally by the present circumstances…I myself have tried everything possible…to make the measures against [my] scientists somewhat bearable…If I can be of any assistance to you somehow, then I am naturally gladly at your disposal.[79]

Haber replied, stating that he saw his retirement as a relief and thanked Bosch for his note. He also indicated that he was looking for another position and hoped that Bosch would help him secure a position.

By 1935, it was obvious that Bosch was wearing down. Farben eventually removed him as director and gave him an honorary position. Bosch retreated to his Heidelberg estate and began drinking heavily. One of Bosch's last public speeches was at the *Deutsches Museum* in Munich in 1939. He appeared on the podium staggeringly drunk, and proceeded to declare that science must be free from government interference. He did not give the Nazi salute and mentioned Hitler only once. Although all the attending Nazis yelled at Bosch and

---

[76] Hager (2008), p. 232.
[77] Hager (2008), p. 244.
[78] Hager (2008), p. 245.
[79] Hager (2008), p. 246.

stomped out of the room, Bosch was not arrested or publicly condemned. It was not worth the headlines. Bosch's depression and drinking both increased, and he fell completely out of public view. In April 1940 he had lost any will to live, burned all his personal correspondence, and declared that he didn't want to go on. He died on April 27, 1940, a broken and forgotten man.

## The Heirs

They were each fighting for Fritz Haber. First Charlotte had the upper hand, then Hermann. You know, it was Hermann who really suffered.[80]

Hermann, the son of Haber and his first wife Clara, married Margarethe Stern in 1926 and the family relocated to Paris. They would eventually have three daughters. To stay in Paris, Hermann applied for citizenship, which was turned down. Since his father had been called a "good German" during his memorial service, the French rather oddly decided that "the loyal son of any good German must certainly be disloyal to France." It seemed that Hermann was doomed to remain in the shadow of his father. When World War II broke out, Hermann had two choices: he could serve in the Foreign Legion or be sent to an internment camp. His family fled to the south of France, and Hermann joined them later. In 1941, friends were able to get the family to the USA. Tragically, Hermann's wife Marga died shortly thereafter of leukemia, and Hermann, who could not shake his depression, committed suicide in 1946. This was the second suicide from the Haber-Clara marriage. Little is known of the three granddaughters. Eva spent many years in Kenya before returning to England, where she died in 1950. Claire, the oldest daughter, committed suicide in 1949, the third suicide in the Clara/Haber legacy.

Haber and his second wife Charlotte had two children: Ludwig (Lutz) and Eva Charlotte. Charlotte and the children moved to England around 1933, and after the war, both children became British citizens. Ludwig (Lutz) became a historian and has written a book about his father's development of poison gases entitled *The Poisonous Cloud*. He passed away in 2015 in England. Haber's daughter Eva Charlotte Lewis was eight years old when her parents divorced, and she had an image of him as "an old man who needs assisting going upstairs, who must never be disturbed…" She of all the children seems to be the one who escaped the shadow of her father. She married, studied agricultural science, worked on a farm in Kenya, then settled in Bath. She died in 2015.

---

[80] Charles (2005), p. 251.

## Dramatizations and Fictionalizations

Fritz Haber's discovery of the fixation of nitrogen has been labeled as perhaps the most impactful and important chemical discovery of the twentieth century. This process has provided fertilizer for the world's farmers and is credited with saving millions of lives. Yet, Haber has been largely forgotten in the popular or scientific press. There are several excellent biographies that have been widely referenced in this chapter, but there are few other dramatizations of Haber's story. A December 2013 BBC radio program underscores this with its title "Why has one of the world's most important scientists been forgotten?" Even in many of the popular books and plays that discuss Haber, it's more often in the context of his relationship with another famous scientist, frequently Einstein, including the play *Einstein's Gift*,[81] and the National Geographic production of *Genius*.[82] Other books and plays focus much of the story on Haber's gas warfare endeavors or his wife's suicide or both, with little reference to the fixation of nitrogen. In the 2012 *Radiolab* podcast entitled "How do you solve a problem like Fritz Haber", the producers address the issue of how one reconciles the idea of a bad person who does great good or a good person who does terrible harm. Finally, two of the most interesting or unusual productions focusing on Fritz Haber include a novel and a song. The first, *The Reunion of Ghosts*,[83] fictionalizes Haber and the suicides within his family. The second is a song entitled *Father*[84] by the Swedish metal band Sabaton that includes references to Haber's gas warfare and nitrogen fixation. Two of the verses from this song follow:

> Father of toxic gas and chemical warfare
> His dark creation has been revealed
> Flow over no man's land, a poisonous nightmare
> A deadly mist on the battlefield
> "Perversions of ideals of science"
> Lost words of alienated wife
> And in the trenches of the western front
> Unknowing soldiers pay the price

---

[81] Thiessen, V. (2015). *Einstein's Gift*. Playwrites Canada Press.

[82] *Genius* (2012). Season 1, episode 7. https://www.imdb.com/title/tt6012932/.

[83] Mitchell, J.C. (2015). *A Reunion of Ghosts*. Harper. https://www.imdb.com/title/tt6012932/.

[84] Sabaton (2022). *Father*. https://www.sabaton.net/historical-facts/fritz-haber-was-born/.

# The Haber Legacy

His life was a stunning mix of triumph and tragedy. We remember him as a great, flawed, and enigmatic figure of 20th-century chemistry and history.

If one simply enters the name Fritz Haber into an internet search, the descriptors that appear overwhelmingly emphasize the duality of Haber: Dr. Fritz and Mr. Haber; the monster who fed the world; experiments in life and death; creator of good and evil; the amoral scientist. Haber was a highly revered scientist who discovered a method to fix ammonia and feed millions, but he also was an extreme German patriot who truly believed that science in war belongs to the Fatherland and in peace to humanity. There are many examples of colleagues who seemed to ignore his use of chemistry to fuel Germany's war effort: he was awarded a Nobel Prize and his memory lives on in both the renamed Fritz Haber Institute as part of the Max Planck Society, and the Fritz Haber Center for Molecular Dynamics in Israel. And yet there are others who still remember him as the Father of Chemical Warfare and perhaps unfairly associate his work with the chemical used to exterminate millions of Jews. How then do we think about this man and his science?

Roald Hoffmann, a Polish American theoretical chemist who won a Nobel Prize in 1981, has also published plays, poetry, and philosophy. In a book entitled *The Same and Not the Same*, Hoffmann stresses the themes of opposites: help or hurt, harm or benefit,[85] what he terms a cauldron of polarities. As Hoffmann states: "the chemist would like the world to leave him alone, but it has its own ways of touching him….In no case…has this fact been …played out with more drama, than in the life of…Fritz Haber."[86] Hoffman underscores the primary goal of Haber and his eventual fixation of nitrogen to produce ammonia for its use as a fertilizer. Most experts would agree that synthetic fertilizer has probably saved millions, maybe hundreds of millions of lives. Then the world intervened. Haber, being the ultimate "good" German, figured out a way to retool the ammonia factory to make nitrates for gunpowder, allowing Germany to extend the war beyond a year, with many more war-related casualties. And to make matters considerably worse for Haber, he decided to create the ultimate horrific weapon (poison gas) to win the war for Germany. In the latter he failed, but he had in fact moved from bread from the air to gunpowder to poison gas. One might ask: is it worse to die from gas than from shrapnel or other traditional weapons? As Hoffman suggests, there

---

[85] Hoffmann, R. (1995). *The same and not the same*. Columbia University Press, pp. xiv–xv.
[86] Hoffman (1995), p. 167.

is "something in the psyche, something deep that associates life with breath; that is the difference. And here we have the "cauldron of polarities."

Haber, perhaps the greatest physical chemist of the twentieth century, used his great intellect to create one giant benefit for mankind and one horrific weapon that would have tragic effects on his personal and professional life. Cornwell describes Haber as the "Janus-faced power for good and evil",[87] and the Museum of the Jewish People describe him as "Dr. Fritz and Mr. Haber."[88] As the quote above points out, his work was a prime example of the scientist in peacetime versus war. As one article states, Haber exemplified "both sides of science … Fritz Haber is the Jekyll and Hyde of the chemistry world."[89] The controversy that surrounds Haber continued into the twentieth century. In 1968, the University of Karlsruhe honored the centenary of his birth. The commemorative ceremony was interrupted by students who unfurled a banner reading[90]:

Celebration for a Murderer
Haber = Father of Gas Warfare

## Bibliography


Charles, D. (2005). *Master Mind: The Rise and Fall of Fritz Haber, the Nobel Laureate Who Launched the Age of Chemical Warfare*. Harper Collins.

Cornwell, J. (2004). *Hitler's Scientists: Science, War and the Devil's Pact*. Penguin Books.

Friedman, R. M. (2004). *The Politics of Excellence: Behind the Nobel Prize in Science*. Henry Holt and Company.

Goran, M. (1967). *The Story of Fritz Haber*. University of Oklahoma Press.

Haber, L. F. (1986). *The Poisonous Cloud. Chemical Warfare in the First World War*. Clarendon Press.

Hager, T. (2008). *The Alchemy of Air*. Harmony Books.

Hoffmann, R. (1995). *The Same and Not the Same*. Columbia University Press.

Stolzenberg, D. (2004). *Fritz Haber: Chemist, Nobel Laureate, German, Jew*. Plunkett Lake Press.


---

[87] Cornwall (2003), p. 47.

[88] Ushi (2024).

[89] *Fritz Haber*. (2012, May 18). Scicommstudios. https://scicommstudios.wordpress.com/2012/05/17/fritz-haber/.

[90] Charles (2005), p. 250.

# 3

## The Atom Is Split

### Otto Hahn—1944 Nobel Prize in Chemistry
*for his discovery of the fission of heavy nuclei*

*Sixteen hours ago an American airplane dropped one bomb on Hiroshima, an important Japanese Army base. That bomb had more power than 20,000 tons of T.N.T. It had more than two thousand times the blast power of the British "Grand Slam" which is the largest bomb ever yet used in the history of warfare… It is an atomic bomb. It is a harnessing of the basic power of the universe. The force from which the sun draws its power has been loosed against those who brought war to the Far East… We are now prepared to obliterate more rapidly and completely every productive enterprise the Japanese have above ground in any city. We shall destroy their docks, their factories, and their communications. Let there be no mistake: we shall completely destroy Japan's power to make war…*

Harry S. Truman, President of the United States

August 6, 1945

Two months after the dropping of the first atomic bomb, with the horrors of war and the images of the mushroom clouds still fresh in everyone's minds, the Swedish Academy announced the awarding of the 1944 Nobel Prize in Chemistry to Otto Hahn, the German scientist who discovered fission in 1938. Insulated from the ravages of war by their neutrality, the Swedes could be perhaps more dispassionate than others in selecting a winner who, in Nobel's words, "during the preceding year, shall have conferred the greatest benefit to humankind." Even for those who believed that "all knowledge is good,"[1] this was a truly remarkable choice: awarding the prize to someone whose discovery had led directly to the practical application of fission—the atom bomb—that killed a hundred thousand Japanese. Beyond that, the decision by the Academy to ignore the enormous contributions to this discovery by Hahn's close colleague Lise Meitner is equally controversial and reflects both the effects of Nazi policies and the politics within the Swedish Academy. What follows is the fascinating story of one of the most impactful scientific collaborations, involving the German chemist Otto Hahn and the Austrian Jewish physicist Lise Meitner. Their story swirls amongst their personal challenges, the rise of Germany's dictatorship, and the emerging age of nuclear physics, leading ultimately to the discovery of fission and one of the most controversial Nobel prizes ever awarded in the sciences.

## The Rise of Germany

Until 1871 Germany was a collection of dozens of small states, remnants of the Holy Roman Empire that had existed for over 900 years. In a series of successful wars with Denmark, France, and Austria, Otto von Bismarck unified Germany, and Wilhelm, King of Prussia, became Kaiser Wilhelm I. Berlin, the capital of Prussia, became the capital of Germany, and under the Kaiser, it grew from its agrarian beginnings to become the center of culture and science in Europe. As one British expert exclaimed in 1906: "Berlin represented the most complete application of science, order, and method of public life; it is a marvel of civic administration, the most modern and most perfectly organized city that there is."[2] And as described by Matthew Stanley: "Germany had become a world empire with an enormous army and a streamlined economy whose intellectual and cultural institutions dominated the continent. There seemed to be no better model for modernity."[3]

Berlin was moving quickly from an agrarian society to an industrial city. In 1902, the U-Bahn subway was completed, and several museums were built. The city was rapidly becoming a major musical center as well as dominating the German theater scene. In 1892, while attending a lavish celebration in Berlin for two scientists (Rudolph Virchow and Hermann von Helmholtz), Mark Twain wrote an intriguing column for the Chicago Daily Tribune

---

[1] Huxley, T.M. *Thomas Henry Huxley quotes on knowledge.* Retrieved September 18, 2021, from https://todayinsci.com/H/Huxley_Thomas/HuxleyThomas-Knowledge-Quotations.htm.

[2] Ashworth, P.A., & Phillips, W.A. (1911). Berlin. In H. Chisholm (Ed.), (11th ed, pp. 785–791). Encyclopedia Britannica. Cambridge University Press.

[3] Stanley, M. (2020). *Einstein's war: How relativity triumphed amid the vicious nationalism of World War I.* Penguin, p. 24.

entitled "The Chicago of Europe."[4] Twain commented that Berlin looked new, it felt spacious, the street grid was straight and rational, and streetcars and subways crisscrossed the city. Like Chicago, Berlin was exploding in population and had grown to over two million people by 1905. Berlin had also become a Mecca of international science. In the first 20 years of the Nobel Prize, years which were dominated by major advances in physics and chemistry, German or German-speaking scientists had won half of all science prizes.[5]

## Atomic Physics Comes of Age

In the early 1900s, the universities in Berlin began to attract students and professors who flocked to the city to attend conferences and seminars, and study with some of the best-known scientists of the era. Both Hahn and Meitner trained in the area of atomic physics and radioactivity and were eager to continue their research in Berlin, the city that was quickly becoming the science center of Europe and perhaps the world. By 1907 when they arrived in Berlin, there had been significant recent progress in understanding the atom, but the details of atomic structure were still a mystery. The concept of atoms was first proposed by Leucippus and his student Democritus in the fifth century B.C.[6] They suggested that matter is composed of atoms separated by empty space. The atoms are solid and indivisible; there are different kinds of atoms that differ in size and shape; and the properties of matter reflect the properties of their atoms. In the eighteenth century, Isaac Newton expressed his version of the atom, which was similar to Democritus' theory, in his book Optiks (Query #31):

> All these things being considered, it seems probable to me that God in the Beginning form'd Matter in solid, massy, hard, impenetrable, moveable Particles, of such Sizes and Figures, and with such other Properties, and in such Proportion to Space, as most conduced to the End for which he form'd them; and that these primitive Particles being Solids, are incomparably harder than any porous Bodies compounded of them; even so very hard, as never to wear or break in pieces; no ordinary Power being able to divide what God himself made one in the first Creation.[7]

---

[4] *Chicago Tribune* (1982, April 3), p. 9. Newspapers.Com. https://www.newspapers.com/image/349852542/?terms=The%20Chicago%20of%20Europe&match=1.

[5] Cornwell, J. (2003). *Hitler's scientists: Science, war, and the devil's pact.* Penguin Books, p. 40.

[6] *Leucippus and Democritus.* (n.d.). http://chemed.chem.purdue.edu/genchem/history/leucippus.html.

[7] *The Newton project: The third book of optics.* https://www.newtonproject.ox.ac.uk/view/texts/normalized/NATP00051.

It wasn't until 1803 that John Dalton, a British chemist, meteorologist, and physicist proposed another significant piece of the atom puzzle. In a series of postulates, he attempted to describe matter in terms of atoms and their properties. In his first postulate, he reiterated Democritus and Newton: all matter is made up of atoms that are indivisible and indestructible building blocks. He next proposed that all the atoms in an element are identical in mass and properties, and compounds are a mix of two or more different kinds of atoms. It wasn't until the groundbreaking work of J.J. Thomson that anyone thought of atoms as anything more than indivisible blocks. In 1897, he discovered the first subatomic particle. In his experiments with electrical discharges, Thomson observed that the cathode rays traveled much faster through air than expected for atom-sized particles. He concluded that the rays were actually extremely light, negatively charged particles that were a new basic building block of atoms. He called these electrons. Based on these findings he proposed a "plum pudding" model for the atom in which the newly discovered negatively charged particles (electrons) were embedded in a sphere of positively charged matter.

Research in the new area of radioactivity further changed how scientists viewed the structure of the atom. In 1896, Henri Becquerel made the serendipitous discovery that when photographic plates were exposed in the dark to uranium salts, a fuzzy image appeared on the plates. He reported to the Academy of Sciences that the uranium salts had emitted "radiation" with no stimulation from sunlight. In Paris, Pierre and Marie Curie began to study this strange new observation and found that several other elements also emitted radiation. Marie Curie coined the term "radioactivity" to describe this new phenomenon. In 1898, Ernest Rutherford examined these new rays in more detail. He distinguished two types of radioactivity—he called them alpha rays and beta rays—that differed in their ability to penetrate ordinary objects. In 1902, Rutherford and his colleague Frederick Soddy demonstrated that atoms of a radioactive element transform into new elements, and published their theory of atomic disintegration. In further experiments, they soon identified these radiations with known particles: beta rays were shown to be electrons by Walter Kaufmann in 1902, and alpha rays were shown to be helium nuclei by Rutherford and Thomas Royds in 1907. By this time, when Hahn and Meitner entered Berlin in 1907, there had been significant progress in understanding the atom, but the details of atomic structure were still a mystery.

# Hahn and Meitner: The Early Years

**Otto Hahn**. Otto Emil Hahn was born in 1879 in Frankfort am Main, Germany, the youngest son of a successful glazier. He attended Klinger High School, where the "entrance exams and requirements were less severe."[8] Until he was 14, Hahn dealt with a multitude of ailments, including asthma, diphtheria, and pneumonia, but he appeared to outgrow these ailments in adult life. He did, however, deal with what might have been depression, suffering nervous breakdowns throughout his adult life. At the age of 15, he began to dabble in chemistry, performing experiments in his home. His father had hoped that Hahn would become an architect, but he had no "talent for drawing, and no artistic imagination whatever."[9] He asked his father's blessing to follow chemistry, not architecture, and his father relented. Chemistry was a rather odd choice for Hahn, since he had not excelled in this subject, and by his own admission, was drawn more to non-science subjects.

Based on a recommendation from a friend, Hahn entered the University of Marburg in 1897 to begin his doctoral studies (comparable in the USA to an undergraduate degree). Hahn related in his autobiography that as a young man (Fig. 3.1) his "attention to science was not very noticeable" and his days were filled with the "duty of drinking beer."[10] He did spend time learning his major (chemistry), but little time on the minor subjects of physics and math. As he later stated: "The end of the closing century was spent in the usual training course of young chemistry students who had no further ambition…Thus my time as a student passed carefree…because [I] believed that for a position in industry it was not necessary to be cultivated in more than…chemistry."[11] At the end of his second year, he changed universities and spent two semesters at Munich. He intended to study higher-level chemistry but spent much of his time on lectures on art and art history. He then returned to Marburg, where he completed his doctorate in 1901. After completing his doctorate, Hahn served the required year of military service in Frankfurt and left the army as a vice sergeant major.

---

[8] Hahn, O. (1966). *Otto Hahn: A scientific autobiography*. Charles Scribner's Sons. p. 5.

[9] Hahn (1966), p. 5.

[10] Hahn (1966), p. 7.

[11] Hoffmann, K. (2001). *Otto Hahn: Achievement and responsibility*. Springer. p. 17.

**Fig. 3.1** Otto Hahn—college years

Hahn's goal at this point was to continue his chemistry studies in preparation for a career in industry; he had no interest in an academic career but spent two more years working with his former professor, Theodor Zincke. At the end of his studies in 1904, Hahn received an offer of employment from Eugen Fischer, the director of Kalle and Co., a major company in the dye industry in Biebrich, Germany. A condition of employment was that he had to live in another country for a short time to gain a reasonable command of another language. With this in mind, Hahn accepted a post at University College in London in 1904, working under Sir William Ramsay. Hahn moved to London in 1904 and entered the famous University College in Gower Street. As a young scientist, Hahn was acutely aware of the honor that he had been given to work with Sir William Ramsay, a giant in the chemistry world who won a Nobel Prize in 1904 for his work on inert gases. Hahn requested a project to work on and was assigned the task of separating radium from barium chloride. He couldn't know in 1904 that this same separation process would be the key to his Nobel Prize work three decades later. Hahn, the organic chemist, knew nothing about radioactivity, nor about the hazards associated with working with radioactive compounds, but he learned the skills and tools needed to complete the project. In the process, he discovered a new radioactive substance which he called radiothorium; this was his first major achievement in the new field of radiochemistry.

With this discovery, Hahn became intrigued by radiochemistry and decided that he needed to know more about the subject. He wrote to the leading expert at the time—Ernest Rutherford in Montreal—and requested a position as an assistant (Hahn's parents paid his expenses). In 1905–1906, Hahn worked in Rutherford's group at McGill University and began a project to study the emission of alpha particles from radiothorium. During these studies, he identified several new elements that he called radium D (later identified as lead $Pb^{210}$), and thorium C (later identified as polonium $Po^{212}$). It must be understood that these new "elements" were present in such exquisitely small amounts that they couldn't be seen or felt or weighed by the most sensitive instruments. They could be detected and identified only by their radioactivity. But as Rutherford later remarked: "Hahn has a special nose for discovering new elements."[12]

In 1906, Hahn moved to Berlin to join the Chemistry Institute at Berlin University, which had opened in 1900 with Emil Fischer as the director. Hahn was appointed as an *Assistent* in the institute with a small salary and no teaching duties. Fischer provided Hahn with a workshop in the back of the Institute to begin his chemistry research, and in 1907, Hahn discovered yet another radioactive substance that he called mesothorium. At this point, Hahn began moving up the traditional academic ladder, preparing for promotion from *Assistent* to *Privatdozent* (instructor); and, after written and oral examinations, to the *Habilitation*. As part of his *Habilitation*, Hahn submitted a total of 11 publications. He conducted research and gave lectures, supporting himself through the fees from his students. In spite of his early disinclination toward academia, by 1907, Hahn had become a well-established young professor in Berlin, focused on advancing his career in academia rather than industry. It is worth noting that some of his colleagues were less than impressed with his credentials for the title of Professor. His mastery of the techniques of radiochemistry was unimpeachable, but Professor Walther Nernst, for example, was not convinced that Hahn had demonstrated "the capability…in conducting original research independently."[13] Another colleague stated: "it is unbelievable what qualifies as a university lecturer nowadays."[14] Nevertheless, as will be seen throughout this discussion, Hahn's career continued to progress, all the way to the top recognition: the Nobel Prize. And in his later life he became what might be described as one of the most revered scientists in post-World War II Germany. Certainly, some of his training prepared him for his ascendance in academia, but perhaps most important in his reaching the Prize was his 30-plus years of collaboration with one of the top physicists in the early twentieth century: Lise Meitner.

---

[12] Hoffman (2001), p. 39.

[13] Hoffmann (2001), p. 45.

[14] Hoffmann (2001), p. 45.

**Lise Meitner.** Elise Meitner (later changed to Lise) was born in Austria on November 7, 1878, the third of eight children in a gifted and liberal Viennese family. Her father was one of the first Jewish lawyers admitted to practice in Austria. All of her siblings ultimately pursued an advanced education. As an adult, she converted to Christianity and was baptized in 1908. Her young adult years from 1892 to 1901 were described by Meitner as her "lost years" as she struggled to get an education (Fig. 3.2). For women, formal education ended at the age of 14; according to Austrian law, that was sufficient for a woman to become a proper wife. By the time girls were admitted to high school in 1899, it was too late for Meitner. In 1899, she earned a certificate to teach French, to make sure that she could support herself. Over the next two years, studying with Arthur Szarvassi, a young assistant at the University of Vienna, she completed eight years of school work and was eager to put her new knowledge of math and physics to work. In 1901, after passing the university entrance exam, Meitner enrolled in the University of Vienna at the age of 22, and in 1906 became the second woman in Austria to earn a physics doctorate. Unfortunately, there still was little hope of obtaining a position in physics, and she began to think that she might never work as a scientist. Following up on her father's suggestion, she signed up as a practice teacher at a local girls' school. By day she would teach, although she disliked teaching, and in the evenings, she would conduct experiments in the relatively new and exciting field of radioactivity.

**Fig. 3.2** Young Lise Meitner

In Vienna, Ludwig Boltzmann, the theoretical physicist who discovered the meaning of entropy, had taught all her courses in physics. He had a lasting impact on Meitner. Later in her life she commented that Boltzmann's lectures were "the most beautiful and stimulating that I have ever heard…one left every lecture with the feeling that a completely new and wonderful world had been revealed."[15] After completing her coursework, she began her doctoral research. In this early research and throughout her career, Meitner, unlike Hahn, tried to understand theoretically what she saw in her experiments. She received her doctorate in 1906, and immediately began working with Stefan Meyer, continuing to teach during the day and conduct research in the evenings in radioactivity. In her first study, she measured the absorption of alpha and beta radiation emitted by thorium and actinium. The objective of her work, as she stated it, was to understand the nature of radiation. She continued these studies by examining whether alpha particles were not only absorbed but also scattered. These experiments anticipated Rutheford's and Soddy's similar experiments in which they discovered the nucleus four years later. She found that alpha particle scattering increased with the atomic mass of the metal atoms. These studies led to her first publication in June 1907.

At this point in her life, Meitner decided it was time to leave Vienna, and go to Berlin, hoping to study with some of the best physicists of the time. She knew no one in Berlin, but she had heard of Max Planck and his discovery of the quantum of light. When she arrived in Berlin, she approached Planck and asked permission to attend his lectures. As Meitner later stated: "He received me very kindly and invited me to his home. The first time I visited…he said to me, "But you are a Doctor already! What more do you want?"[16] He was quite conservative in his views of women in science. He stated that there were certainly rare instances where a woman might possess a special gift for physics, but "in general it cannot be emphasized strongly enough that Nature itself has designated for women her vocation as mother and housewife."[17] He agreed tentatively to allow Meitner to attend the lectures on a trial, revocable basis. Since the lectures occupied only part of her time, she approached Heinrich Rubens, head of the experimental physics institute, about joining his laboratory group. He offered her a place as an unpaid volunteer. Meitner had intended to stay in Berlin for a few terms; she stayed for over 30 years.

---

[15] Sime, R.L. (1996). *Lise Meitner: A life in physics*. University of California Press. p. 5.
[16] Sime (1996), p. 24.
[17] Sime (1996), p. 26.

# The Collaboration Begins: The Hahn-Meitner Team

Hahn and Meitner arrived almost simultaneously to continue their research at the University of Berlin. The university had been founded under Friedrich Wilhelm III in 1809, and by the turn of the century, it ranked as one of the top universities in Europe. The various positions on the academic ladder that will be referred to in this chapter start with *Assistent* (a research position with no teaching duties); *Privatdozent* (an unsalaried research position with teaching, where the students pay fees to the instructor); and finally the level of Professor (the highest level which most academics never reached). Most universities had two professors per discipline. In the early 1900s, for example, the University of Berlin had two physics professors: Max Planck (theoretical) and Heinrich Rubins (experimental). Hahn joined the chemistry department as a professor, and Meitner was an unpaid "volunteer" in Heinrich Ruben's group. Both attended Ruben's regular Wednesday physics colloquium, and it was most likely there that Meitner and Hahn first met. They liked each other from the very beginning, and both realized that each had qualities that would complement the other. Hahn was outgoing and charming; Meitner shy and withdrawn. Meitner knew physics and mathematics, while Hahn's background was in radiochemistry. As Meitner would later recall:

> Rubens added that a Dr. Otto Hahn had indicated that he would be interested in collaborating with me. Hahn himself came in a few minutes later. Hahn was of the same age as myself and very informal in manner, and I had the feeling that I would have no hesitation in asking him all I needed to know.[18]

In 1907, the Chemistry Institute at the University was off limits to women. Meitner could enter the Institute only through the back door. Fortunately, Hahn's small workshop in the back of the University, shown in Fig. 3.3, was perfect for their studies, allowing Meitner to work quite separate from the other researchers. As their collaboration began, Meitner was an unpaid guest with no salary. Her family continued to provide a small stipend for Meitner, which allowed her to stay in Berlin during these first years.

By this time, both Hahn and Meitner had honed their skills in working with radioactivity, and in their first study together they measured the emission characteristics of a number of pure beta sources and mixtures using the electroscope shown in Fig. 3.4. Basically, when a moving charged particle crosses

---

[18] Rife, P. (1990). *Lise Meitner and the dawn of the nuclear age*. Birkhäuser. p. 26.

**Fig. 3.3** The collaboration begins. (Courtesy of Archives of the Max Planck Society, Berlin. VI_001_Meitner_Lise_II_2)

a magnetic field, its path is bent from a straight line to a curve. In this way, the magnetic field separates particles based on their mass and velocity: lighter particles (beta particles) are deflected more than heavier particles (alpha particles), and within each type, slower particles are deflected more than faster ones. To study beta emission, Hahn and Meitner built their own brass electroscope, similar to the one that was used by Rutherford. As shown in Fig. 3.4,[19] the element to be examined was collected on a thin wire (S) placed in a groove at the bottom of the apparatus. About 23 mm above the groove was a slot (F), mounted parallel to the groove. The radiation from the source would pass through the slot and fall on the photographic plate (P) about 17 mm above the slot. Figure 3.5 shows one of the photographic plates representing the spectra from mesothorium-2: the black band in the middle is alpha particles, displaced only slightly by the magnetic field. It darkens over time as more particles accumulate on the plate. The other bands represent beta particles, and more bands appear over time as new beta-emitting species appear.[20] Contrary to their previous experience with alpha particles, the beta particles appeared with many discrete velocities, but at least the spectroscope could sort them out. The supposedly pure specimens that they began with were apparently mixtures of several species, and to confuse matters further, the

---

[19] Hahn (1966), pp. 55–56.

[20] *Band spectra of mesothorium.* (n.d.). Retrieved January 18, 2020, from https://www.europhysicsnews. org/articles/epn/pdf/2019/04/epn2019504p22.pdf.

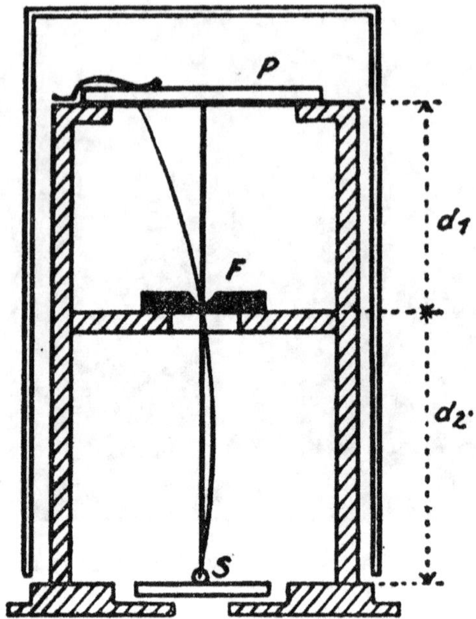

**Fig. 3.4**  Electroscope

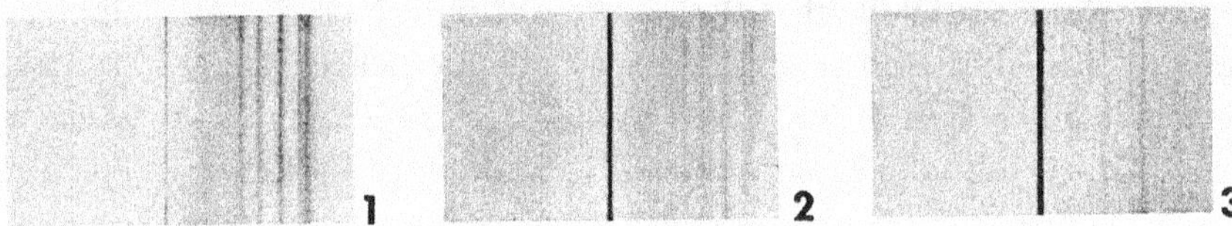


**Fig. 3.5**  Beta spectrum of mesothorium-2: freshly prepared (1); after one day (2); after eight days (3)

radioactive decay of the starting mixtures produced other species that radiated and appeared later in their spectra. Some of the species lived for only a few minutes, and since they could be detected and identified only by their beta radiation, their true nature remained a mystery. From these experiments, they compiled lists of species such as actinium-C (4.76 minutes); thorium-D (3.1 minutes); radium-$C_2$ (1.3 minutes); etc.

In the first two years of their collaboration (1908–1909), they published a total of nine articles together, and over the next several years, the number of publications rose to 20. As Ruth Sime has noted, "Meitner and Hahn had accomplished far more together than either could have done alone…"[21] Their

---

[21] Sime (1996), p. 40.

studies, as in all their future work together, would depend on the combination of Hahn's chemical separation expertise and Meitner's physical measurements and mathematical and graphical skills. But they still didn't know anything about the nature of these new substances, or about the structure of atoms. The nucleus wouldn't be discovered by Rutherford until 1911, and the existence of several "isotopes" of the same element with different radioactive decays wouldn't be discovered by Frederick Soddy until two years after that.

During a trip to a conference in 1910, Hahn's personal life changed dramatically. He met Edith Junghans, a student at the Royal School of Art in Berlin, who would become his wife. Their relationship developed over the next several years. In 1912, Hahn took Edith on a walk around Dahlem where the new Kaiser Wilhem Institute (KWI) for Chemistry had been erected. He told her that he would be working and living there soon and was going to be made a leader of a section. Hahn would have a tenured position and now felt comfortable in asking Edith to marry him. They married in her hometown of Stettin on March 22, 1913 (Fig. 3.6). Meitner never married, although she did apparently have at least one marriage proposal, which she turned down.

**Fig. 3.6**   Otto and Edith Hahn. (Courtesy of Archives of the Max Planck Society, Hahn_Otto_X_15_5.jpg)

She felt that she needed to give her entire life to her physics studies, and, as she stated later, she "just never had time for it."[22]

Hahn regarded Meitner as his closest coworker, and probably his closest friend. Their relationship has been called a kind of work marriage. Hahn was very close to Meitner in some ways, and she was protective of him throughout her life. They enjoyed each other's company, supported by their immediate professional successes. In the laboratory they would sing duets, with Hahn's excellent voice and Meitner humming along. Meitner developed many close friends besides Hahn, mostly in the physics community. They formed an incredible group, including many of the top physicists. One of Meitner's closest friends was Max Planck, and they spent many social evenings at his home. These gatherings might include Planck on the piano, perhaps playing a duet with Einstein on the violin, with Hahn a frequent soloist. Meitner described these social evenings as the "magic musical accompaniment" to her life.[23] Meitner would often visit Hahn's home, and she became fast friends with Hahn's wife Edith. She was also the godmother to Hahn's only child, Hanno. Meitner recalled "We were young, happy, and carefree—perhaps politically too carefree."[24] But as Hahn stressed, there was nothing beyond friendship. They never went out walking together. They would work in the laboratory until eight in the evening, then go home. As their collaboration matured, each realized that they complemented the other and formed a bond deepened by their strengths and differences. Hahn's easy-going manner was probably a big part in helping Meitner overcome her reserve. This chance meeting at a science colloquium between a physicist and chemist led to an amazing personal and professional collaboration that would last for the rest of their lives: "the closest of colleagues, the best of friends."[25]

# World War I (1914–1918)

In 1912, the working conditions for Hahn and Meitner dramatically improved. With great celebration, the Kaiser Wilhelm Institutes (KWIs) were inaugurated on October 23, 1912, in Dahlem, just outside Berlin. The establishment of the KWI reflected the value that Germany placed on science and the stature that Germany had achieved during the early 1900s. Hahn and Meitner moved into the Institute for Physical Chemistry and Electrochemistry in 1913. Hahn

---

[22] Sime (1996), p. 35.

[23] Sime (1996), p. 39.

[24] Hoffmann (2001), p. 54.

[25] Sime (1996), p. 29.

was appointed as a professor with a salary, while Meitner was still an unpaid guest. Later that year, she was appointed as an Assistant for Max Planck, her first paid position. That same year, Hahn and Meitner were appointed as Associates of the Institute, and their radioactivity section officially became the *Laboratorium Hahn-Meitner*. This move for them was incredible. They now had office and laboratory space, with access to grants and students. Perhaps more importantly, their space was pristine, with no contamination from prior studies. Their old laboratory had become so contaminated with radiation that any study of weakly radioactive substances had become impossible. Now they had the chance to begin a search for the radioactive precursor to actinium—the mother substance. By now, they were experts in the field of radioactivity, and as a chemist and physicist, they had built their expertise over the prior period of five years. They were ready for the challenge, but very little was known about actinium. It was extremely scarce, and its chemistry was uncertain. There was some evidence that it was connected to uranium since it was always and only found in uranium ores. The goal therefore was to "find that substance which…forms the starting point for the actinium series, and to determine…[from] which intermediate actinium is derived."[26]

It is worth stepping back at this point to discuss the history and meaning of the periodic table since it plays such an important role in all that Hahn and Meitner were doing, and later in the discovery of fission (Fig. 3.7). In 1869,

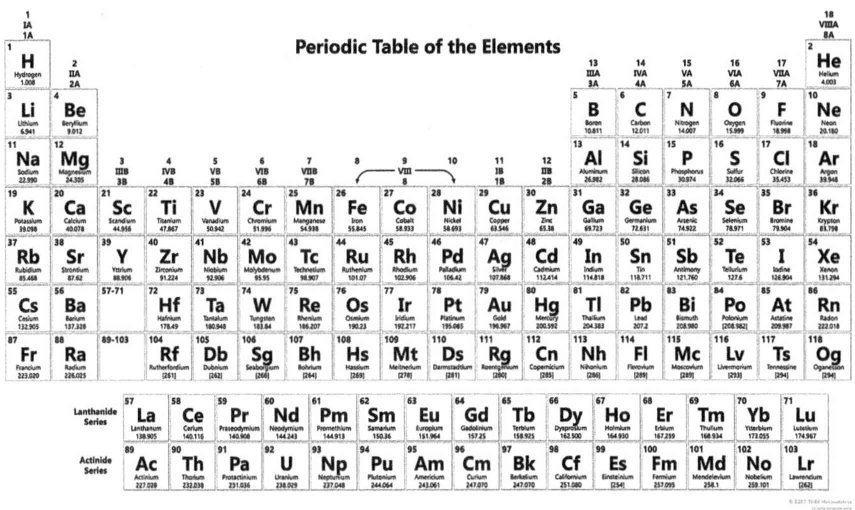


**Fig. 3.7** Periodic table of the elements

---

[26] Hahn (1966), p. 89.

Russian chemist Dmitri Mendeleev wrote down the symbols of the known elements and put them in order according to their atomic weights, resulting in what we now call the periodic table. The elements were grouped according to their known chemical behavior. All the alkali metals were placed in the first column, all the inert gases in the last column, and the others according to their atomic weight in the rows in between. There were missing elements in this first table, but it provided a basic framework that chemists would continue to expand. Although Mendeleev arranged the elements according to atomic mass, from the work of Harry Moseley we now know that the table is ordered by increasing atomic number. Within each column, all elements have essentially the same chemical properties. Mendeleev's original table had gaps but based on the properties of the other elements in the columns, he could predict what properties that unknown element would have. A good example of this comes from Mendeleev's original table that included the element aluminum (Al), but there was an unknown element or empty space directly below it that he called eka-aluminum for lack of a better name. Based on the properties of aluminum and the other elements in this same column, he predicted that the unknown element would have a density of 6 g/cm$^3$; melt at a low temperature; conduct heat well; and be a shiny metal. Sure enough, the element gallium was discovered in 1875 with all the properties that Mendeleev predicted.

Examining the table in a bit more detail, while all the elements in a single column have mostly the same chemical properties, there are small differences. In column 2, for example, barium and radium react with chlorine to form $BaCl_2$ and $RaCl_2$. This similarity makes it difficult to separate and identify elements in the same column, especially when one of the elements is present in only trace amounts. This property is key to the experiments performed by Hahn, Meitner, and Strassmann as they attempted to identify the element produced after bombarding uranium with neutrons. Their chemical separations continued to indicate that barium was produced, which, as we will see, was at that time assumed to be wrong: the nucleus simply could not be split. In this particular case, Strassmann recognized that barium precipitated with chlorine slightly faster than radium does, and he hoped to use this to separate the radium from the barium. But, no matter how he tried to separate the radium and purify the barium, the radioactivity always stayed with the barium, even when he added radium to test his method. Ultimately, it was Strassmann's expertise in recognizing and taking advantage of the exquisitely small differences between the elements in a single column that resulted in the finding that the uranium atom had in fact been split.

In 1913, two major developments by Frederic Soddy provided a clarification for the entire field of radioactivity and opened up additional information for their search. First, Soddy was able to organize the radioactive species into the periodic table for the first time. Soddy assigned actinium to column III in the periodic table, radium to column II, thorium to column IV, and uranium to column VI. This left a space between thorium and uranium. Additionally, Fajans and Soddy observed that an element was displaced two places to the left in the periodic table following alpha emission, and one place to the right following beta emission. Importantly for Hahn and Meitner, this missing element between thorium and uranium would have chemical properties similar to tantalum, located in the same column in the periodic table. Hahn and Meitner considered two approaches to identify the precursor: isolation from uranium salts or from pitchblende. Unfortunately, each source contained a variety of highly radioactive elements such as thorium and polonium, which would obscure the weaker emitting precursor. With the slow rate of alpha emission by the precursor, both knew that the work would be painstakingly slow. It might take years to complete. Since Geiger counters had yet to be discovered, the work would have to be completed by squinting at electroscopes for hours to monitor the radiation.

A major disruption in the work occurred in June 1914: the assassination of Archduke Ferdinand ignited World War I. Many scientists left their laboratories and headed for the front. Hahn was called up in 1914, and by December he was in heavy fighting in Belgium. In January he was recruited for a new project: a special unit on chemical warfare that Fritz Haber was organizing. His unit was transferred to the Eastern front, where the Germans used a mixture of chlorine and phosgene directed against Russian soldiers. Although he was horrified by the agony that the gases caused, Hahn stayed with the unit for the remainder of the war. He later remarked:

> I was very ashamed and deeply agitated. First, we attacked the Russian soldiers with our gas, and then, when we saw the poor chaps lying on the ground and slowing dying, we restored their breathing with our self-rescue equipment. The total insanity of war became obvious to us. First one attempts to eliminate the unknown enemy in his trench, but when one comes face to face with him, one cannot bear it and sets about helping him. Yet often we could no longer save the poor victims.[27]

---

[27] Fitzgerald, G. J. (2008). Chemical warfare and medical response during World War I. *American Journal of Public Health*, *98*(4), 611–625.

As Hahn wrote in his memoirs: "As a result of continuous work with these highly toxic gases, our minds were so numbed that we no longer had any scruples about the whole thing." Hahn also commented in an interview 40 years later that "Haber…explained to me that using gas was the best way of bringing the war to an end quickly."[28] This latter comment would have significant implications for how he and his colleagues viewed America's development and use of the atomic bomb in 1945.

Like Hahn, Meitner also had a desire to help her country, and in July 1915, she left for Vienna to volunteer as an X-ray nurse-technician. Three months later, she returned to Dahlem after convincing herself that she could serve her country better by continuing the research that she and Hahn had started. Upon returning to the institute, she found that Haber had demanded that all laboratories under his control switch their research to military work. With the help of Emil Fischer, the ongoing work of Hahn and Meitner was protected from Haber's military research. She wrote to Hahn on 25 October 1916:

> I hope that the Lion [Haber] will not get his claws into our modest section, especially since our private laboratory with its physical apparatus would hardly be useful for the chemical studies in question. If I am not prevented from doing so, I will try to work.[29]

Back in the laboratory, Meitner continued the search for the mother substance of actinium. The work was arduous. She had to procure the pitchblende and precipitate the desired residue with acid. After several rounds of boiling with concentrated acid, a small amount of solid residue remained. This presumably contained the actinium precursor. Meitner monitored the material regularly; if their calculations were correct, the precursor should emit a "short-range alpha radiation, and decay to products with faster alpha radiation…whose intensity climbed to double the initial intensity in several weeks."[30] With the war going on, equipment became expensive and hard to find; pitchblende was not readily available; food and fuel were scarce; and students, assistants, and technicians had been conscripted into the army. In addition, Hahn had just two leave periods where he was able to return to the laboratory, but only for a few weeks at a time. Meitner regularly wrote to Hahn to keep him abreast of her findings and results.

---

[28] Rife (1990), p. 80.

[29] Sime (1996), p. 62.

[30] Sime (1996), p. 654.

Finally, in a letter to Hahn on January 17, 2018, Meitner opened with "Take a deep breath before you begin reading...."[31] She had finally been able to define the actinium precursor by "measuring the range of its alpha particles and estimating its half-life." She closed the letter with "In any case, we can now think of publishing very soon." The paper was published in 1918, and although Meitner had done essentially all the laboratory work by herself, Hahn was listed as the senior author. As Sime comments: "Twenty years later when the situation was reversed and Meitner was forced to be away from the lab for the last six months of the fission work, she might have expected the same loyalty from Hahn."[32] She didn't get it.

The final issue associated with the discovery of the new element was to decide on the name. Stefan Meyer, Meitner's former mentor in Vienna, suggested that the most appropriate names might be Lisonium or Lisottonium, since Meitner had done most of the work. However, after consulting with others in the field, they decided to name the element protactinium. It sits two places to the right of actinium in the periodic table, next to uranium. With a half-life of 32,700 years, geologists now use it for radiometric dating of sediments.

## The "In-Between" Years: 1918–1933

The German surrender in 1918 left the country bleak and miserable, with widespread poverty and hunger. The value of the mark had declined steadily. By 1923, it plunged by a factor of 50, and later that year, it plummeted by a factor of a billion. Meitner wrote to her mother that "prices had risen fast." She like many others feared that this winter would be hard; "We are already freezing in the institute and at home."[33] All aspects of life were affected, both professionally and personally. Many of the country's scientists were disillusioned and demoralized. Several men, including Otto Hahn, had enlisted into the emergency technical service (*technische Nothilfe*) to keep utility services running. Scientific meetings were poorly attended because participants could not afford the travel. By 1924, the mark stabilized to some extent after austerity measures were introduced. Throughout this time, Meitner and Hahn were able to continue their work, but under the poor conditions their research primarily involved tying up loose ends from the protactinium work.

---

[31] Sime (1996), p. 69.
[32] Sime (1996), p. 71.
[33] Sime (1996), p. 97.

Soon after the protactinium work was completed, Meitner and Hahn realized that their scientific paths were headed in different directions. In January 1917, Meitner was appointed head of her own section, essentially dividing the *Laboratorium* into two parts: one directed by Meitner (physics) and the other by Hahn (chemistry). Hahn was spending more of his time on funding and institute politics, and less on research. They continued to meet daily to discuss projects of mutual interest, even after the institute reorganized into separate Physics and Chemistry sections, but their publication record was evidence of their separate projects. From 1903 to 1925, almost half of their publications were joint. As their paths began to separate, they published almost no articles together between the years 1926–1934.

Meitner's situation began to improve as the country gained more stability and Berlin slowly regained the scientific prominence it once held. Doors were beginning to open for women in academia, with increased opportunities to advance in the German academic ranks. In 1921, Planck offered Meitner the opportunity to teach in the University physics department. To qualify, she had to go through the *Habilitation* that Hahn had completed. The written part of the exam was waived since by this time she had over 40 publications. She gave a stellar oral presentation to her two reviewers, Rubens and Max von Laue, who passed her enthusiastically. She achieved a University Lectureship position, allowing her to teach courses, give lectures, and mentor students, which she did for the next ten years (Fig. 3.8). She eventually rose to Professor in the Institute, the first woman to achieve this status. Given her increased

**Fig. 3.8** Lise Meitner teaching. (Courtesy of the Archives of the Max Planck Society, Berlin, VI. Abt., Rep. 1, Meitner, Lise I/10)

stature in the Institute, she could now look forward to attracting students and obtaining grants. It would eventually afford her some semblance of defense against the coming of the new Nazi regime in the 1930s.

Meitner was gaining recognition as one of the preeminent physicists of her time, establishing herself as a leader in the newer and smaller field of nuclear physics. During this time, she investigated nearly every fundamental problem in experimental nuclear physics, including such areas as radioactive decay, the continuous beta spectrum, alpha scattering, and neutron reactions. In fact, Hahn noted that it was her work, more than his own, that contributed to the international reputation of their Institute. Personally, she considered the physics community as her home. She was accepted by many colleagues of the time and became close friends with both the physicists and their wives. She was invited to give lectures throughout Europe, including Bohr's Institute in Copenhagen in 1921. Notably, she spent several weeks in Sweden in 1921 as a visiting professor. The invitation came from Manne Siegbahn, who was interested in her knowledge of beta spectroscopy. Essentially no radioactivity research was being conducted in Sweden at this time, and she was soon teaching a full course to physicists and chemists there. Meitner also spent some time in Siegbahn's laboratory, where she learned more about X-ray spectroscopy. Although one would assume that the warm reception that she enjoyed by Siegbahn and his colleagues would have extended to her eventual return to that laboratory in 1938, it was not to be.

Hahn continued research through the 1920s despite his increasing administrative roles. What research he could fit into his schedule focused on the refinement of radiochemical techniques. Since the series of natural isotopes appeared to be complete, Hahn turned to the possible applications of radioisotopes in various chemical fields. He developed a method for deriving information about the properties of solids from measurements of their radioactive emanation. Although his emanation method did not lead to any practical applications, the techniques developed by his group became an integral part of radiochemistry. His group also focused on the chemical precipitation of trace amounts of radioactive materials. Throughout this period they developed a number of important precipitation methods that would play a critical role in the discovery of fission.

Hahn's professional advancement had also progressed rapidly after World War I. Prior to the war when the KWI opened in 1913, Hahn was appointed as a professor (in 1913) and promoted quickly to an Associate. When the Director of the KWI retired in 1926, Hahn took over as the provisional leader. His scientific achievements also began to be noticed. In 1924, he was elected to the Prussian Academy of Science. His nomination letters stated that he was being considered for his work in the field of radioactivity, with the discovery of nine new radioactive elements. During the years that he led the Institute, he made the Institute a focal point for students and foreign researchers. It has been suggested that under his leadership, the KWI might have developed into an institute that would rival the one that Rutherford had built in Canada. But the political events during the Nazi regime interrupted the great science community that Berlin had become.

## Transuranes and Nuclear Fission: 1934–1939

Our national policies will not be revoked or modified, even for scientists. If the dismissal of Jewish scientists means the annihilation of contemporary German science, then we shall do without science for a few years.[34]

Adolph Hitler: reply to Max Planck, 1933

Although the first half of the twentieth century saw groundbreaking discoveries in atomic physics throughout Europe, overshadowing these discoveries were the political events that would soon overtake Germany beginning in 1933, and ultimately lead to World War II. Following World War I, the situation in Germany was dreadful, with power struggles and endless elections. Economically, the country was in a deep depression with high unemployment. Amidst all the chaos, on January 30, 1933, Meitner listened to the radio announcement that Hitler was sworn in as Chancellor of the German Reich. Quickly following this announcement, the *Reichstag* was dissolved, and in March, new elections were once again held. Hitler's brownshirts made certain there was no opposition. Worse was yet to come: Jews were openly beaten and jailed, and concentration camps were filled with political prisoners. During this time, Hahn was in the USA, serving as a guest professor at Cornell University. This left Meitner, a Jew, as temporary director of the Institute. Hahn's first reaction to the events in 1933 was a statement on April 8 to the Toronto Star: "I am not a Nazi. But Hitler is the hope, the powerful hope, of German youth…At least 20 million people revere him…In a word: Hitler is an unequivocal Christ."[35]

In April 1933, Hitler announced that "non-Aryans" would be purged from government agencies, including the universities, with the passage of the Law for Restoration of the Civil Service. In the universities, the law created havoc.

---

[34] Cornwell (2003), p. 34.

[35] Sime, R. L. (2006). The politics of memory: Otto Hahn and the Third Reich. *Physics in Perspective*, 8, 6. https://doi.org/10.1007/s00016-004-0248-5.

Over 25 percent of the physics community was Jewish, and Jews from *Privatdozent* to Professor were dismissed. As events got increasingly more frightening, Meitner wrote to Hahn on May 3: "I do think it would be better if you return here…A rather large Nat[ional] Soc[ialist] cell has formed in the institute…."[36] Meitner decided to stay in Berlin, even as many of her colleagues were leaving, including Einstein (1933), James Franck (1933), and Fritz Haber (1933), and others were urging her to consider leaving. For example, Bohr had arranged a Rockefeller grant for her to work in his institute, and Swarthmore College in the USA had offered her a position. But she decided to stay in Berlin, convinced that her Austrian passport and citizenship would protect her. However, she did not totally escape persecution. As an adjunct University Professor, Meitner was required to fill out a questionnaire, including name, birth date and place, citizenship, and, most importantly, the race of her four grandparents. She signed it on April 28, 1933. Four months later, more bad news arrived. Meitner received an ominous notice from the Prussian Minister of Education that referred directly to the questionnaire she had filled out. It read:

> Immediately! For the reason set forth in paragraph 3 of the statute concerning the reinstatement of the professional civil service, I hereby revoke your authorization of professorship at the University of Berlin as of April 7, 1933.
> Signed: The Prussian Minister of Education, Siegel[37]

Both Hahn and Planck protested this decision, but to no avail. Meitner was forced to resign from the University, and she gave her final lecture in the fall of 1933. All future courses and lectures were canceled. Notably, Hahn resigned his position at the University in opposition to the dismissal of his Jewish colleagues, including Meitner. As Meitner's situation became more precarious, her colleagues in Berlin continued to help her however they could. In 1936, Hahn refused an invitation to lecture to the German Chemical Society, stating that the transuranic research was not only his, and he felt "it was not quite right…to take credit for intellectual property which is not mine exclusively."[38] Of course, he was speaking of Meitner, but his attitude of giving her credit for her role in the research would change drastically by 1939 and throughout the rest of his life. Hoping that a Nobel Prize for Meitner might offer protection

---

[36] Sime (1996), p. 145.
[37] Sime (1996), p. 150.
[38] Rife (1990), p. 148.

from the Nazi regime, von Laue, Planck and Heisenberg nominated her in 1936 and 1937, but she was not awarded a prize.

There were other startling discoveries in the early 1930s. In 1932, James Chadwick discovered the neutron, a neutral particle making up roughly half of the particles in the nuclei in all elements except hydrogen. Emitted in some nuclear decays, neutrons could penetrate matter and reach the heaviest nuclei. In the spring of 1934, Enrico Fermi in Italy used neutrons to induce radioactivity, using a mixture of radon and powdered beryllium as the neutron source. The Fermi group started by irradiating hydrogen with neutrons, and systematically worked their way up the periodic table, irradiating one element after another, up to the last element, uranium. In each case, Fermi observed that uncharged neutrons easily attached to the positively charged nucleus, and the nucleus would subsequently undergo beta decay. It had already been shown by Fajans and Soddy in 1913 that when a nucleus emits a beta particle, the nucleus transforms to the next heavier element in the periodic table. Fermi's hope was that nuclei bombarded by neutrons would be unstable and emit a beta particle, transforming to the next heavier element. His goal was to go beyond uranium to reach element 93, the first transuranic element. If this reaction could be induced in uranium, then element 93 would be produced according to the following equation:

$$U^{238} + n \rightarrow U^{239} \rightarrow 93^{239} + \beta^{-}$$

The U in the equation is the element uranium, and the superscript number 238 is its atomic mass. The new transuranic element ($93^{239}$) in the equation above would probably be radioactive, and if it also emitted a beta particle, then an additional transuranic element ($94^{239}$) could be produced:

$$93^{239} \rightarrow 94^{239} + \beta^{-}$$

In support of this theory, Fermi's group reported the production of at least five beta emitters after bombardment of uranium, with half-lives of 10 seconds, 40 seconds, 13 minutes, 40 minutes, and 1 day. However, they knew that the only way to definitively show that new transuranic elements were formed was to isolate these elements using chemical methods and show that they behaved chemically as expected for their proposed location in the periodic table. But Fermi's group did not possess the expertise to conduct the chemical studies.

Fermi's enthusiasm for the transuranic research began to fade as Mussolini's reign of terror in Italy intensified, and several members of his group left to find work in other countries. An interesting event that also excited the Fermi group and moved them in a new direction was the rather serendipitous discovery of slow neutrons. As described by Emilio Segre, who was Fermi's student at the time, uranium seemed to react differently in different parts of the laboratory. Neutron irradiation produced more radioactivity when conducted on a wooden table than on a marble table. In an amazing insight, Fermi surmised that the neutrons were slowed by collisions with the hydrogen atoms in the wood. Fermi's group had discovered how to produce slow neutrons. Fermi dropped the research on transmutation and focused his subsequent work on an investigation of the physics of slow neutrons.

Fermi's exit from the transmutation field opened up this research area for two principal groups: Hahn and Meitner in Berlin and Irene Joliot-Curie and Paul Savitch in Paris. In Berlin, Meitner was immediately excited by Fermi's neutron bombardment experiments. She had correctly realized that it would take the expertise of both a physicist and a chemist to unravel the complexities of radioactivity, and she immediately spoke to Hahn about resuming their collaboration after 12 years of interruption. They also brought in another chemist, Fritz Strassmann, to carry out many of the chemical separations. In Paris, Joliot-Curie and Savitch realized that Fermi's results with neutrons provided another exciting approach to discovering transuranic elements. These two groups—Joliot-Curie/Savitch and Hahn/Meitner/Strassmann—became locked in a competition to identify the products that resulted after bombarding uranium with neutrons. The Berlin group had one great advantage over the Paris group: they had both chemists and a physicist working on the problem. In contrast, in Paris, physicist Frederic Joliot-Curie, who had worked with his chemist wife Irene, was appointed as a Professor at the College of France in 1937. He left Paris to build his laboratory at the Radium Institute, a few miles outside Paris. For the first time in their lives, the Joliot-Curies would be working apart, breaking up one of the most productive teams in radioactivity research. Paul Savitch, a chemist, joined Irene Joliot-Curie to work on the transuranic problem, but now they had no physicist.

Hahn and Meitner began the slow process of bombarding uranium and identifying the products. From the results already reported by Fermi, they knew the work would be difficult. Following neutron bombardment, Fermi had seen many beta emissions, but he had done no chemistry. Since uranium (atomic number 92) was positioned just below tungsten in the same column of the periodic table, then anything beyond uranium (i.e., elements 93 and 94) would be positioned just below rhenium and osmium and should behave chemically like them. To test for an element heavier than uranium, Hahn and

Meitner added rhenium to their mixture, then carried out fractional precipitation. What precipitated was rhenium sulfide, likely to be the "eka-rhenium-93" that Fermi had reported. However, complicating their work was the discovery that there were actually several substances produced when uranium was hit with neutrons as identified by their radioactive emissions. The actual amount of these substances was so exquisitely small that there was no hope of detecting them except by their radioactive emissions. And given the difficulty in chemically identifying these substances, making any claim of where they fit in the periodic table was enormously difficult. By the end of 1938, they had discovered multiple products with varying half-lives, and had published a total of 14 publications, with titles such as "New transformation processes when uranium is irradiated with neutrons" (1936) and "A new long-lived transformation product in the trans-uranium series" (1938). But they were still no closer to chemically identifying these products.

In similar experiments, Joliot-Curie and Savitch reported in May 1938 that they had found a substance with a half-life of 3.5 hours with chemical properties similar to the lanthanides, in addition to the elements found by Hahn and Meitner. In their effort to identify this new substance (which they designated as R), Joliot-Curie and Savitch used fractional precipitation in the presence of lanthanum or lanthanum plus actinium. They consistently found that R precipitated with lanthanum and had properties very different from the expected transuranes. Unable to actually "see" the radioactive element, all they knew was that it precipitated with lanthanum, and not with rhenium or osmium, as they had expected. Unwilling to go further, they commented: "It therefore seems that this body can only be a transuranic element possessing properties very different from those of the other known transuranic elements, a hypothesis which raises difficulties of interpretation."[39,40]

Challenged by the publications from Joliot-Curie and Savitch, the Berlin team hurriedly began experiments to reproduce the Paris findings. But, just as their work on uranium bombardment was heating up, Meitner's situation as a Jew in Germany became perilous. Some colleagues encouraged her to stay, while others thought she should leave immediately. But she believed her Austrian citizenship and passport still protected her, and she felt that as a vital part of the team, she needed to stay. Then, on March 12, 1938, Germany invaded Austria, and overnight, Meitner lost any protection she still held, since her Austrian citizenship was now useless. Threats to her continuing work

---

[39] Curie, I., & Savitch, P. (1938). Radioelement of period 3.5 hours formed from uranium bombarded by neutrons. *Comptes Rendus, 206*, 906–908.

[40] Curie, I., & Savitch, P. (1938). On the nature of a radioelement with a period of 3.5 hours formed in uranium irradiated by neutrons. *Comptes Rendus, 206*, 1643–1645.

at the Institute increased with claims such as "The Jewess endangers the institute." Personally, and professionally, her life in Berlin was suddenly in extreme peril. The KWI treasurer (Heinrich Horlein) suggested that Meitner should resign her position because nothing else could be done for her, and suggested that perhaps she could work unofficially. In her diary, Meitner wrote she was extremely hurt by Hahn when he told her she "must not come to the Institute anymore."[41] She was encouraged to move to the Hotel Adlon where the head of the KWI—Carl Bosch—was now staying. Bosch was still willing to help Meitner. He objected to the suggestion by Hahn that Meitner resign, indicating that it was he (Bosch) and not Hahn who should determine if Meitner should stay. On May 22, Bosch wrote to the Minister of Education requesting permission for "...the famous scientist Lise Meitner to leave for a neutral foreign country..."[42] The response from the Minister was chilling. "It is considered undesirable that renowned Jews should leave Germany for abroad to act there against the interests of Germany...as representatives of German sciences or even with their reputation and experience."[43] Her fate was becoming clearer: she was Jewish, she was banned from working as a scientist, and she could not legally leave Germany. Several of Meitner's colleagues worked furiously to find an option for Meitner to begin at least a temporary position in another institution (in the USA, England, Copenhagen, or Groningen). As doors began to shut from one country to another, Meitner's colleagues became frantic. After many negotiations involving among others Bohr, Peter Debye, Adriaan Fokker, and Dirk Coster, a plan was set in motion: Meitner would leave for a "talk" in the Netherlands, then travel on from there to take a position in Sweden.

The story of Meitner's escape from Germany reads like a spy thriller. She realized—too late—that she should have left Germany months or years earlier. The only possible route at this point was to secure an invitation to speak at a conference in Holland and attempt a train crossing with her now invalid Austrian passport. Meitner and a few colleagues planned her escape. She spent the last day at the institute on July 12. She packed two small suitcases containing items necessary for a short trip and stayed at Hahn's house overnight. Her colleague Dirk Coster met her in the morning to escort her across the German-Dutch border. To alert her colleagues that she had safely crossed the border,

---

[41] Rife (1990), p. 163.
[42] Rife (1990), p. 164.
[43] Rife (1990), p. 166.

> We agreed on a code-telegram in which we would be let known whether the journey ended in success or failure…People trying to leave Germany were always being arrested on the train and brought back…We were shaking with fear whether she would get through or not.[44]

Finally, from Groningen in the Netherlands, Coster sent the coded telegram to Hahn that the "baby" had arrived. Pauli sent a telegram to Coster that spoke for the entire physics community: "You have made yourself as famous for the abduction of Lise Meitner as for [the discovery of] hafnium."[45] After 31 years she left Berlin forever with ten marks in her purse and a diamond ring from Hahn that he had given to her to cover emergency expenses. For Meitner, the ring had a deeper meaning—a symbol of friendship, a gesture of caring. She kept the ring all her life.

After several weeks in Holland and Copenhagen, Meitner proceeded to Sweden, where her colleagues had secured at least a temporary position for her with Manne Siegbahn. Sweden was her choice: she had worked previously with Siegbahn, and their relationship was cordial and respectful. In addition, she chose Sweden because nuclear physics was undeveloped in Sweden, and she believed she could be of use.

In November 1938, after Meitner escaped to Sweden, Hahn was invited to Copenhagen to give a series of talks. While he was there, he met with Meitner, who was also at the conference; he was to discuss the work that he and Strassmann were continuing on the bombardment of uranium. By this time, Hahn and Strassmann, like Curie and Savitch, had found that the radioactive products from bombarded uranium did not behave like rhenium or osmium, but unlike Curie and Savitch they found radioactivity in barium precipitates. In discussions of his current uranium work, Meitner and all the other attending physicists urged Hahn to continue experiments, disbelieving that the uranium atom could be split into smaller nuclei such as barium or lanthanum; there must be another explanation. Hahn returned to Berlin where he and Strassman continued the chemical separation techniques, relying on Meitner to come up with an explanation from the physics perspective of what might actually be happening. None of the scientists, Hahn and Meitner included, were ready to conclude that the atom was being split.

In November 1938, Hahn and Strassmann published an article entitled "On the origin of radium isotopes from uranium by irradiation with fast and

---

[44] Sime (1996), p. 204.
[45] Sime (1996), p. 205.

slow neutrons."[46] The authors claimed that radium (atomic number 88) must have been produced from uranium (atomic number 92) by successive emission of two alpha particles, but every experiment continued to yield a substance that could not be separated from the added barium carrier. Since radium and barium are both in column two of the periodic table, they share similar chemical properties, so radium could explain the radioactivity of the barium precipitates. As Hahn stated: "All chemical elements were excluded except radium and the carrier barium…Since barium…did not come into the question, there was only radium left."[47] They were on the verge of a momentous discovery, but not yet ready to take that step.

It is worth describing the details of the experiments that led to the discovery of fission. Although superficially simple, they describe a heroic effort to discover something that they knew, or thought they knew, couldn't possibly be true. Hahn and Strassmann worked in three rooms in the basement of the KWI. There was Hahn's personal chemistry room at one end of the corridor, a measurement room across the hall, and the irradiation room at the far end of the hallway. The three laboratories were separated to avoid contaminating the non-radioactive rooms with radiation. The irradiation room had a worktable with large paraffin blocks containing the neutron source (radium salts plus beryllium powder). Handmade Geiger counters were surrounded by lead shielding in the measurement room. Hahn's chemistry laboratory had a table set out with the equipment and glassware for the chemical separations. The two men (now without Meitner) had a regular schedule to move throughout the rooms to ensure that they would not contaminate the non-radioactive rooms. The definitive experiment was set up in late December, just before Christmas. In their earlier experiments, they had bombarded uranium with their neutrons, and performed a series of crystallizations to determine which element was created from the uranium. Since they had some reason to believe that radium-88 was being created by the emission of two successive alpha particles, they added barium as a carrier. Referring to the periodic table, since barium-56 is in the same column as radium, the two elements would have similar, if not identical, chemical properties. Radium would initially follow the barium but could then be separated by the crystallization procedure. They could then trace the radioactivity with their Geiger counters. At this point, Strassmann proposed using a form of barium that would have slightly different chemical properties from radium, and with this they would be able at last

---

[46] Sime, R. L. (1989). Lise Meitner and the discovery of fission. *Journal of Chemical Education.* 66, 374. https://doi.org/10.1021/ed066p373.

[47] Sime (1989), p. 374.

to determine if indeed they were producing radium from their uranium bombardment. As Strassmann described it, he "suggested using barium chloride as a carrier rather than barium sulfate because the chloride…forms beautiful little crystals of exceptional purity." Coincidentally this was precisely the method Hahn had used to separate radium in his first experiments in Rutherford's lab decades earlier. When they tried the experiment, the radioactivity refused to separate out during the crystallization; it continued to follow the barium. Just before the annual KWI Christmas party, they concluded that what they had was really barium-56, not radium. But how this could happen was, to a chemist, an impossibility. Or so they thought.

The timeline of what transpired from December to January is outlined in the letters that Hahn and Meitner sent to each other. To fully understand the sequence of events leading to the discovery of fission, and the contributions of Meitner (as well as Strassmann and Frisch) to this discovery, translations from their correspondence as well as important quotes from the original articles are included below.[48,49] One of the most momentous series of events in both scientists' careers started with the letter from Hahn to Meitner dated December 19, 1938, "Monday evening in the lab:"

Actually there is something about the "radium isotopes" that is so remarkable that for now we are telling only you. The half-lives of the three isotopes have been determined quite exactly, they can be separated from all elements except barium, all reactions are consistent [with barium]…Our Ra isotopes act like Ba…I have agreed that for now we shall tell only you. Perhaps you can come up with some sort of fantastic explanation. We know ourselves that it can't actually burst apart into Ba…So please think about whether there is any possibility—perhaps a Ba-isotope with much higher atomic weight than 137? If there is anything you could propose that you could publish, then it would still in a way be work by the three of us.

Letter from Meitner to Hahn, December 21, 1938:

Your radium results are very startling. A reaction with slow neutrons that supposedly leads to barium!…At the moment, the assumption of such a thoroughgoing breakup seems very difficult to me, but in nuclear physics we have experienced so many surprises, that one cannot unconditionally say: it is impossible.

---

[48] Sime (1996), pp. 233–236.
[49] Translation of original Hahn/Meitner letters by authors.

Letter from Hahn to Meitner, December 21, 1938:

How nice and exciting it would have been if we had been able to do our work together as before…we conclude that we as "chemists" must draw the conclusion that the three isotopes studied in detail are not Ra at all, but from the chemist's point of view Ba…We cannot hush up our results, even if they may be physically absurd. You see, you are doing a good work when you find a way out. If we finish tomorrow or the day after, I'll send you a carbon copy of the manuscript.

Letter from Hahn to Meitner, December 28, 1938:

I want to quickly write you something about my Ba fantasies etc. Perhaps Otto Robert [Frisch] is with you in Kungalv and you could discuss it a bit. You will have received the manuscript of our work. I made a few minor changes: most notably, a note to point out that the trans uraniums are not Ma, Ru, Rh, Pd. At least I personally know so little about these bodies that I cannot rule them out chemically. (Strassmann is away). Would it be possible for the uranium 239 to burst into a Ba and a Ma. A Ba 138 and a Ma 101 would result in 239. The exact mutual mass number is not important. It could also be 136 + 103 or something like that…Is that energetically possible? I don't know; I only know that our radiums do not have the same properties that Ba has…Everything else is still untested. I would of course be very interested to hear your frank judgment. Eventually you could calculate something and publish it.

Hahn and Strassmann submitted their first article to *Naturwissenschaften* on December 22, 1938, with publication on January 6, 1939. In this initial publication, the authors were still hesitant about stating that they had witnessed the breakup of the uranium atom.[50] Toward the end of this article, the authors insert the following statement:

As chemists we really ought to revise the decay scheme given above and insert the symbols Ba, La, Ce, in place of Ra, Ac, Th. However as "nuclear chemists," working very close to the field of physics, we cannot bring ourselves yet to take such a drastic step which goes against all previous experience in nuclear physics. There could perhaps be a series of unusual coincidences which has given us false indications.

---

[50] Pearson, J. E. (2015). On the belated discovery of fission. *Physics Today*, 68(6), 40–45. https://doi.org/10.1063/pt.3.2817.

On December 24, Meitner met with Otto Frisch in Sweden. Meitner had Hahn's letter dated December 21, and she continued to go over the possibilities in her mind. The account of working out the theory of fission is a wonderful story, and one related by Frisch in his memoirs[51]:

> The suggestion that they might after all have made a mistake was waved aside by Lise Meitner; Hahn was too good a chemist for that, she assured me…We walked up and down in the snow, I on skis and she on foot…and gradually the idea took shape that this was no chipping or cracking of the nucleus but rather a process to be explained by Bohr's idea tht the nucleus is like a liquid drop; such a drop might elongate and divide itself…[N]uclei differed from ordinary drops in one important way: they were electrically charged, and this was known to diminsh the effect of the surface tension [that holds the drop together]. At this point we both sat down on a tree trunk and started to calculate on scraps of paper. The charge of a uranium nucleus…was indeed large enough to destroy the effect of surface tension almost competely; so the uranium nucleus might indeed be a very wobbly, unstable drop, ready to divide itself at the slightest provocation (such as the impact of a neutron). But there was another problem. When the two drops separated they would be driven apart by their mutual electric repulsion and would acquire a very large energy, about 200 MeV in all; where could that energy come from?…[W]henever mass disappears energy is created, according to Einstein's formula $E=mc^2$, and one-fifth of a proton mass was just equivalent to 200 MeV. So here was the source for that energy; it all fitted.

Frisch returned to Copenhagen to write out their results, which he then finalized during a long-distance phone conversation with Meitner. As Frisch later states, this was a rather expensive way to write a paper. The paper was submitted as a letter to *Nature* on January 16, 1939 and published February 11, 1939. In this paper, Meitner and Frisch provide an explanation for the breaking up of uranium (Frisch introduces the term "fission" for the first time)[52]:

> On account of their close packing and strong energy exchange, the particles in a heavy nucleus would be expected to move in a collective way, which has some resemblance to the movement of a liquid drop. If the movement is made sufficiently violent by adding energy, such a drop may divide itself into two smaller drops. It seems therefore possible that the uranium nucleus has only small stability of form, and may, after neutron capture, divide itself into two nuclei of

---

[51] Sime (1996), p. 236–237.
[52] Meitner, L. & Frisch, O. R. (1939). Disintegration of uranium by neutrons: A new type of nuclear reaction. *Nature*, 143, 239–240.

roughly equal size…These two nuclei will repel each other and should gain a total kinetic energy of c. 200 MeV, as calculated from nuclear radius and charge. This amount of energy may actually be expected to be available from the difference in the packing fraction between uranium and the elements in the middle of the periodic system. The whole 'fission' process can thus be described in an essentially classical way…

The second article by Hahn and Strassmann was submitted to *Naturwissenschaften* on January 18, with "proof" of the barium findings. They cited the Meitner and Frisch *Nature* article, clearly indicating that Hahn had talked with Meitner prior to the publication of their second article. This second paper by Hahn, together with an additional paper by Frisch, also submitted on January 16, absolutely confirmed that Hahn and Strassmann had split the atom. Meitner and Frisch were the ones to confirm and explain the astounding discovery.

From this discussion, it is clear that despite her forced absence, Hahn considered Meitner a key part of the collaboration, up to and including the publication in *Naturwissenschaften* of the fission results. Hahn knew that he could not include Meitner, a Jew, on the publication, but continued to ask for her interpretation of his results. And if she could come up with an explanation that she could publish that would continue their collaboration. As stated in his December 19 letter: "If there is anything you could propose that you could publish, then it would still in a way be work by the three of us." And in fact, Meitner and Frisch had come up with the exact conclusion that ultimately shocked the physics world: the uranium atom could be split, with the release of tremendous energy.

One complication arose before the Meitner and Frisch paper came out in *Nature*. Frisch mentioned to Bohr that he thought that Hahn and Strassman had discovered fission. He made it clear that Bohr should keep this confidential prior to the publication of the results from Meitner and Frisch. However, on the ship traveling to the USA, Bohr unfortunately mentioned the fission discovery to his colleague Leon Rosenfeld, without also mentioning that this was confidential. When Rosenfeld talked about the discovery after reaching the USA, word spread quickly within the physics community. Bohr later admitted that Meitner and Frisch were the first to confirm the fission discovery in a letter to *Nature*,[53] but nothing could now stop the hundreds of researchers who jumped into the exciting new field of atomic fission.

---

[53] Bohr, N. (1939). Disintegration of heavy nuclei. *Nature*, 143, 330.

Isolated in Sweden, Meitner eventually felt that she had been left out of the discovery of fission, perhaps one of the most important discoveries of the twentieth century. The injustice of this realization was bitter. She rightly believed that she had been a part of the team from the beginning of the "transuranes" work that started in 1934. In fact, the Hahn, Meitner, and Strassmann team had published continuously since 1934 (five publications from 1934 to 1938). It was her initiative that had been the impetus for starting their joint work, and she had felt that even in Sweden, she had been a valued member of the team through their constant communication. The situation was even more difficult to accept for Meitner when she compared their teamwork in World War I to the discovery of fission. Their work in 1917 and 1918 that led to the discovery of protactinium was carried out largely by Meitner while Hahn was away from the laboratory at the front, and they continued the collaboration through letters (a situation identical to the fission work). Both were credited with the protactinium discovery, and Meitner placed Hahn's name first on the publication. As Strassmann would later write[54]:

> What difference does it make that Lise Meitner did not directly participate in the "discovery??" Her initiative was the beginning of the joint work with Hahn—4 years later she belonged to our team—and she was bound to us intellectually from Sweden…[She] was the intellectual leader of our team…

## World War II: 1939–1945

**Meitner.** Meitner had been forced to leave Germany with only a few of her possessions. Her writings and all of her scientific equipment and materials had to be left behind. Under pressure from scientific colleagues, Manne Siegbahn, Director of the Nobel Institute for Experimental Physics in Stockholm, offered Meitner a position in his university—the only formal offer at this time that Meitner felt she could take and continue her work. After fleeing Berlin in July 1938, she arrived in Stockholm with only a few summer clothes, and little else. She felt totally isolated, an exile from her physics and her personal life.

She had elected to go to Sweden for several reasons: she had spent a short sabbatical in the 1920s working with Siegbahn and had forged a friendly working relationship. She had several friends in Sweden whom she could rely on for support. And she was excited about Siegbahn's current research focus

---

[54] Sime (1996), p. 241.

on nuclear physics. As an internationally known physicist in this area, she felt that she could contribute immediately. However, when she arrived, she received scant welcome from the Swedish physicists at Siegbahn's institute, and Siegbahn was not inclined to use his limited funding to support Meitner. She eventually turned to the Nobel Academy, which provided a small salary and some research funding. She had gotten other offers following her escape from Germany, including some from England and Holland. But Meitner felt that her reputation in nuclear physics would fill a void in Sweden's fledgling nuclear physics community. As she realized too late, "Siegbahn's reception was cold, her pay very low, she had no assistants or technical support...".[55] Although Siegbahn had given her a position, he was not going to give her any monetary support. In Berlin, she had her own institute and had attained the position of Professor. She had a good salary, with access to students and collaborators. In Stockholm, she was living in a small apartment on borrowed money. She was separated from friends and colleagues, she had no books or journals, and she could not get her possessions out of Germany. She continued to publish when she could, and, although Siegbahn always listed her "apart from the institute's own personnel,"[56] he didn't hesitate to count her publications as part of the Institute. After the discovery of fission was announced, she realized with some horror that as a German Jew, her work before and after fission might not be recognized, and she could not improve her situation in her current role in Sweden. Meitner wrote to Hahn after the fission work: "...the important thing is that I came with such empty hands. Now Siegbahn will gradually believe—especially after your beautiful results— that I never did anything and you...did all the beautiful physics in Dahlem." And, a week later, she wrote: "Siegbahn actually did not want to have me. At the time he said he had no money, he could give me only a place to work, nothing more."[57]

But Meitner continued to work, although as she stated, it was just "bits and pieces" of the tremendous work she had done in Berlin. For example, she measured the cross-section (reaction probability) of thermal neutrons in thorium, lead, and $U^{238}$. She initiated studies of neutron capture in thorium, the radioactivity of scandium, and the primary and secondary electrons in beta spectra. She struggled to continue her physics studies, even as her concerns for friends in German-occupied countries continued to take over much of her life: "I don't take this all very tragically [her struggle to continue her

---

[55] Sime (1996), p. 695.

[56] Sime (1996), p. 696.

[57] Sime (1996), p. 255.

experimental work], because I no longer take my life very seriously. One sees far too much suffering to be overly concerned about oneself."[58]

**Hahn**. Following the publication of the splitting of the atom by Hahn and Strassmann in January 1939, Hahn and his colleagues reported the description of 25 elements of atomic numbers 35–58 produced by fission, and over 100 isotopes in a total of 14 publications between 1939 and 1943. Besides his published public work, Hahn integrated himself and his institute into military research, and cultivated strong connections with the Nazi high command and with industry partners. He participated in most if not all the meetings between the scientists working on the atomic bomb and the military high command, he actively sought funding from the Nazis as well as from industry partners, and he was rewarded for his work with recognitions such as the Fuhrer's War Service Cross First Class (a recognition that Hahn never listed among his many recognitions).

His first military contract focused on gas warfare, but he moved quickly into "military research on nuclear fission" and its practical applications.[59] His first active involvement began in September 1939 when the German Army Weapons Bureau assembled a group of scientists to begin fission research, including at least three of the scientists who would end up as captives of the Allies in 1945 (Erich Bagge, Kurt Diebner, and Hahn). In October 1939, the Weapons Bureau officially took control of the KWI for Physics in Berlin-Dahlem, and Werner Heisenberg joined what was now called the *Uranverein* (the Uranium Club). The Weapons Bureau dropped control of the nuclear fission project in 1942, but Heisenberg and Hahn continued to report on the nuclear project to the Reich Education ministry in 1942, and the Academy of Aerodynamical Research in 1943. It is not clear from the reports how Hahn was involved in the research on weapons and reactors. His research appeared to involve mostly measurements of basic physical properties of nuclear materials. However, as the Director of the KWI for Chemistry, his main job would have been to continue the necessary military work, and to secure as much funding as possible to keep the Institute operating. This was important to keep himself and especially his younger colleagues from being conscripted into the *Wehrmacht*. It is also clear that he attended most if not all the meetings where nuclear fission was discussed, and he and Heisenberg are the most often mentioned in German reports.

---

[58] Sime (1996), p. 696.

[59] Sime, R. L. (2012). The politics of forgetting: Otto Hahn and the German nuclear-fission project in World War II. *Physics in Perspective, 14*(1), 70. https://doi.org/10.1007/s00016-011-0065-6.

**Fig. 3.9**   The KWI for Chemistry destroyed by allied bombing. (Courtesy of the Archives of the Max Planck Society, Berlin, VI_001_KWI_f_Chemie_II_15)

In 1944, Hahn's Institute in Dahlem was destroyed by Allied bombs (Fig. 3.9), and much of Hahn's correspondence and other personal items were lost. The Institute was reestablished in the south of Germany at Tailfingen. On April 25, 1945, the Allied "Alsos" team captured Tailfingen, and drove to the displaced KWI. The GIs informed Hahn, who now looked thin and ill, that he was under arrest. He was put into an armored vehicle and driven to Hechingen where other members of the Uranium Club—Bagge, Horst Korsching, von Weizsäcker, and Karl Wirtz—were captured. While stopping in a small town near Paris they heard the news of the unconditional surrender of Germany. Shortly thereafter, Heisenberg, who had been arrested in Urfeld where he had fled to be with his family, also joined the group. The last to join were Paul Harteck and Walther Gerlach. The group of ten was driven next through Belgium, then finally interned at Farm Hall, a country estate near Cambridge, England. There they were kept completely incommunicado. No one, including their families, knew where they were for the nine months that they were interned at Farm Hall.

The German scientists were surprised when they arrived at Farm Hall. There didn't seem to be any serious security surrounding them, and there didn't seem to be any evidence of hidden microphones. Heisenberg laughed: "Microphones installed? Oh, no, they're not as cute as all that. I don't think they know the real Gestapo methods; they're a bit old fashioned in that respect."[60] So they felt free to talk. What they didn't know was that tiny microphones had been hidden behind the many paintings on the walls of Farm Hall, recording every conversation that they had. After these recordings were declassified, Jeremy Bernstein published a comprehensive translation of many of the available recordings, together with extensive footnotes.[61] The transcripts provide a remarkable glimpse into what these scientists were thinking and feeling before and after the dropping of the first atomic bomb on Hiroshima.

August 6, 1945, seemed like any other day at Farm Hall for the detainees. But later in the afternoon, their usual routine was abruptly altered, and the Germans were ordered to leave their rooms. They would be required to listen to the 6 o'clock BBC News. The news report contained only a few sentences:

> ...President Truman has made known a great achievement of the Allies' scientists: they have created the Atom Bomb. The first has been dropped on a Japanese army base and had as much explosive power as two thousand of our own ten ton bombs.[62]

When they first heard the news, they were stunned. None of them believed that it was possible that the Americans might actually have created this powerful weapon based on Hahn's original discovery of fission. Hahn was completely shattered by the news, and Hahn's colleagues were concerned that he might commit suicide. He said that he felt responsible for the deaths of hundreds of thousands of people. Notes from Hahn's diary suggest that upon hearing the news, he held out hope that the news was not true.

> I want it not to be true, but the Major assured me that it was no reporter's fanciful tale but an official announcement of the President of the United States. I almost fell to pieces at the thought of this new, great misery...[63]

Recorded comments by Hahn at Farm Hall, as well as comments he made prior to 1945, suggest that he probably had only a limited understanding of how one might actually produce a bomb. For example, at Farm Hall he asked Heisenberg how much "235" would be needed to build a bomb, and "how does the bomb explode?"[64] In addition, Hahn believed that building a bomb

---

[60] Bernstein, J. (2013). *Hitler's uranium club: The secret recordings at Farm Hall.* Springer Science and Business Media. p. 78.

[61] Bernstein (2013), p. 82.

[62] Hoffman (2001), p. 3.

[63] Hoffmann (2001), p. 4.

[64] Bernstein (2001), p. 128.

so quickly would be impossible. Harteck disagreed, saying that it would have been possible but would require 100 people to work on it. This was an enormous underestimate given the 120,000 people who worked on the Manhattan Project. In one entry in his diary, Hahn writes:

> I am now glad that we had no way or means of developing a bomb for if it had been possible to build it in Germany during the war one would have been compelled to use it against England. To me that is unthinkable.[65]

von Weizsäcker, another Farm Hall detainee, asserted that the reason Germany did not build a bomb was because German physicists didn't want to. "If we had all wanted Germany to win the war we would have succeeded." Hahn replied: "I don't believe that but I am thankful we didn't succeed."[66]

As more time passed after the bombs had been dropped on Hiroshima and Nagasaki, Hahn, von Weizsäcker, and Heisenberg in particular, worked to come up with the story that they had really not wanted to build a bomb. In the *Farm Hall Transcripts*, Bernstein argues that they really just "couldn't" make a bomb.[67] Nevertheless, the internees produced a "press release" (that they never publicly released) stating categorically that they felt such a bomb was morally wrong, and they actively did not want to produce such a weapon. von Laue, the one captive scientist who was opposed to this statement, later called the missive the "*Lesart*" or "version of events." As stated in this "myth," the German scientists believed that:

> …history would record that the Americans and British made the bomb and the Germans produced a workable engine [which they never did—but that was their goal]. Therefore, the peaceful development of the uranium engine was done in Germany under Hitler, while the Americans developed this 'ghastly weapon of war.'[68]

As discussed in the chapter on Heisenberg, the facts would suggest otherwise. But this "denial" or "myth" of what Hahn and others actually did during both World War I and World War II would be an important part of Hahn's efforts to resurrect science in post-war Germany.

---

[65] Hoffmann (2001), p. 7.
[66] Bernstein (2013), p. 122.
[67] Bernstein (2013), p. 185.
[68] Bernstein (20,130, pp. 258–260.

# The Nobel Prize for Fission

On Friday, November 16, 1945, Hahn and the other scientists were still being held incommunicado at Farm Hall. As the detainees were reading through the daily papers, Heisenberg, startled, announced that the *Daily Telegraph* had published a short article stating that Hahn was being awarded the 1944 Nobel Prize in Chemistry:

> The Swedish Academy of Science announced today Nobel Prize awards for 1944 and 1945. Among the prizes awarded were: Prof. Otto Hahn for splitting heavy atom nuclei. He is reported to be in England.[69]

In a celebratory mood, the Farm Hall detainees honored Hahn at dinner that evening with congratulatory speeches and a song that Diebner and Wirtz sang; the translation of the first stanza and refrain is:

> Detained since more than half a year.
> Are Hahn and we in Farm Hall here.
> If you ask who bears the blame,
> Otto Hahn's the culprit's name.
> The real reason, by the by,
> Is we worked on nuclei.
> They're for the war, the nuclei
> And the general victory.[70]

Hahn expressed doubt that the announcement was authentic. Then there was confirmation from the Swedish news agency, but it was not known if Professor Hahn was alive, and if so, where he was. The Nobel Prize cannot, by its statutes, be awarded to anyone posthumously. The *Sunday Pictorial* on November 18, 1945, reported that: "…the Swedish Academy has…offered a reward for any person who can give useful information as to Otto Hahn's present location."[71] As with all previous winners, Hahn was invited to travel to Stockholm to receive his award and present a lecture in December of 1945. But Sweden and the world still did not know his whereabouts. It's certainly curious that the Academy announced this award at a time when they had no idea where Hahn was, whether he was alive, or if he might be charged with war crimes. When official notification reached Hahn from the Swedish Academy, his captors allowed him to reply that he would be unable to attend

---

[69] *The Daily Telegraph* (1915, Thursday Mar 6), p. 8.
[70] Bernstein (2013), p. 300.
[71] Bernstein (2013), p. 290.

the award ceremony and could not give a date for his attendance due to "certain circumstances"—that is, his captivity at Farm Hall. Once the detainees were returned to Germany on January 3, 1946, Hahn was officially invited to Stockholm to deliver his lecture and receive his prize at the awards ceremony in December 1946, as shown in Fig. 3.10.

There will always be the unanswered question as to why the Swedish Academy decided to award the prize to the discoverer of fission at just this moment in time, when the practical application of Hahn's discovery had led directly to the development of the atomic bomb, the most destructive weapon yet known to mankind. In Alfred Nobel's will it explicitly states that the prize should be awarded annually to "those who have conferred the greatest benefit on mankind." Did the Swedish Academy truly believe that the practical application of fission had conferred a great benefit? Of course, following the rules of the Prize, the nominations had been received early that year, well before the bombs had been dropped. But it was already understood in 1939 that a bomb was possible. One can certainly speculate that the timing of the award was more political than scientific. Perhaps the Academy wanted to highlight the end of the Nazi regime, and through this award promote future German

**Fig. 3.10**  Hahn at the Nobel Prize ceremony. (Courtesy of the Archives of the Max Planck Society, Berlin, VI_001_R_1_1_51_2.jpg)

science that was now out from under Hitler's power, just as they had rewarded Haber at the end of World War I when Kaiser Wilhelm II was ousted.

Since Hahn's whereabouts were unknown at the time of the announcement, the press and public focused their attention on someone who was available; in fact, many scientists and journalists believed that Meitner had been the discoverer of fission. One of her first interviews was with Eleanor Roosevelt on August 9, shortly after the bombing of Hiroshima. In September, she had been invited to spend the winter semester as a visiting professor at the Catholic University of America. In October, she was notified that she had been elected as a foreign member of the Swedish Royal Academy. She would be just the third woman to receive this honor in the 200 years of the Academy's existence; she was finally receiving some recognition from her Swedish colleagues. After arriving in the U.S., plans were made for a lecture tour through a number of prestigious universities in the east. In February, she was presented with the Woman of the Year award by the Women's National Press Club; at the gala event, she was seated with President Truman who exclaimed: "So you're the little lady who got us into this mess!"[72] In addition, numerous newspapers reported that it was Meitner who had been "responsible for the bomb." In a *New York Times* article on August 7, 1945 under the headline "First Atomic Bomb Dropped on Japan," a paragraph buried deep in the article had this statement: "The key component that allowed the Allies to develop the bomb was brought to the Allies by a female, 'non-Aryan' physicist." Newspapers all across the U.S. praised Meitner."[73]

Meitner traveled back to Sweden to meet Hahn on the occasion of his acceptance of the Nobel prize in chemistry. On December 10, 1946, Meitner met Hahn and his wife as they arrived for the awards ceremony. Meitner had decided that she must attend the ceremony. She wrote to her nephew: "…if I don't go, I fear it might be misunderstood."[74] Newspapers carried accounts of the Hahn's arrival, noting the presence of Lise Meitner, Hahn's "former pupil."[75] Meitner's focus in Stockholm was to try to recover their personal relationship. Hahn's sole purpose was quite different. During his visit to Stockholm and during the surrounding ceremonies, he campaigned for Germany, portraying a country struggling to rebuild, oppressed by the Allied occupation. He even went so far as to publicly state that he was "glad Germany

---

[72] *Lise Meitner's fantastic explanation: Nuclear fission.* (n.d.-b). - ANS / Nuclear Newswire. https://www.ans.org/news/article-938/lise-meitners-fantastic-explanation-nuclear-fission/.

[73] First atomic bomb dropped on Japan. *The New York Times*, August 7, 1945.

[74] Sime (1996), p. 340.

[75] Corradini, D.A., Geiger, K., & Mazohl, B. (2021). Lise Meitner (1878–1968): Pioneer in nuclear physis. U. Frevert, E. Ostercamp, & G. Stock (Eds). *Women in European Academies: From Patronae Scientiarum to Path-Breakers*. DeGruyter, p. 173. https://doi.org/10.1515/9783110634259-010

had not built the bomb and caused the needless deaths of thousands."[76] Meitner listened carefully to Hahn's Nobel lecture on December 13. Although he could not ignore Meitner entirely, he described the discovery as involving only radiochemistry, giving no indication that the discovery and importance of fission was absolutely dependent on both chemistry and physics. As Sime notes: "Hahn's behavior in Sweden hurt Meitner personally, damaged her professionally, and contributed to her ongoing isolation in Sweden."[77] It is probable that Hahn never understood or cared.

It is interesting to note that in his autobiography, Hahn mentions his Nobel Prize in just two sentences: "In 1945 the Nobel Prize Committee awarded me the prize in chemistry for the year 1944. Since I was not allowed to leave England at that time, the Prize was finally presented to me in Stockholm on December 10, 1946."[78]

## Why Was Meitner Overlooked for a Nobel Prize?

One of the major controversies surrounding the Nobel Prize awards in the sciences is the issue of scientists that have been overlooked. Many reasons have been cited for the lack of recognition, including that there were more important discoveries made during the years of nomination; the focus of the committee at the time of the selection was either more experimental or more theoretical; or there was political or personal bias toward the scientist or the area of work. At the top of most lists of deserving scientists who were not recognized is the failure of the committee to award a Nobel Prize to Lise Meitner, the Jewish physicist who was intimately involved in the discovery of fission.

Many articles support Meitner for a prize.[79] Some cite her contributions to the team in Berlin (which she initiated in 1934), in which she was the intellectual leader, according to Strassmann.[80] Others cite her theoretical explanation of fission, which she published with Frisch while she was in exile in Sweden. Many of her colleagues felt that she, together with Hahn or Frisch, should have been awarded a prize for the explanation of the energy produced during fission. In fact, she was nominated 29 times after 1939 by several

---

[76] Sime (1996), p. 341.

[77] Sime (1996), p. 342.

[78] Hahn (1966), p. 179.

[79] Sime, R. L. (1998). Lise Meitner and the discovery of nuclear fission. *Scientific American*, 278(1), 80–85. https://doi.org/10.1038/scientificamerican0198-80.

[80] Sime (1996), p. 451.

Nobel laureates, including Bohr, Franck, Planck, and von Laue. As James Franck wrote in his nomination letter:

> In my mind there is no doubt that the finding of fission in radioactivity is the greatest progress made in nuclear physics in the last thirty years. The work which led to this success was done jointly by Hahn and Meitner at the Kaiser Wilhelm Institute…the first proof was offered by Liese [sic] Meitner in working together with Frisch.[81]

Even Carl von Weitzsäcker, a strong supporter of Hahn, acknowledged that "whether Meitner should have gotten the physics prize remains an open question."[82]

Meitner's and Hahn's professional and personal relationship did not end when one or the other was away from the research. During both World War I and World War II, wartime circumstances tested their collaboration. As discussed above, Hahn was at the front much of the time between 1915 and 1918, while Meitner was continuing their research to identify the precursor to actinium. Meitner achieved the identification and wrote a letter to Hahn, declaring with excitement that she had found this precursor and that they could now publish an article together. Although she had done the work and made the protactinium discovery in Hahn's absence, she included both their names as authors. Similar circumstances confronted them when Meitner had to leave Germany in the summer of 1938. Hahn and Strassmann continued their experiments to identify the products of bombardment of uranium with neutrons. This research had begun in 1934 at Meitner's urging, and she was an active, necessary member of the team. After she left, Hahn kept her apprised of the work through letters and a "secret" meeting in Copenhagen in July, 1938, continuing to ask for her expert guidance and suggesting if she could come up with some fantastic explanation for the bizarre findings then they could both be part of the discovery with two separate publications. If one examines all the writings between Meitner and Hahn, it is very obvious that both partners felt that the team was continuing as usual, with one partner (Meitner) outside Germany because of the war. Arguably, the Nazi censors would not have allowed Hahn to publish a paper with a Jewish co-author, but Meitner could have submitted a joint article from Sweden to a British or American journal. Even this might have caused Hahn trouble, but there is no evidence that he ever considered this option.

---

[81] James Franck nomination letter, Nobel Archives (retrieved by authors); 1998.

[82] von Weizsäcker, C. F. F. (1996). Hahn's Nobel was well deserved. *Nature*, 383, 294. https://doi.org/10.1038/383294a0.

# Why Was Meitner Ignored by the Swedish Academy?

Understandably, there was considerable confusion regarding the nominations, reports, and recommendations during the war years. As discussed by Elizabeth Crawford and colleagues, evidence from examination of committee reports suggests that there were several factors that contributed to the decision by the Academy to not include Meitner in a Nobel prize for fission: the committees lacked expertise in theoretical physics; they had difficulty in evaluating an interdisciplinary discovery; and Sweden was politically and scientifically isolated at the time of the discovery, all leading to a lack of understanding of the importance of the contribution of both physics and chemistry, and experiment and theory.[83] It is therefore worth examining in detail the role of the Swedish Academy in excluding Meitner from consideration.

There is no question that the discovery of fission was the most momentous discovery of 1938. Not only did it represent extraordinary new physics, but by early 1939 everyone was aware of the enormous energy—and potential for destruction—revealed by Meitner's and Frisch's calculations. In early 1939, Theodore Svedberg, chair of the Chemistry Committee, recognized the importance of the discovery, and nominated both Hahn and Meitner. However, the report that he prepared for the committee concluded that the discovery was made after Meitner left Germany, and that Hahn alone had made the discovery after the collaboration had allegedly ended.[84] Based on this, Svedberg was prepared to remove Meitner from consideration and propose that Hahn alone should be considered. However, in a further move, Svedberg concluded that based on rapid developments in the field, the committee should wait before awarding a prize for fission.

In 1940, the Physics Committee considered a prize for fission since they had received several nominations for Hahn and Meitner together. One especially important letter from Arthur Compton specifically noted that the discovery involved both Hahn's chemical and Meitner's physical results.[85] However, like the Chemistry Committee in 1939, the 1940 Physics Committee again decided it was "too early to make the award intelligently."[86]

---

[83] Crawford, E., Sime, R. L., & Walker, M. (1997). A Nobel tale of postwar injustice. *Physics Today*, *50*(9), 26–32. https://doi.org/10.1063/1.881933.

[84] Friedman, R. M. (2001). *The politics of excellence: Behind the Nobel Prize in science*. W H Freeman and Company, p. 241.

[85] Arthur Compton nomination letter, Nobel Archives (retrieved by authors); 1008.

[86] Friedman (2001), p. 241.

In 1941, the Physics Committee again took up the issue. James Franck wrote a strong letter stating that the discovery was part of an ongoing collaboration between Meitner and Hahn, and the award should go to both of them. However, the Physics Committee balked on making a decision, and declared that the issue should be taken up by the Chemistry Committee, throwing the debate back to the chemists. Svedberg (still the chair of Chemistry) dug in his heels and continued to maintain that only Hahn deserved the prize. Since the prize had been canceled for 1941, no decision was made. In 1942, an influential member of the chemistry committee, Palmaar, changed his nomination from Hahn alone to Hahn and Meitner, but tragically died before he could write his report. The other members of the committee went along with Svedberg and agreed that Hahn alone should be the prize winner. But, based on the chaotic situation created by the war, the committee decided to reserve the 1942 prize. The situation remained the same in 1943 and 1944: nominations continued for Hahn alone or for Hahn and Meitner together, but the prizes were reserved.

After World War II ended in 1945, the Chemistry and Physics Committees had the opportunity to thoroughly evaluate the significance of the discovery of fission and the persons who played early and important roles. The Committees had access to important wartime publications, as well as the ability to call on international figures to weigh in on who should be included in the prize, as Svedberg had recommended in 1939. In 1945, Klein and Bohr had both recommended that the physics prize be awarded to Meitner and Frisch and the chemistry prize to Hahn and Strassmann; "this would be the fairest distribution of recognition."[87] At this point, the Chemistry Committee hesitated after realizing the complexity of fission research. They decided that it would be wrong to act too quickly, and guidance would likely come from the nominations in the following year. In their original written statement, they admitted that more time might shed light on whether Meitner should share the prize with Hahn. Svedberg, still an outspoken supporter of Hahn alone, crossed out Meitner and added "other researchers" instead.[88] However, in November 1945, the full membership of the Royal Academy voted by a slim margin against the Committee recommendation, and awarded the reserved 1944 prize to Hahn alone. They announced the award on the 16th of that month.

The Physics Committee still had the opportunity to reevaluate the contributions of Meitner and Frisch for a prize in 1946. Meitner alone or Meitner

---

[87] Friedman (2001), p. 246.
[88] Friedman (2001), p. 244.

and Frisch together had received a number of nominations from prominent physicists and former laureates. Letters from all the nominators stressed the importance of physics in the discovery, and the role that Meitner had continued to play after her escape. One of the nominators also stressed that it would be striking if a physics prize was not awarded for fission. Unfortunately, the Committee assigned Erik Hulthen to write the report on Meitner and Frisch. Hulthen was perhaps the worst person to write the report. Not only was he politically and personally biased against Meitner, but his report also clearly underscored how little he understood the science. In his 1946 report Hulthen erroneously stated that the model for fission that Meitner and Frisch reported in 1939 was previously suggested by Bohr, and that based on this information, "Niels Bohr should probably be the one who should come to mind when awarding a Nobel Prize in this context." But, since Bohr was not nominated for the prize that year, no one should therefore be awarded a prize in physics for fission.[89] Based on the committee reports in 1946, "the committee members themselves think that a decision should be postponed…concerning a prize award for Meitner's and Frisch's contribution."[90]

Although Meitner, either alone or together with Frisch, received several nominations after 1946, they knew that it was too late, and the Nobel Committee would never acknowledge their role in the discovery of fission.

## Would Meitner Have Slowed Fission Research If She Had Stayed in Berlin?

In an article published in *Nature* in 1996, von Weizsäcker stated that some members of Hahn's team told him that "If Lise Meitner had stayed in Berlin, it is quite possible that Hahn would not have started his experiment and so would not have been the one who discovered uranium fission."[91] Additionally, when Meitner was included in the Hall of Fame (*Ehrensaal*) in the *Deustches Museum*, von Weizsäcker was asked to give remarks. In his talk he spoke enthusiastically about Meitner as a person but stated that she had not contributed to the discovery of fission, but in fact had impeded it, and that the discovery was totally Hahn's. In other words, he was suggesting that Meitner was actually holding Hahn back by insisting on repeating his experiments and

---

[89] Friedman (2001), p. 248.

[90] Friedman (2001), pp. 248–249.

[91] Von Weizsäcker, C. F. F. (1996). Hahn's Nobel was well deserved. *Nature*. 383, p. 294. https://doi.org/10.1038/383294a0.

stating that splitting the atom was very difficult to believe as a theoretical physicist. Additionally, there is some evidence that Hahn himself felt that Meitner's presence would have negatively affected the discovery of fission. In a personal conversation, Bagge stated that Hahn told him in 1945 that "If Frl. Meitner had still been in the institute in December 1938, she would have talked us out of barium."[92] Additionally, Heisenberg indicated that Hahn stated, "I don't know; I'm afraid Lischen would have forbidden me to split uranium."[93] In contrast, Strassman wrote that it was actually fortunate that Meitner's opinion was so important to Hahn; her concern about their results prompted Hahn and Strassmann to conduct additional experiments which confirmed the breaking up of the atom.[94]

## Why Did Hahn Not Support Meitner After the War?

Hahn's behavior after the war with respect to acknowledging Meitner's role in the discovery of fission is complicated. Although it might have been difficult for Hahn to acknowledge Meitner, a Jew, under the Hitler regime, this was no long a concern after the war. What is clear is that immediately after learning of the dropping of the first atomic bomb by the USA, Hahn's reaction was confusing, toggling between feeling fully responsible for the massive number of Japanese deaths, and taking the moral high ground by stating that he was glad they didn't develop the bomb. As Bernstein points out:

> Hahn was completely shattered by the news and said he felt it was his original discovery which had made the bomb possible. He…had originally contemplated suicide when he realized the terrible potentialities of his discovery…[95]

At some point during the discussion of the dropping of the bomb, the ten scientists at Farm Hall decided to create their version of their military work, developing a memorandum that all ten scientists at Farm Hall signed, now called the *Lesart*. This statement was certainly not factual and was in essence a "self-serving revision of history that would place the Germans on the moral

---

[92] Sime (1996), p. 454.
[93] Sime (1996), p. 454.
[94] Friedman (2001), p. 235.
[95] Bernstein (2013), p. 115.

high ground."[96] Bernstein further states that the German scientists wanted history to treat them kindly and believe that they had "behaved morally in an immoral regime…"[97] Hahn grabbed hold of this version and continued to build upon this myth for the rest of his life. He could take on the role as savior of German science—the "good" German who had never joined the Nazis and only wanted to continue his research.

An important component of the *Lesart* was the idea that the discovery of fission was an inevitable chemical discovery, one that had not involved physics, and that Hahn alone had made the discovery. There was no physics needed; it was discovered by tedious, repetitive chemical separation experiments. Hahn desperately needed the scientific world to believe that he alone discovered fission, a belief that might provide him with the prestige to rebuild German science.

In a strange twist of fate as Hahn was trying to create his Nobel persona, the public had no idea where he was located, and he was therefore unavailable for interviews. But one person was available, Lise Meitner. Her role in the discovery was known from her publication with her nephew Otto Frisch. The public jumped at this opportunity to learn more about the discovery and hounded Meitner in Sweden. She was the only one available. Frisch was still at work at Los Alamos and not available, and Strassmann was in occupied Germany. This complication, as well as the huge number of publications by physicists after the fission discovery (over 100 in 1940 alone) led Hahn to believe that the physicists were hijacking "his" science. To make matters worse, many of these publications did not even reference the Hahn and Strassmann articles but focused instead on the Meitner and Frisch publication. As Gerard Lander and Michael Steiner point out, this "forced silence in the face of these press reports [honoring Meitner]…reinforced his resentment of the physicists in general, and perhaps of Lise Meitner more than she realized."[98]

There was still a possibility of a Nobel prize in physics in 1946, and Meitner and Frisch received a number of nominations for that year. Many in the physics community, including Bohr, still believed that the work by Meitner and Frisch deserved recognition by the Nobel committee. Hahn, as a Nobel laureate, had the opportunity that year to nominate Meitner—but he did not. And, when their names were not announced in 1946, they knew that their chances were over. Hahn did nominate Meitner for the 1948 prize, but it was

---

[96] Lucas, A. A. (2007). Revisiting Farm Hall. *Europhysics, 38*(4), 25–29.

[97] Bernstein (2013), p. 148.

[98] Lander, G. & Steiner, M. (2015). Revisiting the discovery of nuclear fission – 75 years later. *Journal of Neutron Research*. 18, 8. https://doi.org/10.3233/jnr-150018.

too late by that time: fission was old news. If Hahn made some effort to support Meitner, and to include her as a vital member of the fission team, if he had stressed their lifelong professional and personal friendship, it might have made some difference to the committee. But he did not. As Meitner's supporter Oskar Klein wrote to Bohr: "Hahn's…knowledge of the history of the discovery is lacking, to put it mildly."[99]

Hahn also deceived others about Meitner. As early as 1946, when Hahn was in Sweden to receive his Nobel, a Swedish newspaper described Meitner as Hahn's former pupil. He did nothing to correct this image. In 1953, she gave a public talk in Austria on the future of nuclear energy and its peaceful uses. Newspapers considered the talk worthy of coverage, but not Meitner herself. She was described as Hahn's assistant. Hahn did try perhaps to show some acknowledgment—at least to Meitner—when he gave her part of the financial award that came with his prize. Characteristically, in January 1947, she gave it to the Emergency Committee of Atomic Scientists, a committee of scientists who were greatly concerned about the politicization of atomic research. Meitner received a friendly thank you from Einstein, who was chairing the committee at that time:

> Dear Professor Meitner:
> On behalf of my colleagues of the Emergency Committee of Atomic Scientists, I send sincere thanks for your generous help in the great educational task we have undertaken. We value not only the practical support you have sent, but also the good will towards this work and the hope for a reasonable solution of this immense problem which your contribution expresses.
> Cordial greetings, Albert Einstein

## The Post-War Years

**Otto Hahn.** As a result of the war, the situation at many of the KWIs was chaotic. Most of the institutes had been relocated or damaged, and very few of the original personnel were left. Following his internment at Farm Hall, Hahn had been released into the British Zone of Germany at Alswede in January of 1946. A month later, Hahn and Heisenberg were given approval by the Allies to move to Göttingen to attempt to recreate their institutes in the empty rooms of the old Institute for Aeronautical Engineering. Hahn's move

---

[99] Sime (1996), p. 329.

to Göttingen did not lead to much improvement in his personal life. In a letter to Meitner he stated:

> Near me live von Laue and Heisenberg, each in his own room. Not exactly very opulent—palliasse bed, no armchair, no sofa, no pictures on the wall. But I am encouraged. I have running water, even if it is not hot. The fate of our institute, of physics, and of chemistry is not yet clarified. So unfortunately I cannot go to Tailfingen and visit Edith and the Institute.[100]

It was not until July 1946 that Edith was allowed to move to Göttingen.

The Allies had major concerns about the moral integrity of the Kaiser Wilhelm Society (KWS) since many of the institutes had been commandeered for war research. Building on the legacy of the original KWS and recognizing that some scientists had maintained their moral integrity during the war, a new society was established and named the Max Planck Society (MPS), with Max Planck as the first President. Planck served for only one year, with Otto Hahn appointed as his successor in 1946. The Institutes in the British and American zones were added in 1948; the Institutes in the French zone in 1949; and, finally, the Institutes in Berlin in 1953.

Hahn played a major role in rehabilitating German science. He was held in high regard in Germany, known for his Nobel prize for fission and for maintaining his position during the war years as a "decent" German, one who was "politically upright as his scientific research was 'pure'."[101] He was viewed as the grand old man of science in the post-war years, as we see in Fig. 3.11, and it was widely accepted that he would bring German science back to its pre-war eminence. He also was aware of the potential for expansion of atomic weapons, and he became actively involved in supporting peaceful uses of fission. By the 1950s, he had become involved in the public movement devoted to emphasizing the dangers of nuclear warfare. For example, in 1954, he wrote a paper on the exploitation of nuclear energy, a paper that was widely circulated around the world.[102] He assisted other scientists and Nobel laureates in the preparation of the Mainau Declaration in 1955, signed by 51 Nobel prizewinners, that called attention to the dangers of atomic weapons. He later led a group of leading German atomic scientists in releasing the Göttingen Manifesto on April 12, 1957, which renounced the use of atomic

---

[100] Hoffman (2001), p. 206.
[101] Sime (2006), p. 5.
[102] Cornwell (2003), p. 413.

**Fig. 3.11** Hahn in later years. (Courtesy of the Archives of the Max Planck Society, Berlin., Hahn_I_9)

weapons of any kind and protested against the proposed nuclear arming of West Germany.[103]

After the war, Hahn began to have personal issues that made it difficult for him to continue much of his traveling away from home. In 1951 as he was entering his home, he was shot in the back by a disgruntled inventor who was hoping to raise awareness of his work among prominent scientists.[104] Having just recovered from this episode, he was injured in a minor car accident in 1952, followed shortly by a mild heart attack. And in 1951 his wife Edith suffered a nervous breakdown and was hospitalized for several months. Tragically, in 1960, his only son, Hanno, and his wife were killed in an auto accident from which Edith never fully recovered.

Hahn received numerous recognitions and honors for his service in both world wars, his scientific accomplishments, and his work in later life as an advocate for peace. Just before his retirement in 1960, it was announced that the Institute of Nuclear Research in Berlin would be renamed the "Hahn-Meitner Institute", and the Max Planck Institute for Chemistry in Mainz would be renamed the "Otto Hahn Institute" (where Fritz Strassmann had become a Professor). He was awarded a number of medals and prizes; he held numerous honorary memberships in foreign academies and scientific

---

[103] Hoffman (2001). p. 221.

[104] Hoffman (2001). p. 231.

societies; he was awarded several honorary degrees; and he received several war recognitions including the Iron Cross First and Second class (1915) and the General Honor Decoration of the German Empire (1916). Hahn also has a number of unique items named after him, including the only European nuclear-powered civilian ship (NS *Otto Hahn*); a crater on the moon; an asteroid; the 1970 Otto Hahn postage stamp; the Otto Hahn Prize from the German Chemical and Physical Societies; and the Otto Hahn Peace Medal. In 1966, Lyndon Johnson and the Atomic Energy Commission (AEC) awarded Otto Hahn, Lise Meitner, and Fritz Strassmann the Enrico Fermi Award for "pioneering research in the naturally occurring radioactivities and extensive experimental studies culminating in the discovery of fission."

In 1968, Hahn suffered a fall from which he did not recover, and he died a few weeks later on July 28, 1968. As Glenn Seaborg would later state:

> It has been given to very few men to make contributions to science and to humanity of the magnitude of those made by Otto Hahn. He has made those contributions over a span of nearly two generations, beginning with a key role in the earliest days of radiochemistry in investigating and unraveling the complexities of the natural radioactivities and culminating with his tremendous discovery of nuclear fission of uranium. The work of many men has made possible our entrance into this nuclear age. But among them few have made as momentous a contribution as Otto Hahn has. For his special genius the world of science will be forever grateful.[105]

Edith and Otto Hahn are buried in the Göttingen City Cemetery. At the bottom of Hahn's tombstone is the formula for fission (Fig. 3.12).

**Lise Meitner.** After the debacle of the 1944 Nobel prize to Hahn alone, and the failure to win a prize herself, Meitner went on to create a new chapter for herself in Stockholm. Meitner stayed in Stockholm from 1938 to 1949 and became a citizen in 1949. In 1947, she moved to the Royal Institute of Technology, where a new facility for atomic research was starting. She had laboratory space, assistants, and a young colleague in the laboratory next to hers. Her research focused on some theoretical speculation about the shell structure of fission fragments; these would be her last scientific publications. Looking back, Meitner commented "that in Sweden…physics has brought life and fullness into my life."[106]

One difficulty that Meitner faced for the rest of her life was her connection to Otto Hahn. Scientifically, Hahn went to great lengths to continue

---

[105] Seaborg, G. (1967). Bulletin of the Atomic Scientists, 23(4), p. 23.

[106] Meitner, L. (1964). Looking back. Bulletin of the Atomic Scientists, 20(9), p. 7.

**Fig. 3.12** Grave and tombstone of Otto Hahn. (Creative Commons Attribution 4.0 International; photo by Julian Herzog. Photo cropped and changed to B/W)

the myth that fission was his work alone, and belonged to the chemists, not to the physicists. He simply did not believe that she deserved any scientific restitution at all. One episode in particular underscores the place that Meitner held in Germany. A new building had been erected at the renovated KWI for Chemistry, which is now being used by the Free University of Berlin. It was decided to name this building after Otto Hahn. When von Laue proposed that the new institute for nuclear research near Berlin be named for Meitner, there was some resistance to using just Meitner's name, so the institute became the Hahn-Meitner-*Institut* in Mainz. At least in this one instance, Meitner still had her name on a building in Germany, to remind the scientific community of the role she had played in pre-war physics in Germany. Hahn, however, continued to overshadow any contribution from Meitner, and she would in a sense always be considered his *Mitarbeiterin,* or employee, which in fact was used in public reports of their fission work. Even close colleagues, such as Heisenberg, referred to her as Hahn's *Mitarbeiterin.* Finally, in 1953, she reached a breaking point and wrote to Hahn:

Now I want to write something personal, which disturbs me and which I ask you to read with our more than 40-year friendship in mind…I read in an article…by Heisenberg about the relationship between physics and chemistry in the last 75 years, where the only mention of me…is as 'Hahn's longtime

*Mitarbeiterin…*'. What would you say if you were only characterized as the "longtime *Mitarbeiterin*" of me?[107]

There is no record that Hahn replied.

One other example is particularly worth noting. Many papers and books written about the discovery of fission and collaborative work by Meitner and Hahn include the now famous picture of the workbench that the team used for their radioactive studies. Sime describes the workbench as "Meitner's physical apparatus" that was used by the Berlin team in their discovery of fission.[108] However, as time elapsed, the workbench became known as Otto Hahn's, and eventually ended up on display in the *Deutsches Museum* in Munich in 1953 as *Arbeitstisch von Otto Hahn*—the worktable of Otto Hahn (Fig. 3.13). The wording above the workbench simply states that Hahn had worked at this table, with a recording by Hahn explaining the components accompanying the display.[109] It's important to understand however, that the workbench is, in fact, a mere representation of some of the instruments used

**Fig. 3.13** Original display of Hahn/Meitner workbench. (Courtesy of the Deutsches Museum, Munich, Archive, BN24876)

---

[107] Sime (1996), p. 370.

[108] Sime (1996), p. 258o.

[109] Sime, R. L. (2010). An inconvenient history: The Nuclear-Fission Display in the Deutsches Museum. *Physics in Perspective*. 12(2), 204. https://doi.org/10.1007/s00016-009-0013-x.

by the Hahn-Meitner team. Their experiments were actually carried out in three different rooms in Meitner's physics section on the ground floor of the KWI, and Meitner had been responsible for designing the experiments used for the fission research.[110] At some point, the museum recognized the contribution of Fritz Strassmann (not Meitner) and attached a plaque above the display that read:

> Otto Hahn discovered in 1938 together with F. Strassmann the splitting of uranium by neutrons and with it laid the foundations for the technical utilization of energy from the atomic nucleus.

In truth, it was the team—Hahn, Meitner, and Strassmann—that ultimately discovered fission and explained the physics. But nowhere was Meitner given any credit for the work or the equipment display. Shortly after Meitner's and Hahn's deaths in 1968, voices rang out, demanding that Meitner get the credit she was due. After many protests, and years during which the museum simply refused to diminish Hahn's memory by including his coworkers, the display was altered in 1990 to reflect more accurately the contributions of all to the fission discovery (Fig. 3.14). As described by Sime[111]:

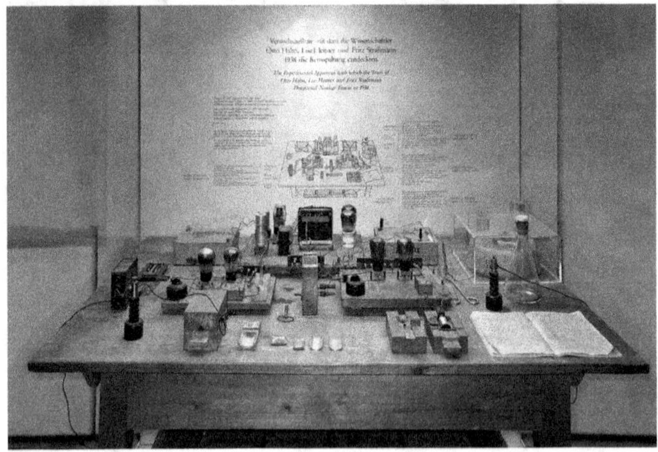

**Fig. 3.14** Hahn/Meitner/Strassman workbench, updated. (Courtesy of Versuchsaufbau Hahn Deutsches Museum.jpg. License: Creative Commons Attribution-Share Alike 3.0 Unported license)

---

[110] Herrmann (1990), p. 484.
[111] Sime (2010), p. 208.

The glass case was new, the Arbeitstisch von Otto Hahn sign was gone, and Hahn's recorded voice was updated with a new introduction. Behind the display was a much larger printed text, in German and in English… designating this as the experimental table of Otto Hahn, Lise Meitner, and Fritz Strassmann and explaining the apparatus. At the side was another panel with explanation of the history and youthful photographs of the scientists, including Otto Robert Frisch.

Heisenberg summed up the Hahn and Meitner relationship in his address at the 1962 ceremony for the presentation of the *Orden pour le Merite*:

…Hahn owed his success primarily to his qualities of character. His untiring energy, his unbendable industry in acquiring new knowledge allowed him to work even more accurately and more conscientiously, to reflect on his own experiments with even more self-criticism, and to carry out more controls than most other scientists who entered the new territory of radioactivity…Meitner's relation to science was somewhat different. She not only asked 'What' but also 'Why.' She wanted to understand…she wanted to [understand] the laws of nature…in the new field. Consequently, her strength lay in the asking of questions and the interpretation of the experiment.[112]

Even though Hahn had redefined their working relationship to diminish Meitner's contributions, their personal friendship survived, as is evident in Fig. 3.15. She stayed in touch with the family, and was informed of every major family event, including the death of Hahn's son and daughter-in-law in a car accident. Meitner also shared her life events, such as the death of her relatives, as well as those lost during the Nazi occupation. She thanked Hahn for his warmhearted and affectionate letters to her, emphasizing their lasting relationship with the notation: "There are no people here who would have known me or my [family] to whom one could say: Do you remember? Or do you still know?"[113]

Although Meitner was not honored with a Nobel Prize for her work on either protactinium or fission, she was recognized with a number of prestigious honors, both in years preceding the discovery of fission, and in the postwar years (Fig. 3.16) right up until her death. The many recognitions include the Lieven Prize (1925), the Ellen Richards Prize (1928), Woman of the Year by the National Women's Press Club (1964 on the occasion of her 80th

---

[112] Rife (1990), p. 263.
[113] Sime (1996), p. 369.

**Fig. 3.15** The friendship endures. (Courtesy of the Archives of the Max Planck Society, Berlin, VI_001_Meitner_Lise_II_13.jpg. Original image cropped)

**Fig. 3.16** Portrait of Lise Meitner at Bryn Mawr College. (Photo by Robert R. Davis, courtesy of AIP Emilio Segrè Visual Archives, Physics Today Collection)

birthday); the Prize for Science and Art (1947); the Planck Medal (1949, with Otto Hahn); the first Otto Hahn Prize in 1954; and the Enrico Fermi Prize in 1965 together with Hahn and Strassmann. At this point in her life, she was too ill to attend the ceremony; she was later presented the award by Glenn Seaborg in the house of Max Perutz in Cambridge where she was staying. Meitner's award bore the words, "For pioneering research in the naturally occurring radioactivities and extensive experimental studies."[114] She was

---

[114] AEC's Fermi award. (1966). *Physics Today*, 19(11), 95. https://doi.org/10.1063/1.3047835.

elected to Academies of Science in several countries and received numerous honorary degrees. Several awards and recognitions stand out in particular. In 1924, she was awarded the Liebniz Medal following a nomination by Einstein, von Laue, and Planck, with this description:

> As a woman, she is excluded from many kinds of academic recognitions; exactly for this reason, it seems fitting to express such recognition by conferring the Leibniz Medal.[115]

In 1948, she was the first woman to be elected to the Austrian Academy of Sciences with these words:

> It brings me great pleasure that with the election of your person in particular, esteemed Professor, the first woman has been elected to the ranks of the Academy's membership since its founding.[116]

Awards and recognitions continued to occur after Meitner's death. Fourteen years after her death, she was honored with a place in the periodic table with the naming of element 109 as Meitnerium with the symbol Mt. More recently, the Institute of Physics established the Lise Meitner Medal and Prize for distinguished contributions to public engagement within physics since 2017.[117] In 1991, Meitner was the first woman to be honored with a bust in the *Deutsches Museum* (among 39 men), and in 2014 a statue of Meitner was erected along with other notable scientists at Humboldt University.[118] In 2003, the physics department of Humboldt University moved into its new building named the Lise-Meitner-Haus.[119] Finally, on October 29, 2016, over 70 physicists attended the inauguration of the Uddmanska House in Kungalv, the second European Physical Society historic site in Sweden. The Uddmanska house in Kungalv is where Meitner was staying when she and her nephew first understood that it was possible to split an atomic nucleus. There is now a plaque recognizing that achievement. As the organizers stated:

---

[115] Rife (1990), p. 96.

[116] Corradini et al. (2021).

[117] Lise Meitner Medal and Prize Winners (2023). *Institute of Physics*.

[118] Corradini et al. (2021).

[119] *The Lise-Meitner-Haus (Department of Physics)*. (n.d.). Institut für Physik. https://www.physik.hu-berlin.de/en/department/about/the-lise-meitner-haus-department-of-physics/das-institutsgebaeude.

It took the scientific community a long time to recognize the scientific achievements of Lise Meitner. She did not share the Nobel Prize in chemistry for the discovery of fission with her colleague, Otto Hahn, in 1944. Today she is justly considered to be one of the most important physicists of the 20th century.[120]

In 1960 Meitner suffered a broken hip, and moved to Cambridge, UK near her nephew Otto Frisch and his family. She suffered a series of small strokes, and died on October 27, 1968, at the age of 89; she was buried in the churchyard of St. James Church in the UK near her youngest brother. She survived her lifelong friend Otto Hahn by only three months. The epitaph on Meitner's gravestone, written by her nephew Otto Frisch, reads[121]:

Lise Meitner: a physicist who never lost her humanity.

## The Conscience of Otto Hahn

Science without conscience is the ruin of the soul—Rabelais

Meitner felt strongly about her fellow scientists' contributions to the war effort (on all sides). Shortly after the dropping of the atomic bomb, James Franck emphasized Meitner's pacifist feelings: "Never have I seen such an ardent pacifist as Lise Meitner, who worked out the mathematical formula that made the atomic bomb possible."[122] Meitner knew that Hahn was suppressing or even rewriting his role and that of others during the war. As she wrote to a friend: "He suppresses the past with all his might, even though he always truly hated and despised the Nazis. As one of his main motives is to gain international respect for Germany once again…he deceives himself about the facts."[123] She continued:

Hahn is without a doubt a decent man with many good traits. He only lacks thoughtfulness and perhaps also a certain strength of character, things that in normal times are minor flaws, but in the complicated times of today have deeper implications.[124]

---

[120] Bidragsgivare till Wikimedia-projekten. (2021). *Uddmanska huset.* https://sv.wikipedia.org/wiki/Uddmanska_huset.

[121] Rife (1990), p. 267.

[122] *Boston Globe*, August 11, 1945.

[123] Sime (1996), p. 345.

[124] Hafner, K. & Gupta, A. (2023). How antisemitism and professional betrayal marred Lise Meitner's scientific legacy. *The lost women of science initiative.* J. DelViscio (Ed). https://www.scientificamerican.com/article/how-antisemitism-and-professional-betrayal-marred-lise-meitners-scientific-legacy/.

She felt particularly strongly that Hahn had done little to take responsibility for contributing to the Nazi war machine. Meitner stated her feelings in a letter that she wrote to Hahn on Jun 27, 1945[125]:

> I have written many letters to you in my thoughts in the last few months, because it was clear to me that even people like you and Laue have not comprehended the reality of the situation…You all worked for Nazi Germany and you did not even try passive resistance. Granted to absolve your consciences you helped some oppressed person here and there, but millions of innocent people were murdered and there was no protest. I must write this to you, as so much depends upon your understanding of what you have permitted to take place…I and many others are of the opinion that one path for you would be to deliver an open statement that you are aware that through your passivity you share responsibility for what has happened…But many think it is too late for that. These people say that you first betrayed your friends, then your men and your children in that you let them give their lives in a criminal war, and finally you betrayed Germany itself, because even when the war was completely hopeless, you never once spoke out against the meaningless destruction of Germany…In the last few days one has heard of the unbelievably gruesome things in the concentration camps; it overwhelms everything one previously feared. When I heard on English radio a very detailed report by the English and Americans about Belsen and Buchenwald, I began to cry out loud and lay awake all night. And if you had seen these people who were brought here from the camps…Perhaps you will remember that while I was still in Germany (and now I know that it was not only stupid but very wrong that I did not leave at once) I often said to you: as long as only we have the sleepless nights and not you, things will not get better in Germany. But you had no sleepless nights, you did not want to see, it was too uncomfortable…I beg you to believe that everything I write here is an attempt to help you.

Hahn thanked her "for trying to make us understand, for guiding us with remarkable tact" toward understanding how the German physicists had worked under, and in many ways for, the monstrous Nazi regime.[126] Hahn continued: "We all knew that injustice was taking place, but we didn't want to see it, we deceived ourselves… Come the year 1933 I followed a flag that we should have torn down immediately. I did not do so, and now I must bear responsibility for it."

---

[125] Sime (1996), p. 310.
[126] Ball, P. (2020, April 8). *Lise Meitner: the nuclear pioneer who escaped the Nazis.* BBC Science Focus Magazine. https://www.sciencefocus.com/science/lise-meitner-the-nuclear-pioneer-who-escaped-the-nazis/.

In the end, one can say that Meitner never lost her moral compass, and she continued throughout her life to work for peace. Alternatively, "Hahn lost his moral compass over and over again—in World War I when he worked on chemical weapons and in World War II when he allowed the Nazis to take over his lab. He lost it again when he accepted his Prize and refused to share credit with a Jew."[127]

## Bibliography

Bernstein, J. (2013). *Hitler's Uranium Club: The Secret Recordings at Farm Hall.* Springer.

Friedman, R. M. (2001). *The Politics of Excellence: Behind the Nobel Prize in Science.* W H Freeman and Company.

Hahn, O. (1966). *Otto Hahn: A Scientific Autobiography.* Charles Scribner's Sons.

Hoffmann, K. (2001). *Otto Hahn: Achievement and Responsibility.* Springer.

Rife, P. (1990). *Lise Meitner and the Dawn of the Nuclear Age.* Birkhäuser.

Sime, R. L. (1996). *Lise Meitner: A Life in Physics.* Univ of California Press.

---

[127] Hafner et al. (2024).

# 4

## The Atomic Bomb That Wasn't

### Werner Heisenberg—1932 Nobel Prize in Physics *for the creation of quantum mechanics, the application of which has, inter alia, led to the discovery of the allotropic forms of hydrogen*

*The gaunt face hardened to grimness, and with both hands the bomb-thrower lifted the big atomic bomb from the box and steadied it against the side. It was a black sphere two feet in diameter. Between its handles was a little celluloid stud, and to this he bent his head until his lips touched it. Then he had to bite in order to let the air in upon the inductive. Sure of its accessibility, he craned his neck over the side of the aeroplane and judged his pace and distance. Then very quickly he bent forward, bit the stud, and hoisted the bomb over the side.*
*"Round," he whispered inaudibly.*
*The bomb flashed blinding scarlet in mid-air, and fell, a descending column of blaze eddying spirally in the midst of a whirlwind. Both the aeroplanes were tossed like shuttlecocks, hurled high and sideways and the steersman, with gleaming eyes and set teeth, fought in great banking curves for a balance. The gaunt man clung tight with hand and knees; his nostrils dilated, his teeth biting his lips.*

This terrifying vision of future war was described by H. G. Wells in 1913 in his novel *The World Set Free*, soon after Ernest Rutherford discovered the atomic nucleus and its vast energy.[1] Rutherford himself, while World War I was raging, was quoted as saying

Talk softly, please. I have been engaged in experiments which suggest that the atom can be artificially disintegrated. If it is true, it is of far greater significance than a war.[2]

---

[1] Wells, H.G. (1924). *The World Set Free*. W. Collins Sons. pp. 106–107.

[2] Cornwell, J. (2003). *Hitler's scientists: Science, war, and the devil's pact.* Penguin Books, p. 94.

© The Author(s), under exclusive license to Springer Nature Switzerland AG 2025
V. Shepherd, C. Brau, *Beyond the Genius*, Copernicus Books,
https://doi.org/10.1007/978-3-032-02735-1_4

Ultimately, Werner Heisenberg and his team in Nazi Germany and an Allied team in England and America inherited the task to make Wells' vision of an atomic bomb a reality. Despite having a head start, Heisenberg, one of the most brilliant physicists of the twentieth century, failed utterly while the Allies succeeded. How did that happen?

This chapter is about Heisenberg the man and Heisenberg the scientist, and his life before, during, and after the Nazi regime. Scientifically, Heisenberg, together with Albert Einstein, was responsible for the most profound changes in our understanding of physical reality in the history of science. Einstein, with his Theory of Relativity, changed our understanding of the very nature of time and space, showing that both are parts of a 4-dimensional reality, and there is no absolute meaning to the concept of simultaneity. Heisenberg, with his "uncertainty principle," fundamentally changed our understanding of the nature of matter itself, showing that the position and motion of matter can never be known precisely, and the future exists only as probabilities. But when Hitler came to power, Einstein left Germany for the United States while Heisenberg stayed. In fact, so many of the leading scientists in Europe—including Einstein, Erwin Schrödinger, Hans Bethe, and Enrico Fermi—left Europe that by the beginning of World War II Heisenberg had become the leading theoretical physicist on the Continent, surpassing even Niels Bohr. His extreme nationalism compelled him to remain in his beloved home country, where he continued his research and teaching so that after the war he would be positioned to help rebuild German science. Throughout the war years he spread the word of German greatness in science to occupied countries, traveling with Nazi support, though he was never a Nazi. When fission was discovered in 1938, Heisenberg became the leader of the German project to develop an atomic weapon. But the project failed even its limited objective to create a self-sustaining chain reaction, a milestone the Americans achieved less than a year after they entered the war. Why did the German project fail so utterly? Was Heisenberg, the brilliant scientist, not competent for this position? Did he lack the resources? Or, despite being a strong German nationalist and wanting Germany to win the war, did he deliberately try not to build a weapon for the Nazis? Who, really, was Werner Heisenberg? These are the questions that form the focus of this story.


## Germany Under the Kaiser

Werner Heisenberg grew up in pre-World War I Germany, at a time when the newly united country was exuberantly nationalist. It was a time of war and conflict, and competing ideologies. Although he grew up in Bavaria,

Heisenberg's family was Prussian, in a country dominated by Prussia, and the values he learned as a youth would stay with him all his life.

The German nation emerged as a country in 1871 when Napoleon III was overthrown at the end of the Franco-Prussian War. Prior to that, Germany comprised 22 principalities, duchies, grand duchies, and bishoprics, three free Hanseatic cities, and one imperial territory. The Treaties of Versailles and Frankfurt in 1871 established the German Empire as the dominant country on the European continent. The new Germany was dominated by Prussia, and Otto von Bismarck, Minister President of Prussia, became Chancellor. Wilhelm II, King of Prussia, was proclaimed *Kaiser* (Emperor). Before World War I, Germany was prosperous, energetic, proud, nationalistic, and arrogant. The Prussian military tradition, which only a generation earlier had defeated France and united the new nation, was strong. H. G. Wells called it "the monstrous vanity begotten in 1870."[3] This was the Germany in which Heisenberg began his life.

Werner Karl Heisenberg was born on December 5, 1901, in Würzburg, Germany, a small, medieval city in northern Bavaria, a predominantly Catholic region. Heisenberg grew up as a Lutheran, and in his later years he thought and wrote about science and religion. His family was bourgeois and academic, with roots in Prussia. The Prussian values of rectitude, German nationalism, and military culture were strong in his family. His father, Kaspar Ernst August Heisenberg, was a secondary school teacher of classical languages and later a professor of medieval and modern Greek studies in the university system. While not college educated, Heisenberg's mother Annie came from an academic family and translated papers from Russian for August. Social standing and advancement were important to the Heisenberg family. Heisenberg's grandfather was a Prussian locksmith, and August Heisenberg's rise to a prestigious academic position reflected this drive for advancement. Throughout his life, prestigious appointments and recognition were important to Werner Heisenberg.

When Heisenberg was ten, his father got a university chair in Munich and the family moved to an upscale neighborhood there. Heisenberg attended the *Maximilians-Gymnasium*, one of the best schools in Munich, from 1911 to 1920. Typical for that time, half the curriculum was in Greek and Latin, with some math but little physics. When Heisenberg asked his father for more math texts to read, his father gave him math texts in Latin. By the time he reached the University of Munich in the fall of 1920, Heisenberg's physics education was mostly self-taught.

Heisenberg also joined, and later led, youth groups somewhat similar to our Boy Scouts. Like many Bavarians, he loved hiking in the mountains as a

---

[3] Tuckman, B.W. (1962). *The guns of August.* Ballentine, p. 312.

means of releasing tension and he continued to lead youth groups until the Nazis took over youth programs. These outings carried him through the chaos that followed World War I, the Great War, which would be the time of his greatest discoveries.

On June 28, 1914, while Heisenberg was a *Gymnasium* student, the Serbian nationalist Gavrilo Princip assassinated the heir to the Austro-Hungarian Empire in Sarajevo, and Austria declared war on Serbia. Arrogant and confident of victory, Kaiser Wilhelm supported Austria, while Russia, France, and Great Britain opposed them, and the Great War began. Faced with enemies on two fronts, the Germans first attacked and defeated the Russians. They then turned west, brutally smashing through Belgium before they got bogged down in France. They were finally stopped at the Marne, and remained in their trenches until they were defeated in November, 1918.

Heisenberg's life was only slightly impacted by World War I. Strategic bombing was in its infancy, and the tragic effects of the war were largely confined to the trenches at the front. Heisenberg's father volunteered to go to the front briefly in 1914–15, then later served as liaison to some Greek prisoners, using his knowledge of their language (Fig. 4.1). Toward the end of the war, food and other necessities became scarce in Germany, owing in part to the allied blockade of German ports and to the diversion of ammonia production from fertilizer to explosives. Heisenberg and the other students worked

**Fig. 4.1**  Heisenberg family during World War I. Left to right, Werner, Ann (mother), August (father) and Erwin (brother). (Max-Planck-Institute, courtesy AIP Emilio Segrè Visual Archives)

summers in the agriculture service. But aside from the prisoners and the casualties returning from the front, the war was visible in Germany mostly through the desperate shortages. Finally, the failure of the German Spring Offensive in 1918 and the loss of their allies in Austria, Bulgaria and the Ottoman Empire forced Germany to surrender.

After World War I, Germany was exhausted and chaos reigned. Kaiser Wilhelm was replaced by the Weimar Republic, Germany's first constitutional democracy. Economically, Germany faced ruinous reparations demanded by the victorious Allies, an amount equal to two-thirds of all the gold reserves in the world. Hyperinflation destroyed the German currency: a loaf of bread that sold for half a mark in 1918 cost 200 billion marks in 1923. Politically, extremist groups on the right and left openly tried to overthrow the government. On the left was the Bavarian Soviet Republic, a communist group that briefly took over the government of Munich and instituted a rather repressive regime. Military indoctrination was part of *Gymnasium* education, and *Gymnasium* students were encouraged to join military groups. In 1919, in the political vacuum that followed Germany's defeat in World War I and the Kaiser's abdication, Heisenberg joined a *Freikorps*, a paramilitary group formed to combat the Bavarian Soviet Republic. Bolshevism, and Soviet influence in general, were feared and hated by most Germans. The *Freikorps*— there were many in Germany, at this time—were archly right-wing and had a reputation for brutality and antisemitism. Many future leaders of the Nazi Party served in the *Freikorps,* including Heinrich Himmler, the future head of the SS [*Schutzstaffel,* Hitler's personal terror organization]. Decades later Heisenberg recalled that this time for him was just youthful games, with occasional guard duty. He stayed out of the conflict as much as he could, and focused his attention on his studies. He was still a student in the *Gymnasium* and had to pass his final exams. On the right was the Beer Hall *Putsch* of 1923 led by Adolf Hitler, which failed and ended with Hitler in jail. By 1924, things had stabilized somewhat and Germany began to prosper. But the worldwide economic collapse in 1929 ended this, making room for radical political change. In 1933 Hitler, now out of jail, was appointed Chancellor of Germany, and in 1934 he became the *Führer*. This ended, in practical terms, the Weimar Republic. Throughout these times, academics, including Heisenberg, tried to stay above the chaos and partisan politics and focus on their scholarly work and professional advancement. Politics was "dirty business," no place for academics. This attitude stayed with Heisenberg all his life. He never became politically active, and never joined the Nazi party, but he willingly served Germany, even under Hitler, as a scientist and as a good-will ambassador for the Nazis in occupied countries.

# University of Munich

When Heisenberg began his career, physics was a mess. The classical physics of Newton and Maxwell that worked so well in the macroscopic world failed completely at the atomic level. Max Planck and Einstein introduced what they called "quanta," tiny units of energy, to explain some of properties of light and its interaction with matter, but when Niels Bohr showed that inside atoms quanta restricted Newton's laws of motion, nobody understood what was happening. Heisenberg would turn everything on its head, but the path wasn't straightforward.

Amidst this chaos, Heisenberg entered the University of Munich in 1920. As a student in *Gymnasium*, Heisenberg had excelled in mathematics, and he wanted to pursue number theory, a branch of pure mathematics. Upon acceptance, students were expected to identify a mentor and begin research, so Heisenberg first approached the mathematician Ferdinand von Lindemann. Lindemann, who was close to retirement, declined to accept him, but Arnold Sommerfeld, a brilliant theoretical physicist, saw something in Heisenberg and accepted him as his student. In Heisenberg's second semester, Sommerfeld assigned him a research project on the effect of magnetic fields on the light emitted by atoms, called the "Zeeman effect." Heisenberg published his first paper on an obscure aspect of this called the "anomalous Zeeman effect" in 1921, in just his third semester at Munich.

It's worthwhile stepping back briefly to understand what was known of physics when Heisenberg began his research. Physics, as we think of it now, may be said to have begun when Sir Isaac Newton invented calculus and discovered the laws that govern the movement of bodies acted upon by gravity and other forces. On July 5, 1687 he published these laws in his *magnum opus* entitled *Philosophiae Naturalis Principia Mathematica*. Written in Latin, which was the language of science at that time, it explained everything from falling apples to the motions of the planets around the sun. Arguably, no work was more seminal for physics and astronomy than Newton's *Principia*. His laws of motion are called classical mechanics. They remained unchanged for over 200 years … until Einstein and Heisenberg came along.

Newton also did experiments with light, famously splitting white light into colors with a prism and then reassembling them into white light with a second prism. He conjectured that light consists of particles, which he called corpuscles, and the different colors of light were carried by different corpuscles. Newton's idea of corpuscles prevailed for over a century until Thomas Young discovered diffraction and proved that light actually consists of waves, not particles. In his experiments Young showed that the wavelength of light is less than a millionth of a meter, so perhaps Newton can be forgiven for missing it.

The laws describing electric and magnetic fields and forces took longer to develop. The ancient Chinese knew about natural magnetism and used compasses for navigation centuries before they were introduced in the West. The ancient Greeks knew about static electricity from rubbing amber with cat's fur, and to this day we call it electricity from the Greek word for amber. But other than Benjamin Franklin's experiments with kites and lightning, little happened until early in the nineteenth century when Hans Christian Oersted discovered that an electric current can deflect a magnetic compass.[4] This first connection between electricity and magnetism was followed up quickly by David Faraday and Andre Ampere. Finally, in 1862 James Clerk Maxwell, in a stroke of genius, recognized the last piece of the puzzle and completed what are now called Maxwell's equations. With these equations Maxwell was able to show that light is just a travelling wave of electric and magnetic fields, and he correctly predicted the speed of light. Maxwell's theory is called classical electrodynamics. Taken together, the classical mechanics of Newton and the classical electrodynamics of Maxwell comprise what we call classical physics.

By 1900, Maxwell's equations explained everything from electric motors to radios, but they failed completely at predicting the light from incandescent bodies like the sun or a hot-filament light bulb. In fact, Maxwell's equations predicted an infinite amount of ultraviolet light from such bodies. Instead of a pleasant day at the beach, we would be sunburned to a crisp in seconds. As he thought about this, Max Planck had an idea. Recognizing a similarity between the spectrum of ultraviolet light from the sun and the spectrum of low-energy and high-energy molecules in a hot gas, he asked himself "what if light behaved like particles?" With this in mind, he divided the light into tiny bits of energy that he called "quanta" and gave infrared quanta the lowest energy and ultraviolet quanta the highest energy. When he worked out the math, he got just the amount of ultraviolet light that we see coming from the sun. But what were these "quanta?" Nobody believed in Newton's "corpuscles," they were just his conjecture. Young's experiments and Maxwell's theory of waves were convincing. But Planck's quanta predicted the right amount of ultraviolet light. How could this be reconciled?

---

[4] Actually, mariners had long known that lightning would cause their compasses to spin, but no one thought to ask them.

Then, in 1905, Einstein had an idea. In 1887, Heinrich Hertz observed that ultraviolet light from a spark could eject electrons from the surface of a metal, but infrared light, even intense infrared light, could not. Called the photoelectric effect, this phenomenon is now used to make night-vision equipment. What if Planck's quanta of light could act like real particles, like tiny billiard balls? Then, when the high-energy ultraviolet quanta collided with the electrons in the metal, they would eject them from the surface. The exact agreement of Einstein's theory with the experiments provided the first direct, physical support for Planck's hypothetical quanta. This was also our first glimpse into the strange world of atomic physics, where waves behave like particles and particles behave like waves.

Both Planck and Einstein got Nobel Prizes for their discoveries, but it took more than a decade for the Nobel Committee to accept the radical idea that light is simultaneously both waves, characterized by their color, and particles, characterized by their energy. In some way, although they couldn't have understood it then, both Newton and Young were right. Later in life, Einstein himself confessed

> All these fifty years of conscious brooding have brought me no nearer to the answer to the question 'What are light quanta?' Nowadays every Tom, Dick, and Harry thinks he knows it, but he is mistaken.[5]

Meanwhile, at about the same time, other scientists were learning about atoms. Back in the fifth century BCE, the Greek philosopher Democritus argued that matter cannot be infinitely divided, but consists of indestructible, indivisible units called atoms. Then, in 1897, J. J. Thomson was experimenting with electrical discharges. In his experiments he discovered electrons, tiny particles carrying a negative charge that came out of one of the electrodes in his discharge. Democritus' atoms weren't indivisible after all. Instead of a structureless atom, Thomson proposed his "plum pudding" model: an atom was a solid body of positive charge with the negatively-charged electrons imbedded in it like the raisins in a plum pudding, or the bits of carrot in grandma's Jell-O salad.

In 1896, Henri Becquerel discovered that certain elements, among them uranium and radium, gave off strange emissions that fogged his photographic film. In 1907, following up on Becquerel's discovery, Rutherford showed that two types of particles were emitted, and that the "alpha particles" emitted by radium were really positively charged helium atoms stripped of their electrons. The "beta particles" were Thomson's electrons. Beginning around 1908,

---

[5] French, A.P. (1951). *Einstein: A centenary volume*. Harvard University Press, p. 138.

Rutherford and Hans Geiger directed alpha particles from a radium source toward an exquisitely thin (less than a millionth of a meter thick) gold foil, "Like gold to ayery thinnesse beate," as John Donne would say,[6] about a thousand gold atoms thick. They hoped that by studying the deflection of the charged alpha particles as they traveled through the gold atoms, they might learn something about the structure of the atoms. After passing through the foil, the alpha particles were allowed to strike a fluorescent screen where they made a tiny flash of light. As Rutherford himself said,

> Geiger is a good man and worked like a slave. I could never have found time for the drudgery before we got things going in good style. Finally, all went well, but the scattering is the devil…Geiger is a demon at the work of counting scintillations and could count at intervals for a whole night without disturbing his equanimity. I [swore] vigorously and retired after two minutes.[7]

Most of the alpha particles went through the gold foil only slightly deflected, but almost as an afterthought Rutherford had Geiger check for particles that might be reflected backward. To his amazement, a few alpha particles bounced off the gold foil straight back toward the radium source.

> I remember…later Geiger coming to me in great excitement and saying, 'We have been able to get some of the [alpha] particles coming backwards…' It was quite the most incredible event that has ever happened to me in my life. It was almost as incredible as if you fired a 15-inch shell at a piece of tissue paper and it came back and hit you.[8]

The alpha particles were being deflected by something small, heavy, and positively charged inside the atom. Rutherford concluded that far from being like Thomson's Jell-O filling the atom, virtually all the mass (and all the positive charge) of the atom, must actually be concentrated in the center of the atom with a size too small to detect in his experiments. He had, in fact, discovered the atomic nucleus. Atoms, far from being solid, are mostly just empty space with the electrons orbiting around the nucleus like the planets orbit around the sun, just like a miniature solar system.

But if atoms are mostly empty space, what gives atoms their shape and size? Since the negative electrons are attracted to the positive nucleus, why don't

---

[6] Louthan, D. (1951). *The poetry of John Donne, a study in explication.* Bookman Associates, p. 46.

[7] Eve, A.S. (1939). *Rutherford: Being the life and letters of the Right Honorable Lord Rutherford, O.M.* Cambridge University Press, p. 180. yyyy.

[8] Rutherford, E. (1938). Forty years of physics. In J. Needham & W. Pagel (Eds.), *Background to Modern Science,* MacMillan Company and Cambridge University Press.

atoms just squish down to nothing? And why do the atoms give off light of only certain colors, like the different colors we see from neon signs? In 1913, Neils Bohr came up with an explanation: inside an atom, the electrons can exist only in certain "quantized" orbits, with integer "quantum numbers" 1, 2, 3… proportional to the orbital momentum of the electrons. The smallest orbit (quantum number 1) then has a finite radius. This "explains," in some sense, why atoms have a certain finite size. Bohr further argued that when an electron drops from a higher-energy orbit to a lower one, a quantum of light, or "photon" as we would call it now, is emitted with just the energy lost by the electron, as illustrated in Fig. 4.2, in perfect agreement with Planck's theory of quanta. Bohr's theory correctly predicted the simplest spectra, like that of the hydrogen atom, but atoms with more than one electron had anomalies. Importantly, in Bohr's theory, electrons were still classical particles and Newton's equations of motion were still valid. Only the "quantization rule" was new. This strange mixture of classical physics with quantum rules was physics as it was understood when Heisenberg began his research. In the coming decade Heisenberg and a few others would give us new ways of thinking about matter.

By 1921, when Heisenberg began his research, his first task under Sommerfeld was to explain certain features of the Zeeman effect that didn't seem to fit Bohr's rules. As a brash young student with nothing to lose, Heisenberg defied the rules, introduced orbits with half-integral quantum numbers, and correctly predicted the effect he was assigned to explain. Heisenberg's theory was criticized for having no physical justification: what was the meaning of the half-integral quantum numbers? Heisenberg replied simply "success sanctifies the means."[9] It wasn't until 1925, four years later,

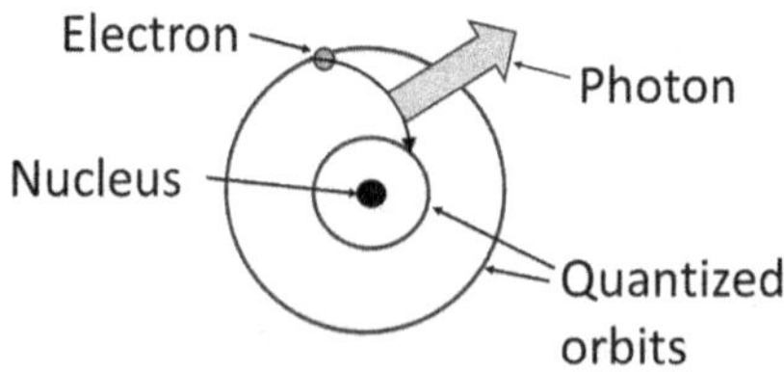


**Fig. 4.2**    Quantized Bohr orbits with an electron emitting a photon

---

[9] Cassidy, D.C. (2010). *Beyond uncertainty: Heisenberg, quantum physics, and the bomb.* Bellevue Literary Press, p. 98.

that Samuel Goudsmit and George Uhlenbeck discovered that Heisenberg's extra half integer was due to spinning of the electron itself.[10]

While Heisenberg was still his student at Munich, Sommerfeld was in Wisconsin for the academic year 1922–1923, so Heisenberg spent the year with Max Born in Göttingen. While he was there, Born sent him to a conference in Leipzig so he could meet the "important people," like Einstein and others. Although Einstein himself never made it to the conference, Heisenberg experienced antisemitism for the first time when Philipp Lenard objected to Einstein's theory of relativity because Einstein was a Jew. Since Heisenberg was very junior at this time, he wanted to stay "above politics" and just stay focused on his scholarship. But the confrontation in Leipzig presaged years of trouble. Heisenberg himself would later be attacked as a "white Jew" for some of the same reasons that Lenard had attacked Einstein.

Typically, at the end of an "undergraduate" program, a student in Germany would submit a research thesis and then take a final oral exam for a doctorate. This would allow him or her to teach at the *Gymnasium* level. Upon his return to Munich in 1923, Heisenberg submitted a theoretical thesis on turbulence in fluids, since his analysis of the Zeeman effect was too controversial. He then prepared for his final oral exam. He did well on the theoretical part, but Wilhelm Wein was also on the examining committee, and he failed Heisenberg on the experimental part of the exam. Previously, Heisenberg had taken Wein's course in experimental physics and performed poorly, even negligently, so Wein was ready to pounce. Heisenberg was unable to answer Wein's simplest questions, questions that a freshman physics student might be expected to answer. In the end, Heisenberg was given the average of Wein's failing grade and Sommerfeld's top grade, which allowed him to pass his doctorate exam, but just barely. Embarrassed, he left the celebration party early and took the night train to Göttingen to begin work on his *Habilitation* (the next higher academic degree) under Born. But Heisenberg's inability to grasp even the most basic experimental physics would become an important piece of the puzzle as to why the German nuclear program failed.

Shortly after he got to Göttingen, Heisenberg left for Finland with the youth group that he led for years. Throughout much of his life Heisenberg had the rather boyish demeanor that we see in Fig. 4.3, and his youth groups were important to him. His youthful enthusiasm and optimism would carry him through difficult times.

---

[10] Electrons don't actually spin, but they have some properties that make them seem as though they do.

**Fig. 4.3** Werner Heisenberg in 1926, at the peak of his scientific creativity. (Creative Commons Attribution 3.0 Unported license. Photo by Friedrich Hund)

## Epiphany at Heligoland

Back in Göttingen, working with Born, and occasionally with Wolfgang Pauli and Niels Bohr, Heisenberg extended his analysis of the Zeeman effect to explain some additional features. It was enough to earn Heisenberg his *Habilitation*, qualifying him to teach at the university level. It also got Heisenberg an invitation to spend a year at Bohr's institute in Copenhagen, beginning in 1924 (Fig. 4.4). There, the breakthrough happened: quantum mechanics was born.

Generations of writers—historians, scientists, biographers—have tried to explain how Heisenberg conceived his radical departure from all earlier theories.[11] His new theory was so abstract, so divorced from all previous ideas, and even from ordinary experience, that it was a different way of thinking even for physicists. Newton was a genius; he invented a new branch of mathematics (calculus) for his theories, but he described what everyone could see. Einstein was more abstract than Newton, but even when he united time and space in his theory of relativity he worked with the particles and waves of classical physics, and he could explain his theory in familiar terms using bullets, trains,

---

[11] MacKinnon, E. (1977). Heisenberg, models, and the rise of matrix mechanics. In *Historical studies in the physical sciences, 8,* pp. 137–188. https://doi.org/10.2307/27757370.

**Fig. 4.4** Niels Bohr Institute, Copenhagen. (Courtesy of the University of Copenhagen Niels Bohr Institute)

and clocks.[12] Heisenberg's level of abstraction was even more profound, and may have been inspired by his reading of philosophy, for he remarked later

> I think that modern physics has definitely decided in favor of Plato. In fact, the smallest units of matter are not physical objects in the ordinary sense; they are forms, ideas which can be expressed unambiguously only in mathematical language.[13]

Heisenberg's discovery of quantum mechanics seemed to come out of the blue, and his "uncertainty principle" was totally unexpected until it emerged from his equations two years later. Until then, everyone had believed in Newton's laws and the absolute predictability of the future. The uncertainty principle showed that there is a fundamental limit to what can be known about the position and velocity of an electron, and that the future exists only as probabilities. It took years for the physics community to understand and accept the new theory.

In Bohr's theory of the atom, it was assumed that the electrons traveled in well-defined orbits around the nucleus of the atom. Aside from the fact that only certain orbits were allowed by Bohr's ad hoc quantum conditions, this

---

[12] These (and other) ideas are amusingly explained by George Gamow in his wonderful little book *Mr Tompkins in Paperback*, Cambridge University Press, Cambridge (1965}.

[13] As quoted in The New York Times Book Review (8 March 1992).

picture was quite classical: the electrons are particles that go around the nucleus just like the planets go around the sun, except that only certain orbits are allowed. In a sense, Bohr's theory was half classical and half quantum-mechanical. But there were problems with this picture: the electrons didn't radiate as would be expected classically. They radiated only as they jumped from one allowed orbit to another, as shown in Fig. 4.2, and there was no theoretical understanding of this process.

In his new theory Heisenberg discarded the notion that one could describe the motions of electrons in their orbits since the electrons could never be seen. Instead, he focused on the so-called observables, which were the photons emitted in the quantum jumps, and he opened his 1925 paper with the remarkable words:

> It is well known that the formal rules which are used in quantum theory for calculating observable quantities such as the energy of the hydrogen atom may be seriously criticized on the grounds that they contain, as basic element, relationships between quantities that are apparently unobservable in principle, e.g. position and period of revolution of the electron.[14]

In more familiar terms, Heisenberg might have said that since a guitar string vibrates too fast to see, we must ignore its motion and describe the string mathematically in terms of its musical tone and all its overtones, which we can hear. So, in his first paper, Heisenberg rewrote Newton's equations of motion and Bohr's quantization rules in terms of what we could see: the optical frequencies of the light emitted by the atoms, which are equivalent, in a sense, to the "tones" of the guitar string. The departure from classical theory to quantum theory could hardly have been more profound. But Heisenberg was able to solve the complicated equations that he found (some equations took as many as three lines to write out) and predict the energy levels of some simple, artificial systems. After sending these results back to Born in Göttingen, he published his new theory in a paper so terse that it's difficult to read and understand even today.[15] Most physicists were unfamiliar with Heisenberg's mathematics, and they had a difficult time accepting the new theory.[16]

---

[14] Heisenberg, W. (1925). *Zeitschrift f. Physik*, 33, 879.

[15] Aitchison, I. J. R., MacManus, D. G., & Snyder, T. D. (2004). Understanding Heisenberg's "magical" paper of July 1925: A new look at the calculational details. *American Journal of Physics*, 72(11), 1370–1379.

[16] MacKinnon (1977), p. 137.

Of course, all this didn't happen in a straightforward, linear fashion. The four brilliant, competitive, and ambitious physicists in Copenhagen at that time—Bohr, H. A. Kramers, Pauli, and Heisenberg—each had their own, sometimes irascible and incompatible personalities. And while Heisenberg had a one-year grant to work in Copenhagen, Born wanted him back in Göttingen for the 1925 spring semester, so Heisenberg had to finish his commitment to Bohr during semester breaks. It was during one of these breaks that Heisenberg had his "epiphany at Heligoland." On June 7, 1925, plagued by hay fever, Heisenberg took a vacation to Heligoland, an island off the coast of Germany where there was less pollen. There he discovered the strange and cumbersome mathematics that describes quantum-mechanical quantities. He later wrote:

> It was about three o' clock at night when the final result of the calculation lay before me. At first I was deeply shaken. I was so excited that I could not think of sleep. So I left the house and awaited the sunrise on the top of a rock.[17]

Back at Göttingen, Born and his former student Pascual Jordan recognized that the quantum mechanical quantities that appeared in Heisenberg's equations formed arrays, like numbers or letters arranged on the squares of a checkerboard. Mathematicians already knew about such arrays, which they called matrices, and they had developed elaborate methods to deal with matrices.[18] Using this mathematical machinery, Born and Jordan, working together with Heisenberg, were able to give Heisenberg's theory the simplicity and elegance that it deserved. Owing to its use of matrices, the new theory came to be called matrix mechanics.

Things moved quickly after that. Two years later, in 1927, Born and Heisenberg realized that Heisenberg's abstract formulas lead directly to the "uncertainty principle," for which Heisenberg was awarded the 1932 Nobel Prize in Physics. According to this principle we cannot know precisely both the position and momentum of an electron, or any other particle, and the future of the particle cannot be precisely predicted. The product of the uncertainty in the position and the uncertainty in the momentum turns out to be just the same hypothetical constant that Planck used to calculate the energy of his quanta and Bohr used to calculate his orbits. Like Einstein, who often used amusing *gedanken* experiments (thought experiments) with trains and clocks to explain his theories, Heisenberg used a fictitious experiment to

---

[17] Heisenberg, W. (1969). *Der Teil und das Ganze*. Piper.
[18] Cassidy (2010), p. 141.

demonstrate his uncertainty principle.[19] In this experiment, a single photon of light is used to locate an electron by bouncing the light off the electron. But when the photon bounces off the electron to mark its position, it changes the velocity of the electron so we no longer know the electron's momentum. If we use a more energetic photon to gain more precise information about the position, we know less about the momentum. It's impossible to escape the joint uncertainty in the position and momentum of the electron. It's important to recognize here that this uncertainty isn't a fault of the experiment. Any experiment would lead to the same uncertainty. It's fundamental to the theory that the precise position and momentum of a particle don't exist. In the analogy with the guitar string, even if we know everything about the tones and overtones of the guitar string, we still don't know anything about the precise position and speed of the string at any time. But whereas a sufficiently fast camera will reveal the position and momentum of the guitar string, the precise position and momentum of a particle in an atom don't exist. And what does it mean to describe a particle with a matrix? But his equations gave the correct values for the energy levels of an electron in an atom, and as he said of his half-integral quantum numbers, success sanctifies the means.

## Wave Mechanics

But all this abstraction didn't satisfy most people. Soon, an alternative theory appeared that competed with Heisenberg's theory, challenging his primacy in atomic physics.

The controversy arose in 1926, a year after Heisenberg discovered matrix mechanics, when Erwin Schrödinger introduced his "wave equation."[20] This theory, quite distinct from Heisenberg's abstract matrices, described electrons as waves. A few years earlier, Louis de Broglie had proposed that electrons and other particles sometimes behave as waves, just as light sometimes behaves as waves and sometimes as particles (quanta). Adopting this point of view, it follows that in the same way that sound waves confined to an organ pipe give rise to specific tones and overtones, the electron waves confined in an atom give rise to specific energy levels. Schrödinger's theory, called "wave mechanics," correctly predicted the observed values for the energy of electrons in hydrogen atoms. But for most physicists, waves—light waves, water waves, sound

---

[19] Boughn, S.P., & Reginatto, M. (2017). Another look through Heisenberg's microscope. *European Journal of Physics*. 39(3), 035402. https://doi.org/10.1088/1361-6404/aaa33f.

[20] *Schrödinger, E. (1982) Collected papers on wave mechanics (3rd ed.).* AMS Chelsea Publishing.

waves—were a familiar concept. Everybody has seen waves, and physicists were familiar with the elegant mathematics that had been developed to describe waves. On the other hand, most physicists (even Heisenberg when he started) didn't understand matrices, or how to use them, and no one understood the physical meaning of Heisenberg's matrices.

When Schrödinger introduced his theory, the intuitive nature of wave mechanics made it more popular than matrix mechanics. Whereas Heisenberg's uncertainty principle appeared as an abstract result, the uncertainty principle was easily understood from the well-known properties of waves: since a wave is not a point, you can't precisely define its position or momentum, and since a wave spreads out in time you can't precisely predict its future. Even Sommerfeld, Heisenberg's thesis advisor at Munich, said "although the truth of matrix mechanics is indubitable, its handling is extremely intricate and frighteningly abstract. Schrödinger has now come to our rescue."[21] Heisenberg's colleague Pauli wrote Jordan that Schrödinger's paper introducing wave mechanics "numbers among the most significant that have been written lately," and Born, Heisenberg's mentor and co-author, said "I would regard [wave mechanics] as the deepest form of the quantum laws."[22] All this upset Heisenberg deeply, and he became quite defensive of his theory. Soon, Schrödinger (and others) showed that both theories are mathematically equivalent, but even this didn't settle the dispute. Heisenberg's approach treated electrons as particles, while Schrödinger's approach treated them as waves. There was a great deal at stake here: which approach was more fundamental, or "deepest," as Born would say? Which approach was more powerful? And behind the physics were the competitive and ambitious physicists themselves. Who would get the most acclaim, and who would get the prestigious chairs (professorships) that Heisenberg and Schrödinger both coveted? When Schrödinger was invited to describe his theory in some lectures in Munich in July 1926, Heisenberg travelled from Copenhagen to hear the lectures. There, he challenged Schrödinger so sharply that Wien, who had invited Schrödinger to lecture there, had to tell Heisenberg to sit down! In the end, both physicists would get prestigious chairs: Heisenberg would get the chair in Leipzig, and Schrödinger would succeed Max Planck at the Friedrich Wilhelm University in Berlin. This relaxed some of the tension between them, and today they are both equally recognized. Wave mechanics and matrix mechanics together are called quantum mechanics.

---

[21] Cassidy (2010), p. 149.
[22] Cassidy (2010), p. 150.

Arguments about the meaning of quantum mechanics continue to this day. As Richard Feynman once remarked, "I think I can safely say that nobody really understands quantum mechanics."[23] Einstein never fully accepted quantum mechanics. Faced with Heisenberg's uncertainty principle, he famously wrote Born

Quantum mechanics is certainly imposing. But an inner voice tells me that it is not yet the real thing. The theory says a lot, but does not really bring us any closer to the secret of the 'old one.' I, at any rate, am convinced that He is not playing at dice.[24]

Einstein believed that underneath the uncertainties in quantum mechanics lay hidden information that, when we learn it, will make the future precisely known, as it was in Newton's classical mechanics. But we are certain about uncertainty. Remarkably, there is a theorem, called Bell's inequality, that shows that there are experiments whose outcome can distinguish the existence or nonexistence of hidden information, even if we don't know what that information might be.[25] Recent experiments to test Bell's inequality were awarded the Nobel Prize in Physics in 2022.[26] The results show that there really is no hidden information; the future cannot be predicted precisely, with all the philosophical and religious implications that contains. Which begs the question "Why don't we see this uncertainty in our everyday experience?" The answer lies in the size of things. For an atom, uncertainty is everything. The very size of an atom is the limit of uncertainty: no more precise information can be learned about an atom without adding so much energy that it bursts apart. On the other hand, we can predict the path of a planet or a baseball to incredible precision; the uncertainty represented by Planck's constant is incredibly small. Fortunately, even a molecule of DNA is large enough to remain stable throughout our lifetime, or even for generations.

With his *Habilitation* in hand, Heisenberg was ready for a university position. He coveted a position back at Munich, to replace Sommerfeld upon his

---

[23] Feynman, R. (2017). *The character of physical law*. The MIT Press.

[24] Newton-John, I.B. (2005). Albert Einstein to Max Born. *Physics Today*, 58(6), 16.

[25] Maccone, L. (2013). A simple proof of Bell's inequality. *American Journal of Physics*. 81, 854–859.

   Bell, J. S. (1971). *Introduction to the hidden variable question. Foundations of Quantum Mechanics*, Enrico Fermi Course IL. Proceedings of the International School of Physics, pp. 171–181.

[26] Garisto, D. (2022). The universe is not locally real, and the physics Nobel Prize winners proved it. *Scientific American*, October 6.

retirement, but this was not going to happen. In 1927, when he was just 26 years old, Heisenberg was appointed professor of theoretical physics and head of the Department of Physics at the University of Leipzig. During the next few years, he worked on a variety of problems. Using quantum mechanics, he was able to explain what makes iron magnetic, and he published several papers on further developments of Paul A. M. Dirac's relativistic version of matrix mechanics. He also attracted a wealth of brilliant students to Leipzig including Erich Bagge, Felix Bloch, Robert Mulliken, Rudolf Peierls, George Placzek, Isidor Rabi, John Slater, Edward Teller, John Van Vleck, Victor Weisskopf, Carl Friedrich von Weizsäcker, and Gregor Wentzel, shown in Fig. 4.5. Many of these later worked on the development of the atomic bomb, some for the Allies and some for the Nazis, and several received Nobel Prizes in their own right.

In December, 1933, Heisenberg was finally awarded the 1932 Nobel Prize in Physics for "the creation of quantum mechanics." Even five years after the discovery of the uncertainty principle, the Nobel Committee still was still hesitant to give the Prize to such a radical idea, so they delayed the award for a year, as allowed by the statutes of the Prize. But sinister forces in Germany were opposed to the radical new physics of Einstein and Heisenberg. This led to problems later, especially after the Nazis came to power.

Fig. 4.5 Heisenberg with students at Leipzig. Front row (L–R): Rudolf Peierls and W. Heisenberg. Back row (L–R): G. Placzek, G. Gentile, G. Wick, F. Bloch, V. Weisskopf and F. Sauter. (Courtesy of the AIP Emilio Segrè Visual Archives, Rudolf Peierls Collection)

## The White Jew

Heisenberg had first encountered antisemitism while he was a student at Munich, when at a conference he observed Lenard attacking Einstein's theory of relativity because Einstein was a Jew. Things would get much worse.

On January 30, 1933, Hitler became Chancellor of Germany. Antisemitism became rampant, culminating in November, 1938, when at least 91 Jews, probably many more, were murdered in a nationwide pogrom called *Kristallnacht* (the night of broken glass). The Law for the Restoration of the Professional Civil Service was passed on April 1, 1933, just a few weeks after Hitler's rise to power. It forced all non-Aryan civil servants (including university faculty) to retire. About 1000 teachers were dismissed. Their positions were filled by non-Jews such as Paul Harteck, who later played an important role in the German atomic weapons program. Harteck filled the position at the University of Hamburg vacated by Otto Stern, who found refuge in the USA. In response to Hitler's rise to power, Einstein resigned from the Prussian Academy while he was traveling in the United States, renounced his German citizenship, and never returned to Germany. Fritz Haber and James Franck, both Jews but protected by their previous military service, resigned their positions, and Born, also a Jew, was forced to leave Göttingen. Heisenberg tried to keep Born at Göttingen, and in June 1933 he wrote to Born

> Since…only the very least are affected by the law—you and Franck certainly not, nor Courant—the political revolution could take place without any damage to Göttingen physics…Certainly in the course of time the splendid will separate from the hateful…Therefore, I would like to ask you not to make any decisions yet, but wait and see how our country looks in the fall.[27]

A month later Born finally responded "One cannot serve a state that treats one as a second-class citizen and that treats children even worse."[28] He departed for Cambridge and became a British citizen the day before World War II broke out. Faced with the loss of so many leading scientists, Max Planck pleaded with Hitler not to remove the Jews. Hitler replied "If science cannot do without Jews, then we will have to do without science for a few years."[29]

---

[27] Cornwell (2003), p. 134.

[28] Cassidy (2010), p. 213.

[29] Cornwell (2003), p. 34.

Things were getting worse at the universities as more and more Jewish faculty were being replaced. Philipp Lenard and Johannes Stark, who had supported Hitler, attempted to take over the German Physical Society, and would have succeeded if not for the efforts of Max von Laue. Throughout the 1930s, Lenard and Stark tried to purify "*Deutsche Physik*" (German physics) by eliminating "Jewish physics." Stark's four-volume work on the subject begins:

> 'German physics?' people will ask. I could have also said Aryan physics or physics of Nordic-natured persons, physics of the reality-founders, of the truth-seekers, physics of those who have founded natural research.[30]

Stark attacked Einstein's theory of relativity as Jewish science, and the abstract quantum theories of Heisenberg and Schrödinger were lumped together with relativity as "not Aryan." In 1934, the Nazi Party formed the Reich Education Ministry to take control of the universities. It was headed by Bernhard Rust, a zealous Nazi who exhibited psychological problems stemming from a World War I injury. He was intent on purifying German education according to Nazi ideals, including the suppression of "Jewish physics." In August 1934 all civil servants were required to sign an oath of allegiance to Hitler. Heisenberg delayed signing until the following year, which probably didn't help in his difficulties later. With the departure of Born from his prestigious position at Göttingen, Heisenberg was his logical successor to continue the development of quantum mechanics that began there. But Rust, together with the Nazi University Teachers League, blocked Heisenberg's appointment. Even so, in a letter to his mother in the autumn of 1935, Heisenberg, always the loyal Prussian, wrote:

> I must be satisfied to oversee in the small field of science the values that must become important for the future. That is the only clear thing left for me to do in this general chaos. The world out there is really ugly, but the work is beautiful.[31]

As late as October, that year, Heisenberg told his mother "Much that is good is also being tried [by the Nazis] and one should recognize good intentions."[32]

In April, 1935, Sommerfeld was forced into emeritus status by a new rule promulgated by Minister Rust. Since Heisenberg had won the Nobel Prize, he

---

[30] Cornwell (2003), p. 178.

[31] Eckert, M. Werner Heisenberg, controversial scientist. *Physics World,* November 30, 2001.
   Cornwell (2003), p. 180.

[32] Cassidy (2003), p. 208.

could hardly be denied his coveted chair at Munich, but Lenard and Stark and others in the *Deutsche Physik* movement marshalled their forces to prevent the appointment. Stark complained publicly that although Einstein had left Germany for America, there were still physicists acting in Einstein's spirit. He protested that "the theoretical formalist Heisenberg, spirit of Einstein's spirit, is now even to be rewarded with a call to a chair."[33] In a vicious 1937 article in *Das Schwartze Korps*, the weekly newspaper of the SS, Stark attacked Heisenberg as a "white Jew:"

> For it is not the racial Jew in himself who is a threat to us, but rather the spirit that he spreads. And if the carrier of this spirit is not a Jew but a German, then he should be considered doubly worthy of being combatted as the racial Jew, who cannot hide the origin of his spirit. Common slang has coined a phrase for such bacteria carriers, the 'white Jew.'[34]

This threat could not be ignored, even by the loyal and apolitical Heisenberg. In these violent times, his life, his career, and his honor were in danger. Heisenberg was saved only by a letter from Himmler himself. When he saw the letter in *Das Schwartze Korps*, Heisenberg wrote the *Reichsfürer-SS* asking that he be exonerated from the accusations in the article, or he would resign. The letter was brought to Himmler through Heisenberg's mother, who was a friend of Himmler's mother. Himmler called for an investigation that lasted nine months, and Heisenberg had to defend himself.[35] In the end, Heisenberg was acquitted by Himmler, with a compromise: Heisenberg could teach and use the new physics, so long as no Jewish scientists were mentioned.

Heisenberg's science also had difficulties during this period. When James Chadwick discovered the neutron in 1932, Heisenberg turned his attention to the atomic nucleus. In nuclei, which consist of uncharged neutrons and positively charged protons, the repulsion between the protons is enormous. What kind of forces could hold the nucleus together? Within this question Heisenberg hoped that he would discover new laws of physics, going beyond his matrix mechanics that worked so well for electrons. In particular, he thought that the answer might be revealed in the showers of particles that were generated by cosmic rays striking the atmosphere, for only in these showers was it possible to observe events with enough energy to probe the forces that might be involved in the nucleus. In laboratory experiments using

---

[33] Cornwell (2003), p. 181.
[34] Cassidy (2010), p. 268.
[35] Cornwell (2003), p. 184.

cyclotrons, E. O. Lawrence's cyclotrons could reach about 80 million "electron volts" of energy. Nowadays, giant "atom-smashing machines" like that at CERN, in Switzerland, reach 100,000 times more energy than was available to Lawrence, but cosmic rays have even higher energy than that, sometimes as much as 40 million times higher. Looking for a clue to the new physics somewhere in the cosmic ray phenomena, Heisenberg proposed a new theory of explosive showers, but in the end other theories proved correct. He also tried to extend his matrix mechanics to higher energy events by introducing his $S$-matrix theory. This, too, didn't produce the breakthrough he was hoping for.

The bright spot in all this was in his personal life. In 1937, Heisenberg met Elisabeth Schumacher at a musical evening in the home of a friend. Fourteen years his junior, she lifted Heisenberg's spirit at a lonely time. It helped that she was from an academic family (her father was a professor of economics at the University of Berlin) and Heisenberg was a Nobel Prize winner. He proposed to her two weeks after they met. In March he wrote her

> Dear Elisabeth!
>
> It is strange to think that this is the first letter I am writing to you. For it actually seems to me as though, for many years already, we have been close acquainted, and the present state of being alone is only a painful interruption in an ever-beautiful, already almost accustomed shared life. I am indebted to you for bringing me so much peace and security and am looking forward with my every thought to the time when, together, we can enjoy the daily changes between the serious and the beautiful. Thank you for everything![36]

They were married soon after, on April 29, 1937. Fraternal twins were born nine months later, and five more children followed over the next 12 years. Heisenberg wasn't a "hands-on" type of father. He wasn't present for any of the births, and never fed or diapered the children. Physics and his career were a priority for Heisenberg, but he and Elisabeth remained devoted to one another for the rest of their lives.

Little did they know of what lay ahead. It wasn't long after they were married that Heisenberg was attacked in *Das Schwartze Korps*. When his appointment to Munich was blocked, the couple settled in Leipzig. The campaign against Heisenberg finally ended in the fall of 1939, just as war was breaking out, when Wilhelm Müller was named as Sommerfeld's successor at Munich.

---

[36] Heisenberg, W., and Heisenberg, E. (2016). *My Dear Li: Correspondence 1937–1946*. Yale University Press.

Müller, an aerodynamicist, was branded a "complete idiot" by Sommerfeld.[37]
A year later Heisenberg wrote to Elisabeth

> Then I met [Walther] Gerlach from Munich. He told me that this certain Mr.
> Müller, who cannot take over Sommerfeld's lectures since he does not know the
> relevant field, has now taken on as an assistant a gentleman by the name of
> Glaser to give his lectures for him. This Glaser used to be in Würtburg and, as
> Döpel told me, has been mentally disturbed for some time. This Mr. Glaser,
> P.G. [member of the Nazi Party], has now written an essay and published it stat-
> ing that there will not be any purity in German physics until all Jewish quantum
> and atomic theories (and their representatives), are forever expunged.[38]

In spite of all this, Heisenberg remained loyal to Germany. With Himmler's
protection and a cottage in Urfeld, in southern Germany, where Elisabeth and
the children might be safe, Heisenberg was willing to face the coming storms.

## The Uranium Club

Although they had a head start at the beginning of World War II, the German
program to develop an atomic bomb never succeeded in achieving even the
smallest steps toward building the bomb. Set in the chaos of the war, the program was beset with infighting, inept management, uncertain support, and
scientific missteps, some attributable to Heisenberg himself. Since they were
certain that if the Germans couldn't do it, then no one else could do it either,
the German program lacked the sense of urgency that energized the Allied
program. While they were interned at Farm Hall after the war, the German
scientists were shocked and disbelieving when they heard about the bombs
dropped on Japan. This is how it happened.

When Rutherford discovered the nucleus in 1911, it was immediately
obvious that the nucleus must contain enormous energy to hold all the positive charges together in a such a tight volume, but no one thought that this
energy would ever be available. More than 20 years later, on September 11,
1933, while he was addressing a meeting of the British Association for the
Advancement of Science, *The Times (London)* quoted Rutherford as saying
that "anyone who looked for a source of power in the transformation of atoms
was talking moonshine."[39] In his address, Rutherford was describing his

---

[37] Eckert (2001).

[38] Heisenberg and Heisenberg (2016), p. 132.

[39] Rhodes, R. (1986). *The making of the atomic bomb.* Simon and Schuster, p. 27.

discovery that bombarding nuclei with artificially accelerated protons could split light elements like lithium into smaller particles, in which case more energy was released than was provided by the proton. Due to electrical repulsion, however, the positively charged protons Rutherford was using couldn't reach nuclei heavier than lithium, which is just the third element in the periodic table. In any event, proton bombardment would be a slow and inefficient way to produce nuclear energy.

The next morning, reacting to the article in *The Times*, Leo Szilard, a Hungarian Jew who had fled to London when Hitler became Chancellor, realized that since neutrons, which had been discovered by Chadwick only the year before, have no charge, they would not be repelled by the positive charge of a nucleus, and unlike Rutherford's protons, neutrons could reach, and potentially react with, any nuclei, even heavy nuclei. Since nuclei contain both protons and neutrons, he conceived the idea that if further neutrons were liberated by the neutron bombardment, this would cause a self-sustaining chain reaction. If, for example, two neutrons are liberated by the first collision, these two neutrons will cause reactions that liberate four neutrons, and so on, releasing energy at each step. Energy, potentially enormous energy, would be released with no need for any initial energy input. If all this takes place in microseconds, the result will be a tremendous explosion, a bomb. Szilard had, of course, no idea that a heavy nucleus could break into two smaller nuclei, since fission hadn't been discovered yet, but any nuclear process that released energy and neutrons would work. He immediately applied for a patent on the idea of a nuclear chain reaction. It was granted in 1936, but recognizing the possibility of a monstrous bomb, Szilard assigned the patent to the British Admiralty and kept it secret until 1949. It's interesting that Szilard conceived these ideas shortly after reading H. G. Wells' futuristic novel *The World Set Free*, in which nuclear power and atomic bombs were first mentioned, as described in the opening passages of this chapter. But it would take 12 years, and the largest scientific undertaking in history, to create the first successful atomic bomb.

Meanwhile, in Rome, Enrico Fermi also recognized that neutrons could reach any nucleus. In 1934 he began a series of experiments in which he irradiated increasingly heavy elements with neutrons from a radioactive source. In each case, Fermi found that the irradiated nucleus absorbed the neutron and underwent radioactive decay to form the next-higher element in the periodic table. When he bombarded uranium, he thought that he had created transuranic elements, that is, elements beyond uranium in the periodic table that are not found in nature. For this work, Fermi was awarded the Nobel prize in 1938. Rather than return to Italy under Mussolini, when he received his prize, he left with his Jewish wife directly from Stockholm to the USA, where he took a position at Columbia University. He was soon to discover that the irradiation of uranium by neutrons is more complicated, and consequential, than he thought.

Stimulated by Fermi's early results, in 1934 Lise Meitner revived her close collaboration with Otto Hahn and began similar neutron bombardment experiments. With her physics background and Hahn's chemistry background, they could carry out experiments that Fermi could not, especially after he lost his chemist Oscar De Agostino amidst Mussolini's terror. When they irradiated uranium with neutrons, Hahn and Meitner kept seeing products that behaved chemically like barium, but they dismissed the possibility that the uranium had split into two smaller elements. At the same time, Paul Savitch, a chemist, joined Irene Joliot-Curie in Paris and they began similar experiments. Upon bombarding uranium with neutrons, they found a substance that behaved like lanthanum, but like Hahn and Meitner they couldn't make sense of their results.[40] Unfortunately, when Hitler invaded Austria on March 13, 1938, Meitner, a Jew, lost the protection of her Austrian passport and had to flee to Sweden, where she continued to collaborate with Hahn by mail. Then, in December, 1938, in an exquisite series of tedious but careful experiments, the chemists Otto Hahn and Fritz Strassman were forced to conclude that when exposed to neutron irradiation, uranium nuclei underwent some sort of reaction that produced barium nuclei.[41] This represented a complete departure from all previous experience since barium nuclei are roughly half the size of uranium. They came to this conclusion reluctantly, only after repeated checks on their results. Within days of hearing of this amazing discovery, Meitner and her nephew Otto Robert Frisch used Bohr's "liquid drop" model of the nucleus to explain this result. Since it involved the splitting of the uranium nucleus into two roughly equal parts, they called it "fission," in analogy to the process of cell division in bacteria.[42] Due to their positive charges, the two new nuclei repel each other with enormous energy, which Meitner and Frisch estimated to be about 200 MeV (million electron volts) in each event. Per pound of uranium, this is about 20 million times the energy of TNT. On January 3, 1939, Frisch met with Bohr, who was on his way to America, and told him of this discovery, swearing him to secrecy until

[40] Curie, I., & Savitch, P. (1938). Radioelement of period 3.5 hours formed from uranium bombarded by neutrons. *Comptes Rendus*, 206, 906–908.

[41] Hahn, O., & Strassmann, F. (1939). Uber den Nachweis und das Verhalten der bei der Besstrahlung des Urans mittels Neutronen entstehenden Erdalkalimetalle. *Naturwissenschaften*. 27, 11–15.

[42] Meitner, L. & Frisch, O. R. (1939). Disintegration of uranium by neutrons: A new type of nuclear reaction. *Nature*, 143, 239–240.

the results could be published. But while on board his ship, Bohr told his colleague Leon Rosenfeld, who was traveling with him. Not understanding the need for secrecy, Rosenfeld told Fermi and others when he got to New York. The secret was out. This immediately attracted the interest of physicists around the world, but it wasn't obvious how to get this energy in useful quantities. Where would the neutrons come from to cause the reaction and release the energy?

The answer came just a few months later. The neutrons would come from the fission reaction itself. Experiments by Frédéric Joliot-Curie, Hans von Halban and Lew Kowarski in Paris showed that when uranium fissions, several neutrons (about 2.5 on average) are released.[43] At the same time, Szilard and Fermi working at Columbia University obtained the same result. These results demonstrated that the neutrons released in the fission reaction can propagate a chain reaction in which the neutrons from one fission create the next fissions, and so on, just as conceived and patented by Szilard in 1933. In other words, a bomb is possible.

That same year, on March 15, 1939, Hitler violated the Munich agreement and invaded Czechoslovakia. The previous fall, Hitler had promised no more territorial demands beyond the Sudetenland, and British Prime Minister Neville Chamberlain had returned from Munich proclaiming "peace for our time." But the peace was to prove tense and short-lived. In spite of this, over Szilard's objections, both Joliot-Curie and Fermi published their neutron results in April, 1939; Joliot-Curie published in the British journal *Nature*[44] and Fermi in the American journal *Physical Review*.[45] Upon seeing the paper in *Nature*, Harteck, who by this time had replaced Stern as a professor in Hamburg, notified Erich Schumann, head of weapons research in the *Heereswaffenamt* (HWA, Army Ordnance Office) of potential military applications of nuclear energy. He wanted to

> draw attention to the latest development in nuclear physics, which is that it is probably possible to produce an explosive which is many orders of magnitude more effective than presently available.[46]

---

[43] Cassidy (2010), p. 298.
[44] Cornwell (2003), p. 223.
[45] Rhodes (1986), p. 293–295.
[46] Hoffmann (2001), p. 157. Bernstein, J. (2013). *Hitler's uranium club: The secret recordings at Farm Hall.* Springer, Science and Business Media, p. xxii. Cornwell (2003), p. 224.

The letter concluded with the ominous warning "The country which makes the first use of it will, in comparison with others, possess a superiority which cannot be caught up with."

On April 29, 1939, Professor Abraham Esau, who was on the Reich Research Council, convened a secret meeting of Hahn and a few other experts on the uranium question. Following this meeting, Esau took the first concrete step in the German nuclear weapon program: he embargoed all uranium exports.

The possibility of making an atomic bomb was now apparent to everyone. Heated discussions took place at the spring meeting of the American Physical Society in Washington "with arguments over the probability of some scientist blowing up a sizeable portion of the earth with a tiny bit of uranium," as described in *The New York Times* on April 30, 1939.[47] But it would not be straightforward. The first obstacle would appear just a few months later.

The following summer, June to August, 1939, Heisenberg travelled to the USA where he visited Columbia, Princeton, Purdue, the University of Chicago, and the University of Michigan. Between the Midwest summer heat and the hectic schedule of visits and lectures it was a tiring trip, but mostly enjoyable for him. By the end of this trip he had attractive offers of positions from Columbia, Princeton, and the University of Chicago.[48]

Along the way there were many discussions, both scientific and non-scientific. Some were heated. Heisenberg's colleagues in America were skeptical of his proposals for explosive cosmic ray showers; the old theories worked just fine.[49] But inevitably, there were discussions of Hitler and Nazi Germany. People couldn't understand Heisenberg's determination to remain in Germany. Recollections vary. Rudolf Peierls and Neville Mott recalled Heisenberg expecting a German victory[50]; Germany would need him when the Hitler regime ended. But in his memoir *Physics and Beyond*,[51] Heisenberg remembers telling Fermi in Chicago that he must return to his students:

If I abandoned them now, I would feel like a traitor. …People must learn to prevent catastrophes not to run away from them. Perhaps we ought even to insist that everyone brave what storms there are in his own country.[52,53]

---

[47] Cornwell (2003), p. 225.

[48] Cassidy (2010), p. 293.

[49] Cassidy (2010), p. 293.

[50] Cassidy (2010), p. 294.

[51] Heisenberg, W. (1971). *Physics and Beyond: Encounters and Conversations*. Harper and Rowe.

[52] Cornwell (2003), p. 230.

[53] Cassidy (2010), p. 295.

In spite of everything, even after the *Deutsche Physik* attack in 1937, even after *Kristallnacht* in 1938 and the invasion of Czechoslovakia in early 1939 with its possible consequences, Heisenberg remained a dedicated German nationalist and would return to Germany. While in New York on June 25 he wrote Elisabeth[54] "By the way, so far, I have not encountered anti-German sentiments; all the Americans have treated me very nicely." Regarding the imminence of war, on July 10, 1939, Heisenberg wrote to Elisabeth from Chicago "I am reading in the papers that, at least for the next weeks, there is no danger of war."[55]

When he returned to Germany in early August, Heisenberg and his family moved into the cottage on the Walchensee, in Urfeld. Three weeks later, World War II began. On September 1, 1939, Hitler invaded Poland, and the Nazi *Blitzkrieg* [lightning war] crushed Poland in six weeks. In response, England and France declared war, but did almost nothing. The next seven months were mocked by the press as the *Sitzkrieg* [Sitting war] in Germany and the "phony war" in England. During this time, German submarines attacked British ships, sinking the aircraft carrier HMS Courageous and the old battleship HMS Royal Oak, while the British began bombing military targets on the coast of Germany. But there was no serious fighting on land.

Shortly after the invasion of Poland, Heisenberg moved back to Leipzig, leaving Elisabeth and the children in Urfeld. Wartime shortages began, worse in Leipzig than in Urfeld, and on September 15, 1939 Heisenberg wrote from Leipzig:

> It is so good that you in Urfeld are far from the war. If you had to look after the three little ones here in Leipzig as well as stand in line for food, I do not know how you could manage it.[56]

On September 23 he added "There also seems to be a great shortage of coal…Nobody seems to believe in a brief war anymore."[57]

The first scientific obstacle that confronted any effort to make an atomic bomb appeared on September 1, 1939, the same day that Hitler invaded Poland, when Niels Bohr and John Archibald Wheeler published the result that only certain uranium nuclei fission with neutrons.[58] All uranium atoms

---

[54] Heisenberg (2016), p. 35.

[55] Heisenberg and Heisenberg (2016), p. 63.

[56] Heisenberg and Heisenberg (2016), p. 76.

[57] Heisenberg and Heisenberg (2016), p. 83.

[58] Bohr, N., & Wheeler, J. A. (1939). The mechanism of nuclear fission. *Physical Review, 56*(5), 426–450. https://doi.org/10.1103/physrev.56.426. Cornwell (2003), p. 232.

contain the same number of protons, 93, but different isotopes of uranium contain different numbers of neutrons. $U^{238}$ contains 93 protons and 145 neutrons (238 particles in all), while $U^{235}$ has only 142 neutrons (235 particles in all). All the promising fission results reported by Hahn, Joliot-Curie, and Fermi were caused by $U^{235}$, but this rare isotope forms only 0.7% of natural uranium. The bulk of the uranium, $U^{238}$, absorbs the neutrons without fissioning and stops the chain reaction. But Bohr and Wheeler also showed that the probability (called the cross section) for a neutron to hit a $U^{235}$ nucleus and cause fission increases enormously if the neutron has a very low energy. This opens up two possibilities: a bomb or a nuclear power reactor.

To make an atomic bomb, the neutrons must multiply rapidly enough to create an explosion. When the uranium nucleus fissions, the neutrons that are produced have high energy, so they're called fast neutrons. Since $U^{238}$ absorbs fast neutrons without fissioning, it stops the chain reaction. The rare isotope $U^{235}$ must be used to make a bomb, and this has to be separated from the common isotope $U^{238}$. Unfortunately, the two isotopes of uranium are chemically identical and have nearly the same mass, making it exquisitely difficult to separate the isotopes. Separation of the $U^{235}$ from the $U^{238}$, called enrichment, was the most expensive part of the Allied bomb effort, and the Germans never achieved it on a large scale.

But there is another possibility. Natural uranium can be used to sustain a chain reaction if the neutrons are slowed down quickly, so they react with the $U^{235}$ before they are absorbed by the $U^{238}$. To sustain a chain reaction with natural uranium, the uranium is imbedded in a substance called a moderator that slows down the neutrons without absorbing them. The good news is that with a satisfactory moderator, the neutrons can be slowed down quickly enough that the fission of $U^{235}$ exceeds the absorption of neutrons by $U^{238}$. The best moderators are "heavy water," in which the hydrogen in $H_2O$ is replaced by deuterium (an isotope of hydrogen with one proton and one neutron in the nucleus) to make $D_2O$, or carbon in the form of graphite. Because it requires a moderator to slow the neutrons, which slows down the overall rate of reaction, a nuclear reactor can overheat and "melt down," releasing dangerous radiation (think Chernobyl), but everything takes place too slowly to produce an explosion. A reactor can't be a bomb. Unable to separate isotopes on a large scale to get pure $U^{235}$, the Germans chose to make a reactor. It was cheaper than trying to enrich uranium, and was an important first step in any event. When they first began, the Allies did the same.

With this information, everybody on both sides of the Atlantic now understood both the possibilities and the obstacles confronting the development of nuclear reactors and atomic bombs. In response to Harteck's April 1939 letter,

Schumann passed the information to Kurt Diebner, Director of the Nuclear Research Council of the HWA, who consolidated all fission research in the HWA. Together with Erich Bagge, a former student of Heisenberg, Diebner convened the first meeting of the so-called *Uranverein* (Uranium Club) on September 16, 1939, just two weeks after the Bohr/Wheeler paper and the invasion of Poland. Following this meeting, rather than being drafted into the army, Heisenberg was conscripted into the Uranium Club. On September 21 he wrote Elisabeth

> The summons yesterday was, as it were, only about "readiness" and did not come from the place I had expected but from the scientific office. … Most likely, now, however, is that I should simply remain in Leipzig and only go to Berlin occasionally. … Then it might be irresponsible to bring the family here. It is better for the children to grow up normally in the family; but they would get less to eat here. … It is also quite likely that the air war will become worse. The Führer's speech points toward that.[59]

Except for brief periods, Elisabeth and the children remained in Urfeld for the duration of the war. Until he moved to Berlin in June, 1942, Heisenberg stayed in Leipzig and commuted to Berlin, about a two-hour train trip, to oversee the work going on there.

On September 26, 1939, Heisenberg, von Weizsäcker, and others attended the second meeting of the Uranium Club where they discussed Bohr's results and their implications for atomic bombs and nuclear reactors. Heisenberg reported that "in principle atomic bombs could be made … it would take years … not before five," owing principally to the difficulty of separating the uranium isotopes.[60] If the war were to end quickly, as Hitler expected in 1939, this time scale would not produce a weapon in time to help. As things worked out, had they been able to develop a bomb in five years it might have completely changed the outcome of World War II.

On October 5, HWA took over the *Kaiser Wilhelm Institut für Physik* (KWIP) in Berlin-Dahlem with Diebner as head, replacing the Dutch physicist Peter Debye. Debye promptly decamped to the United States where he contributed to the Allied war effort. Diebner and Bagge organized the work as follows[61]: Heisenberg was assigned theory; Bagge and Klaus Clusius were

---

[59] Heisenberg and Heisenberg (2016), p. 81.

[60] Wikipedia contributors. (2023, March 8). *German nuclear weapons program*. Wikipedia. https://en.wikipedia.org/wiki/German_nuclear_weapons_program.

[61] Heisenberg, W. (1947). Research in Germany on the technical application of atomic energy. *Nature.* https://doi.org/10.1038/160211a0.

assigned uranium enrichment; Walther Bothe, later a Nobel Prize winner himself, was assigned to measure the absorption of neutrons by heavy water and by carbon (graphite) to test their suitability as moderators; and Harteck was assigned to work on the separation of deuterium from hydrogen to make heavy water. Diebner himself began reactor experiments at the KWIP and at the HWA test site (*Versuchsstelle*) in Gottow, about an hour south of Berlin.[62] With Heisenberg in Leipzig, Bothe in Heidelberg, Bagge in Berlin, Harteck in Hamburg, and Diebner in Berlin and in Gottow, the work was spread out, making communication and cooperation difficult. It also didn't help that Bothe and Heisenberg had been attacked by the Nazis in the *Deutsche Physik* movement, whereas Diebner was a Nazi. Infighting characterized Nazi organization throughout Germany, and the Uranium Club was no different.

Following the second meeting of the Uranium Club, Heisenberg worked intensely on the theory of atomic bombs and nuclear reactors and in two months he produced two secret reports that summarized all that was known at that time. In his first report (December 6, 1939), he concluded that a controlled fission reactor was possible using natural uranium with a suitable moderator, and that uranium vastly enriched in $U^{235}$ could be used to make a new explosive "which surpasses the explosive power of the strongest explosive materials by several orders of magnitude."[63] In his second secret report (February 29, 1940), Heisenberg analyzed the *Uranmachine* (controlled nuclear reactor) using natural uranium.[64] Working with the limited data available at the time, he found that the reactor would be about 1 m in size. It would require at least 600 L of heavy water as a moderator and two to three metric tons of uranium oxide in a layered configuration with the moderator. The Germans explored both graphite and heavy water as moderators, but in June 1940, measurements by Bothe on the purest commercial graphite available from Siemens showed that the absorption of neutrons by impurities in the graphite was too large, making graphite unsuitable as a moderator. Heisenberg abandoned carbon, and focused on heavy water. Fortunately for the German nuclear program, Hitler overran Norway in May 1940, so heavy water was available from Norsk Hydro in occupied Norway. This was the only source of large quantities of heavy water in the world. There was little use for heavy water, and this production capability was left over from an ammonia

---

[62] Cassidy (2010), p. 306.

[63] Rhodes (1986), p. 296.
   Bernstein (2001), p. xxiii.
   Cornwell (2003), p. 232.

[64] Cornwell (2003), p. 232.
   Bernstein (2001), p. xxiv.

production process that had become economically uncompetitive after the discovery of the Haber-Bosh process. Prior to the German occupation, Norsk Hydro had been producing about 11 L of heavy water per month.[65] After the occupation, pressed by the Nazis, Norsk Hydro, which was controlled by I. G. Farben, increased this to 110 L/month. Then Harteck added a catalytic process that increased the production rate to 200–400 L/month by the summer of 1942,[66] but heavy water would always be in short supply. Norsk Hydro remained in production until the plant was damaged by Norwegian commandos in February 1943.[67] The plant was completely destroyed by bombing in November 1943, and commandos sank a ferry carrying about 500 L of heavy water back to Germany. But by this time about 2000 L of heavy water were available, enough for a single nuclear reactor.[68] But this would require a well-run program, which the Germans never achieved.

Illustrating the dysfunctional nature of the Nazi nuclear program, there were three parallel projects constructing nuclear reactors in Germany, plus smaller efforts by Harteck in Hamburg and Bothe in Heidelberg that never got very far. As Director of the KWI/Physics in Berlin-Dahlem, Diebner initiated a reactor effort there. In June, 1940, work began on a reactor on the site of the KWI/Biology, nearby the KWI/Physics, in a building called the "virus house" to keep people away. But Diebner wasn't respected by the staff of the Institute, so Heisenberg became the de facto leader of the program and spent half his time in Berlin. The first experiment was carried out in the fall of 1940 comprising layers of uranium oxide separated by layers of ordinary paraffin with a neutron source in the center. The measurements of the neutron count at the outside of the reactor showed no neutron gain due to uranium fission. Follow-on experiments used heavy water in place of ordinary paraffin. As the nominal head of the German nuclear program, Diebner also ran what would become the most successful reactor project at the *Versuchsstelle* in Gottow. Diebner's first experiment at Gottow in 1941–42 used cubes of uranium oxide stacked together with a moderator of ordinary paraffin, with a neutron source placed at the center of the pile. Like the experiments in the "virus house," these measurements showed no neutron gain. Subsequently, Karl-Heinz Höcker showed theoretically that Diebner's cubes of uranium were better than Heisenberg's layers, but Heisenberg continued to favor layers until

---

[65] Heisenberg, (1947).

  Bernstein (2001), p. 27.

[66] Heisenberg (1947).

  Bernstein (2001), p. 27.

[67] Mears, R. (2004). *The real heroes of Telemark*. Coronet. BBC television documentary.

[68] Heisenberg (1947).

near the end of the war because they were easier for him to analyze theoretically.

Although Heisenberg's role assigned by Diebner was theory, owing to his great scientific prestige he was able to secure funds for a third reactor project, which was carried out in Leipzig by Robert Döpel under Heisenberg's direction. On July 22, 1940, Heisenberg wrote to Elisabeth

> Unfortunately, at the moment our work here in Leipzig is going rather poorly, it is taking much longer than we anticipated … in the lab in a white coat I learned from Mrs. Döppel how to make metal tubes airtight. I enjoy the opportunity to learn the basics in experimental physics.[69]

It seems a bit late for Heisenberg to start to learn something about experimental physics. As a theoretical physicist, he didn't appreciate how slowly experimental physics goes, and on October 10 he wrote Elisabeth from Berlin

> The people in the institute [KWI/Physics, Berlin-Dahlem] are so terribly slow, and when I arrive here on Wednesday they are usually exactly where I left them the previous Saturday. I am now developing a certain respect for Döpel, who does not usually appear to be an especially good physicist. But he is working infinitely more precisely and faster than the people here.[70]

Heisenberg's experiment in Leipzig comprised spherical shells of uranium oxide (difficult for an experimenter to fabricate and assemble but easy for a theorist to analyze!) alternating with heavy water as a moderator. Again, a neutron source was placed at the center. Only 150 L of heavy water were available in the summer of 1941, and the neutron gain was very small. In February, 1942, the uranium oxide shells were replaced by uranium metal, which produced a measurable neutron gain. This was the first positive result in Germany, and showed that controlled atomic energy was possible. All that was needed was to enlarge the reactor with more uranium and heavy water. But with three nuclear reactor experiments underway, there was never enough uranium and heavy water to go around, and the three projects fought each other for the available supplies.

A fourth experiment, one that didn't require heavy water, was attempted by Harteck in Hamburg. As a very clever alternative, Harteck proposed to use dry ice (frozen $CO_2$) as the moderator. In this form, the impurities in the carbon could be reduced to one part per million, and oxygen doesn't absorb

---

[69] Heisenberg and Heisenberg (2016), p. 110.
[70] Heisenberg and Heisenberg (2016), p. 136.

neutrons, making dry ice a suitable moderator. In the spring of 1940 Harteck procured the dry ice from I. G. Farben, but he couldn't get the uranium he needed. Heisenberg wanted it for his experiments, and his prestige gave him priority. Had he gotten his uranium, Harteck might have developed the world's first self-sustaining nuclear reactor, even before Fermi's reactor in the United States. Under Heisenberg's leadership it was not to be. As Heisenberg pointed out in the second meeting of the Uranium Club and in his early reports, building an atomic bomb would take years. Even with pure $U^{235}$, there are several obstacles to building an atomic bomb. The most fundamental problem is the loss of neutrons at the surface of the uranium. If the block of uranium is too small, the neutrons escape through the surface before they cause another fission, and the chain reaction stops. This problem was already recognized by Szilard in his 1934 patent. The amount of fissile material needed to reduce the surface losses to an acceptable level is called the critical mass. At just this time, Rudolf Peierls, who had emigrated to England in 1936 because of his Jewish background, actually published the theory of the critical mass in the November 1939 issue of *Mathematical Proceedings of the Cambridge Philosophical Society*, a well-known journal of which the German team should have been aware.[71] In this article, Peierls used natural uranium for his calculation and concluded that a critical mass would be impossibly large. By 1940, Frisch, also of Jewish descent, had escaped to Britain where he began working with Peierls at the University of Birmingham. As aliens, they couldn't work on the important but highly classified program to develop radar, but nuclear physics was still unclassified. Frisch suggested that Peierls' calculation of the critical mass should be redone using pure $U^{235}$. This changed everything. With no experimental data to work with, Frisch and Peierls used somewhat overoptimistic estimates of the fission cross section in $U^{235}$. They found a critical mass of just 0.6 kg, which corresponds to a sphere with a diameter of 4 cm, about the size of a golf ball.[72]

Frisch and Peierls also considered the problem of separating $U^{235}$ from $U^{238}$ and concluded that the best method would be thermal diffusion, a process discovered by the German scientist Clusius. Clusius was by this time already a member of the *Uranverein* and was working on isotope separation for the German program. Frisch and Peierls estimated that one hundred thousand thermal diffusion columns would be enough to enrich 100 g of $U^{235}$ per day, which they felt was doable. They summarized their findings in a report in

---

[71] Pierls, R. (1939). Critical conditions in neutron multiplication. *Cambridge Philosophical Society*, *35*, 610–615.

[72] Bernstein, J. (2011). A memorandum that changed the world. *American Journal of Physics*, *79*(5), 440–446. https://doi.org/10.1119/1.3533426.

March 1940, now called the Frisch-Peierls memorandum.[73] This optimistic result kick-started the British nuclear weapons program. The actual critical mass for uranium is larger than this, about 52 kg, which corresponds to a sphere with a diameter of about 17 cm, roughly the size of a pineapple. This was 90 times more enriched uranium than Frisch and Peierls estimated, but, at least for the Allies, it was still doable.

The historical record doesn't make clear whether Heisenberg and his colleagues ever correctly calculated the critical mass needed to make a bomb, though they should have been aware of Peierls' published theory. Without this calculation, they couldn't know how much enriched uranium was actually needed to make a bomb. It seems that from the beginning Heisenberg believed that the atomic bomb couldn't be developed in time to affect the war, so he paid little attention to its construction. Clusius did some work on enriching uranium using thermal diffusion columns, and in 1943 Harteck, with his colleague Hans Jensen, did some work on a double centrifuge to enrich uranium, but neither of these efforts was pursued seriously. In any event, the discovery of plutonium as an alternative explosive material in the summer of 1940 made uranium isotope separation unnecessary, provided that a working nuclear reactor could be built to produce the plutonium.

Sometimes, when a nucleus of $U^{238}$ is struck by a neutron, it emits an electron and becomes a new element, neptunium, $Np^{239}$. Then, in just a couple of days, the $Np^{239}$ emits another electron and becomes plutonium, $Pu^{239}$, which is stable enough to last for thousands of years. These decay products are actually the transuranic elements that Fermi had been hoping to create in his experiments leading up to his Nobel Prize but was unable to identify at that time. Crucially, plutonium is an alternative fissionable material that can be separated from uranium by chemical processes. These are much simpler than the elaborate processes needed to separate the chemically identical isotopes of uranium. It was quickly realized on both sides of the Atlantic[74] that this was an alternate route to building an atomic bomb that didn't need the complex, expensive, isotope separation processes that Germany couldn't afford and that would be vulnerable to Allied bombing. But it did require a working nuclear reactor to transform the $U^{238}$ into plutonium. This was described by von Weizsäcker as the "open road to the bomb" in a document he sent to the HWA on July 17, 1940, with a copy to Heisenberg.[75] Nuclear reactors became

---

[73] https://web.stanford.edu/class/history5n/FPmemo.pdf.

[74] McMillan, E. M., & Abelson, P. H. (1940). Radioactive element 93. *Physical Review, 57*(12), 1185–1186. Bernstein (2001), p. 30.

[75] Bernstein (2001), p. 30.

the focus of the German team both as a source of nuclear power, which would be useful for ships and submarines, and to produce plutonium for an atomic bomb. But they never succeeded in building a working nuclear reactor before Germany surrendered five years later.

To make a useful amount of energy or plutonium in a nuclear reactor, it is necessary for the chain reaction to proceed at a rapid—but not too rapid—rate. To keep the reaction going, every neutron lost by various processes must be replaced by the neutrons emitted in a fission reaction. This is called criticality. But if more neutrons are created than are lost, the reaction will accelerate and lead, eventually, to overheating and a meltdown. Fortunately, as shown by Bohr and Wheeler, faster neutrons have less likelihood of causing a fission reaction with the $U^{235}$ than do slow neutrons, so if the reactor heats up, this slows down the rate of reaction, making the reactor stable. Usually... but if things get too out of hand, a runaway reactor will overheat and self-destruct, releasing dangerous radioactivity. While Heisenberg recognized this problem, there is no evidence that he appreciated the seriousness of this issue. He doesn't seem to have been aware of the "delayed neutrons" that are emitted a short time later by unstable fission products. Confident that his reactor would be stable, Heisenberg never took any precautions. Since the delayed neutrons appear on a slower time scale, giving reactor operators time to respond, all nuclear reactors use control rods made of neutron-absorbing materials that the operators can use to control the rate of reaction. Even Fermi's first reactor experiment used control rods. Since he never achieved a self-sustained chain reaction, Heisenberg never had to confront a meltdown. For Heisenberg, "success" might have proved a disaster.

For the first couple of years of the program, 1940 and 1941, the impact of the war on the Uranium Club and its progress wasn't too serious. The Germans invaded Denmark and Norway on April 9, 1940. All resistance ended on June 10 and the occupation of Norway began, providing the Nazis with a source of heavy water. Meanwhile, on May 10, 1940, the Nazis invaded the Netherlands, Belgium and France. The Allies were quickly crushed by the *Blitzkrieg*, and British forces barely escaped from the Continent. The French surrendered on June 25, 1940. All was going well for the *Wehrmacht* (the German army), but not so well for the Allies. After France fell, von Halban and Kowarski from the nuclear research group in Paris fled to Britain and joined Allied fission workers there, while Frédéric Joliot-Curie remained in France and joined the resistance. When Belgium was overrun by the Nazis, the uranium ore from the mines in the Congo became available, but throughout the war Heisenberg and Diebner fought over the limited supplies of refined and processed material.[76]

Except for the casualty lists, the early years of the war had been mostly good news and propaganda for the people at home in Germany. The swift victories in Czechoslovakia, Poland, Denmark, Norway, and even the *Blitzkrieg* in France gave the Germans much to cheer about. The dirty side of the war, the death and destruction, the extermination of Jews and others, was hard to see from the comforts of home, and the cafes in Germany were full. Following the fall of France, the next step for Hitler was to defeat Britain, but since it was protected by the English Channel, Britain couldn't be invaded until the Nazis controlled the skies. From June to September, 1940, the Battle of Britain raged as the *Luftwaffe* (the German Air Force) tried to destroy the RAF. The attempt failed, but it led directly to the German *Blitz* of London and, in reply, the British bombing of Germany. The first bombs fell on Berlin on June 7. The damage was slight, and work on the uranium program went on mostly undisturbed. But by July 15, Heisenberg wrote Elisabeth from Hamburg, where he was visiting:

> I could not but watch the happenings from the window. The shooting of anti-aircraft weapons reflected in the clouds like a nighttime thunderstorm, just the noise is briefer and more mechanical....I was awakened again about 3 o'clock by more sirens...The Hamburg people told us the next day that it is usually much worse.[77]

Soon, strikes on Berlin occurred almost daily. On September 3, 1940, Heisenberg wrote to Elisabeth from Berlin "Around midnight then, there was a pretty intense air strike, from the basement one could distinctly make out bomb explosions...".[78]

But the Luftwaffe suffered major losses throughout the summer and on September 17 Hitler postponed preparations for the invasion of Britain. Nevertheless, the *Blitz* of London by Germany and the bombing of Berlin by the Allies continued.

Now Hitler turned his attention to the East. Obsessed with the defeat of Communism and coveting the "*Lebensraum*" (living space) offered by Ukraine and western Russia, Hitler invaded Russia on June 22, 1941, commencing Operation Barbarossa. On the home front, the effects of the war were still

---

[76] Heisenberg and Heisenberg (2016), p. 239.

[77] Heisenberg and Heisenberg (2016), p. 106.

[78] Heisenberg and Heisenberg (2016), p. 114.

hardly visible as the German army swept across western Russia, and Heisenberg and Diebner continued their experiments mostly uninterrupted. With the rapid success of the German army, there didn't seem to be any urgent need for an atomic bomb. This would change in the next six months.

Throughout the war years, perhaps in part to keep in Himmler's good graces after the *Deutsche Physik* affair, Heisenberg frequently traveled to other countries as an ambassador for German culture and science.[79] Sometimes the trips had a personal as well as a scientific purpose. This was the case when, on September 15, 1941, Heisenberg and von Weizsäcker visited Bohr in Copenhagen. Back in 1924–25, when Heisenberg was developing his matrix mechanics, he spent a year with Bohr in Copenhagen. During that time, Bohr, who was older than Heisenberg, became not just his scientific advisor but also his friend and mentor. By the time of his visit, Denmark had been occupied by Germany for a year, and in addition to wanting to spread the prestige of German physics, Heisenberg was concerned for his erstwhile mentor, who was partly Jewish. While in Copenhagen, Heisenberg gave a lecture on cosmic rays at the German Cultural Institute, which had been set up by the occupying forces. The Danish scientists from Bohr's institute boycotted the talk; they refused to go to the German Cultural Institute.[80] The next day, Heisenberg and von Weizsäcker visited Bohr at his institute. Much has been written about the discussions that Heisenberg had with Bohr during that visit, including a famous play "Copenhagen," written by Michael Frayn.[81] In his talks with Bohr, Heisenberg called the war a "biological necessity."[82] This would seem to be a well-known quote from the influential book *Deutschland und der Nächste Krieg* [Germany and the Next War], written in 1911 by the militarist and historian General Friedrich von Bernhardi.[83] Bernhardi makes the case that a nation that is not at war to expand must inevitably decline, and argues that this is "the natural law, upon which all the laws of Nature rest, the law of the struggle for existence." Evidently, Heisenberg shared this view. With the German forces advancing rapidly across Russia, and every reason to believe that the Germans would swiftly achieve victory there, Heisenberg advised Bohr that the Danes should cooperate with the Nazis. This

---

[79] Walker, M. (1995). *Nazi science*. Cambridge, MA: Perseus Publishing.

[80] Walker (1995), p. 148.

[81] Frayn, M. (n.d.). *Postscript to Copenhagen*. https://nba-old.nbi.dk/files/sem/CopenhagenPostscript.pdf.

[82] Walker (1995), p. 149. Crowther, J.G. (1949). *Science in liberated Europe*. Pilot Press, London, p. 108.

[83] von Bernhardi, F. (2001). *Germany and the next war*. Translated by Allen H. Powles, University Press of the Pacific.

understandably upset Bohr and his Danish colleagues. In a letter to Heisenberg written after the war, but never sent, Bohr wrote:

> In particular, it made a strong impression on both Margrethe [Bohr's wife] and me, and on everyone that the two of you spoke to, that you and Weizsäcker expressed your definite conviction that Germany would win and that it was therefore quite foolish for us to maintain the hope of a different outcome of the war and to be reticent as regards all German offers of cooperation.[84]

Further, Bohr wrote:

> I also remember quite clearly our conversation in my room at the Institute, where in vague terms you spoke in a manner that could only give me the impression that under your leadership, everything was being done in Germany to develop atomic weapons and that you said that there was no need to talk about details since you were completely familiar with them and that you had spent the last two years working more or less exclusively on such preparations. I listened to this without speaking since [a] great matter for mankind was at issue in which, despite our personal friendship, we had to be regarded as representatives of two sides engaged in mortal combat.[85]

Assuming that Bohr's memory of what Heisenberg said is correct, there has been much discussion of why Heisenberg would have made such a sensitive statement at that time.[86] Was he trying to warn Bohr about the German atomic weapon program? Was he trying to enlist Bohr? Although there are various accounts of the conversation, it seems that Heisenberg asked Bohr if he considered it right for scientists to do research on uranium in time of war, to which Bohr responded "Do you really think that uranium fission could be utilized for the construction of weapons?" To this, Heisenberg says that he responded "I know that this is in principle possible, but it would require a terrific technical effort, which, one can only hope, cannot be realized in this war."[87]

Of more personal concern to Bohr, the indications were already clear from the Nazi rhetoric, the treatment of Jews under the Law for the Restoration of the Professional Civil Service, and the murder of Jews on *Kristallnacht*, that

---

[84] Neils Bohr Archives, released February 6, 2002. https://www.aip.org/library

[85] Neils Bohr Archives, released February 6, 2002. https://www.aip.org/library

[86] Bernstein, J. (1995). What did Heisenberg tell Bohr about the bomb? *Scientific American*, 272(5), 92–97.

[87] Powers, T. (1993). *Heisenberg's War*. Boston, MA: Da Capo Press. Cornwell (2003), p. 302.

Bohr, whose mother was Jewish, would be in danger for his life. Heisenberg's complete lack of understanding is evident in his letter to Elisabeth written on September 16, while he was in Copenhagen:

> Tomorrow begin the talks in the German Scientific Institute…Sadly, the members of Bohr's institute will not attend for political reasons. It is amazing, given that the Danes are living totally unrestricted and are doing so well, how much hatred or fear has been galvanized here, so that even a rapprochement in the cultural arena—where it used to be automatic in earlier times—has become almost impossible. …nobody wants to go to the German Institute on principle, because during and after its founding a number of brisk militarist speeches on the New Order in Europe were given. [Saturday night] Today I was once more, with Weizsäcker, at Bohr's. In many ways this was especially nice, the conversation revolved for a large part of the evening around purely human concerns…[88]

When he returned to Leipzig a few days later Heisenberg wrote Elisabeth "The return to Germany from the free Denmark made me quite gloomy."[89] It seems incredibly insensitive when Heisenberg speaks of "Danes living totally unrestricted" in "free Denmark," since Denmark had surrendered to German forces more than a year earlier and had been under occupation since then. At the conclusion of the visit, the official reports from the Copenhagen German Cultural Institute rated Heisenberg's visit a success.

## Failure

Later in 1941, the military situation became more dire for Germany. Everything had gone well as the *Wehrmacht* advanced across Russia, until the German offensive stalled outside Moscow in December. The autumn rain turned the ground to mud, and the German panzers became bogged down. Then the Russian winter struck, with the same bitter cold that defeated Napoleon in 1812. The winter of 1941–42 was the coldest of the twentieth century, and the *Wehrmacht* wasn't prepared. The panzers wouldn't start, and the troops froze in their summer uniforms. On December 5, the Soviets launched a massive counterattack that drove the *Wehrmacht* back as much as 250 km in places. It became clear that the war would not be short, and German victory was no longer assured. Two days later, on December 7, the Japanese attacked Pearl Harbor. The United States responded by declaring

---

[88] Heisenberg and Heisenberg (2016), pp. 169–170.
[89] Heisenberg and Heisenberg (2016), p. 172.

war on Japan the next day. Since he had a mutual aid treaty with Japan, Hitler declared war on the United States on December 11, 1941, and the USA responded by declaring war on Germany the same day.

These setbacks caused major changes in German planning and resource allocation, with disastrous effects on the nuclear weapons program. On December 5, 1941 Schumann, as Chief of the Research Department of the HWA, informed all the leaders of the uranium work that further work "can only be taken responsibility for if the certainty exists that a [militarily interesting] application can be achieved in a foreseeable time."[90] A "foreseeable time" seems to have meant six to nine months. Schumann ordered a review of all research programs, and a progress report compiled by the scientists contains both promises and a warning:

> The enormous significance that [atomic energy] has for the energy economy in general and the Wehrmacht in particular justifies such preliminary research, all the more in that this problem is also being worked on intensively in the enemy nations, especially in America.[91]

A conference was called for February 26–27, 1942, to discuss the status of nuclear research. Himmler, Göring, and other high officials were invited but declined to attend. At this meeting Heisenberg described the potential for making an atomic bomb from enriched uranium. He also discussed nuclear reactors for, among other things, submarines. Aware, by then, of the plutonium alternative for a bomb, he added that the reactors would generate "a new substance (element 94) … which in all probability is an explosive with the same unimaginable effectiveness as pure uranium 235."[92] One high-ranking Nazi official who did attend was Bernard Rust. He was so impressed that following the meeting he took the program over and placed it in the Reich Research Council under Abraham Esau. It seems that Heisenberg had no qualms about offering nuclear weapons to the Nazis.

But progress was slow. After repeated failures, late in 1941 the experiments in Leipzig produced a neutron gain of 13%. This was small, but it was the first neutron gain observed by the Germans. In the first half of 1942, Döppel's results in Leipzig indicated that the spherical geometry, simply enlarged, could sustain a fission reaction. But on June 23, 1942, the Leipzig pile burned in a fire caused by the reaction of its powdered uranium with air. This was the

---

[90] Hoffmann (2001), p. 165.

[91] Cornwell (2003), p. 310.

[92] Bernstein (2001), p. xxviii.
   Cornwell (2003), p. 311.

end of reactor experiments in Leipzig. Meanwhile, Diebner's first Gottow experiment, G-I, produced no useful results in 1942.

Following up on the poorly attended meeting in February, Albert Speer (Reich Minister of Armaments and War Production) called another meeting to get a decision on the nuclear program. The meeting was held on June 4, 1942, at the KWI with Speer, General Field Marshall Erhard Milch (State Secretary in the Reich Ministry of Aviation and Inspector General of the *Luftwaffe*) and other high-ranking military officials in attendance. There, Heisenberg described the enormous potential of an atomic bomb. In response to a question from Milch about the size of such a weapon, Heisenberg himself referred to a critical mass about the size of a pineapple. Hearing this, Milch, Speer and the others immediately sat up and became interested. Heisenberg was forced to tell them that it would be expensive and couldn't be completed in the six to nine months they demanded; it would probably take years. Asked by Speer how much it would cost, von Weizsäcker gave a number on the order of 40,000 marks (about $16,000 at that time). Expecting an estimate on the order of hundreds of millions of marks, Milch later recalled "It was such a laughably small amount that Speer looked at me and we both shook our heads at the unworldliness and naïvety of these people."[93] Instead, the money— more than one billion marks—went to Werner von Braun for the development and production of the V-1 and V-2 missiles.

On the train back to Leipzig, June 5, Heisenberg wrote Elisabeth a letter that reveals much about him:

> Those were eventful days in Berlin. In the Harnack House truly every preeminent person of the Reich was assembled—essentially the entire armaments council and many others in power. It went far better than expected, my own and Bothe's presentations left, apparently, a strong impression; at any rate, we were treated fabulously afterward both personally and in terms of the facts. To me this is exceedingly strange and unnerving; suddenly, I will no longer have to concern myself at all with the whole Lenard-Stark clique, and can push through almost anything I find important. I led the secretary through our institute; strangely, all the scheduled experiments were functioning without a glitch, and afterward one sat together at dinner and on the terrace of the Harnack House until midnight with some wine. At dinner I sat next to General Field Marshal Milch, with whom I had an excellent conversation. ... I was very interested in the human level of the whole assembly: all of them were unusual people, of course, many really unusually smart, I suppose, all of them good diplomats, used to prevailing in a hostile situation of intrigues with decorum (and, I sup-

---

[93] Hoffmann (2001), p. 168.

pose, occasionally with stronger methods as well). Intermingled were a few strange characters who at first repel you, and during conversation exude something disarming, which can leave you quite confused. … I do not envy [Speer] the fact that his words decide the work of many millions of people, the care for the hundred thousand homeless from Cologne [the first Allied 1,000-bomber raid had hit Cologne the week before] and the whole human misery.[94]

Heisenberg's naïveté, apparent in this letter, is astonishing.

In spite of von Weizsäcker's remarks, the uranium program did get one million marks per year (about $400,000 at that time). Heisenberg and the others were afraid to ask for more, fearing the repercussions that might come if they failed. But one million marks was a comfortable budget for the *Uranverein*. It would keep them from being conscripted into the *Wehrmacht* and sent to the front, and it would allow them to continue research on nuclear reactors. These might be useful late in a long war, and might support their careers after the war. Later, while they were interned in England after the war, just after the uranium bomb had been dropped on Hiroshima but before the plutonium weapon had been dropped on Nagasaki, Heisenberg remarked

> If the Americans had not gotten so far with the engine [nuclear reactor] as we did—and that's what it looks like—then we are in luck. There is a possibility of making money.[95]

Just a few days later they learned that the Americans had already developed nuclear reactors and used them to produce the plutonium for the bomb dropped on Nagasaki. Even in nuclear reactor research the Americans were way ahead.

Shortly after the meeting with Speer and others, Hitler placed the Reich Research Council (RFR) under Göring and Speer, and HWA control of the nuclear program was passed to the RFR in July. On July 1, 1942, one week after his reactor experiment in Leipzig caught fire, Heisenberg took over as Director of the KWIP in Berlin-Dahlem (Fig. 4.6) and began a series of reactor experiments using heavy water and uranium plates. Unfortunately, plates were the least efficient way to structure a reactor; it would take Heisenberg another year and half to admit that Diebner's cubes of uranium were better than his plates. Anticipating this move to Berlin, on June 5 Heisenberg wrote Elisabeth: "Our relocation to Berlin is still troubling me, mostly because of

---

[94] Heisenberg and Heisenberg (2016), p. 175–176.
[95] Bernstein (2001), p. 139.

**Fig. 4.6**  Kaiser Wilhelm Institute for Physics, in Berlin. (Courtesy of the AIP Emilio Segrè Visual Archives, Gift of David Cassidy)

the expected air raids. …How nice that you are in Urfeld and safe."[96] By this time conditions in Berlin were getting worse, and on September 13 he wrote:

> It is striking these days how everybody becomes thinner; several people…whom I had last seen before vacation, shocked me when I saw them again.…And the state of war does not contribute to making people more optimistic.[97]

Along with the Directorship of the KWIP, Heisenberg was given a professorship at the University of Berlin, where he spent time teaching and working on his theories. Given the deteriorating conditions, the general lack of optimism, the arrogance that "if they couldn't do it nobody could," and the comfortable level of funding that protected him and the others from military service at the front, Heisenberg seemed to lose the sense of urgency and focus necessary to succeed with either a reactor or a weapon. For the rest of the war Heisenberg spent much of his time on his new mathematical theories, his writing on philosophy, and his music. On September 22, 1942, he wrote Elisabeth:

> I am working on my physics all day long, almost for all of this week already, and I am making wonderful progress. Mathematics is a glorious science; a problem that seemed to me almost insolvably difficult eight days ago I now have fought through to the solution completely.[98]

---

[96] Heisenberg and Heisenberg (2016), p. 176.

[97] Heisenberg and Heisenberg (2016), p. 180.

[98] Heisenberg and Heisenberg (2016), pp. 180–181.

Heisenberg also continued his travels to other countries as an ambassador for German culture and science with Nazi support. On November 17, he wrote from Zurich:

> ...you had feared that, as a German, I would encounter much hatred. But the first impression, which is so deeply moving, is just the opposite: a measure of friendliness we have not seen in years....However...I might mention that newspapers are formally loyal, but their sympathies seem to be on the side of our adversaries.[99]

But things continued to deteriorate in Berlin, and on December 9, he wrote Elisabeth:

> I was in the city today, it looks disastrous. But in between the rubble there is an occasional store with a few things available for sale....Hoffmann's and Döpel's homes burned out, the entire Bonhoeffer institute as well.[100]

In December 1942, Abraham Esau was appointed plenipotentiary for nuclear physics research in Germany, yet another change in the Nazi organization. Then, just a year later, on January 1, 1944, Gerlach replaced Esau as plenipotentiary of fission research sponsored by RFR. Uncertain organization and inadequate support were constant problems for the German nuclear program. For the members of the *Uranverein*, the fear of promising more than they could deliver and the refusal to ask for more support reflected the weak leadership of the German program.

In 1943, Diebner's results in Gottow were better than Heisenberg's in Berlin or Leipzig. Recognizing that small blocks were better than plates,[101] Diebner used small blocks of uranium suspended in heavy water, and Fermi, in Chicago, used blocks stacked in tubes imbedded in graphite. Diebner's G-II experiment (actually carried out in the *Chemisch-Technische Reichsanstalt* in Berlin) in 1943 had a neutron yield significantly higher than Heisenberg's previous experiments in Leipzig.[102] Uranium plates (and sometimes spherical shells) were easier to analyze theoretically, so Heisenberg insisted on these until late in the war, when it was too late. But in the winter of 1943–44, he

---

[99] Heisenberg and Heisenberg (2016), pp. 181–182.

[100] Heisenberg and Heisenberg (2016), p. 184.

[101] Interestingly, at about the same time, William Shockley and James Fisk, at Bell Telephone Laboratories, came to the same conclusion and calculated the optimum size of the uranium blocks. But since they were at a private facility, their work was shut down.

[102] *Deutsches Museum - Online-Sammlung.* (n.d.). https://digital.deutsches-museum.de/de/digital-catalogue/hitlist/?searchTerm=diebner%2C+Kurt.

assembled a model reactor comprising 1.5 tons of heavy water and a like amount of uranium in the form of plates at the KWIP and achieved a three-fold neutron gain.[103] This was very encouraging.

Following the "success" of Heisenberg's 1941 visit to Neils Bohr's institute in Copenhagen, the Reich Education Ministry sent Heisenberg on further missions as an ambassador for German culture. In October 1943, Heisenberg visited his colleague H. A. Kramers, who had been his close collaborator during his amazing year in Copenhagen when he discovered quantum mechanics. Kramers was now at the Kammerlingh-Onnes Laboratory in Leiden, the Netherlands, where things were going very badly. There had been protests that summer at the Dutch universities, and Jews in the Netherlands were being deported to the Mauthausen-Gusen concentration camp. The Dutch laboratories were closed, and the occupation authorities were stripping the laboratories of equipment. While Heisenberg was there, the conversation inevitably turned to political matters, and in a conversation with his Dutch colleague Hendrik Casimir, Heisenberg admitted that he was aware of the concentration camps and looting, but he wanted Germany to rule. Democracy was too weak he said:

> There are, therefore, only two possibilities: Germany and Russia. And then a Europe under German leadership would perhaps be the lesser evil.[104]

Heisenberg was never able to understand why this alienated his Dutch colleagues. However, he was able to lift the ban on research and save some equipment from looting, which helped the Dutch.

Beginning in 1943 with the development of the long-range P-51 Mustang fighter to escort the bombers, U.S. aircraft bombed Germany during the day while British aircraft continued bombing at night. The devastation was most severe in the north, but it extended to Berlin, Dresden, Leipzig, and the southern borders of Germany. In Berlin, the bombing began to interrupt the research and affected Heisenberg's health. On July 21 he wrote Elisabeth:

> You can tell—all my thoughts are about the war again, I find it difficult to free my inner self from all the horrors that are happening. I am too depleted to concentrate on scientific work anymore.[105]

---

[103] Heisenberg (1947).
[104] Bernstein (2001), p. 43.
   Cassidy (2010), p. 346.
[105] Heisenberg and Heisenberg (2016), p. 192.

And on July 26 he wrote:

> This morning, then, the news came of Mussolini's stepping down which, of course, was very exciting news. … One other thing concerns me too: for about eight days now, my state of health has noticeably become worse, the neuritis and the accompanying rash on my hands have become much more intense.…It may have been a mistake that I have taken on the institute here in the first place. One needs to be much more solidly constructed for it than I am.[106]

Diebner continued his experiments in Gottow. By 1943, test G IIIb, with 564 kg of uranium cubes and almost 600 L of heavy water, showed a 106% increase in neutrons. Diebner's reactor experiments had verified Höcker's theory and proved that cubes work better than Heisenberg's plates, but a self-sustaining chain reaction was never observed. In any event, these results came too late.

In December 1943 Heisenberg went to Krakow, Poland. By this time, he must have been aware of the razing of the ghettos in Krakow, Warsaw, and Lotz. In Krakow Heisenberg gave a lecture on physics to which only Germans were admitted. Polish scientists were turned away at the door. The lecture was given at the Institute for German Work in the East. The university had been closed by this time, and most of the faculty sent to the concentration camps; the Polish elite were to be exterminated. Heisenberg stayed at the villa occupied at the time by Hans Frank, his friend from *Gymnasium* days. Frank at this time was the Nazi-installed governor of Poland who presided over the razing of the ghettos and obliteration of Polish culture. In December 1941 he had told his cabinet: "As far as Jews are concerned, I want to tell you frankly that they must be done away with one way or another."[107] During the war, 800,000 Poles were sent to Germany as slave laborers and thousands were transported to Treblinka. At the Nuremburg trials, after the war, Frank was convicted of war crimes and hanged. Although there is no evidence that Heisenberg was an antisemite or a Nazi, his insensitivity, his "pact with the devil," is breathtaking.

Satisfied with his 1941 visit, Heisenberg made several more trips to Copenhagen. In early 1944, by which time Bohr and nearly all the other Jews in Denmark had escaped to Sweden, Heisenberg again visited Bohr's institute. The German occupiers had taken over the institute, imprisoned the acting head, Jørgen Bøggild, and were deciding whether to strip the labs of

---

[106] Heisenberg and Heisenberg (2016), p. 193.
[107] Cornwell (2003), p. 325.

equipment that the Germans wanted, some for the nuclear weapons program. Heisenberg convinced the authorities that the equipment was too difficult to move, and got Bøggild released, which pleased the scientists who still remained at the institute. A few months later he returned to give some lectures at the German Cultural Institute, but the Danes once again boycotted the talk.

By late 1943, Allied bombing of Berlin had forced Heisenberg to move portions of his effort to Hechingen, a small town on the eastern edge of the Black Forest in southwestern Germany. The move to Hechingen was good for Heisenberg. Things were better there than in Berlin. He even had time for his music and to work on his book on philosophy, which was published after his death as *Reality and its Order*. On February 26, 1944, Heisenberg wrote Elisabeth:

I am very much enjoying the relief from air raids here; my work goes well; besides, I am practicing [piano] regularly for one to one and a half hours a day. The philosophy is now finished…Even though there are often enemy plane formations flying above Hechingen, they do us no harm…[108]

And on March 5 he wrote Elisabeth:

It really is true, one lives here once again, for a few weeks, as in former, better times; you are surrounded again by everything you knew so well: a good living space, nice people. You sit together for music, and talk about the good books somebody just wrote—and the war only seems like a bad dream.[109]

Meanwhile, the Theoretical Physics Institute and Döpel's laboratory in Leipzig were destroyed. Back in Berlin, where he spent much of his time, things continued to deteriorate. On May 8 he wrote Elisabeth:

Today I almost went to the opera…The state opera is still standing,…free tickets to Lohengrin, scheduled for tonight. [after yesterday's attack] I was told an unexploded bomb was still in front of the theater…it was definitely canceled. …This is what life has become now. 'Keep smiling,' an American would say.[110]

And on June13, 1944:

---

[108] Heisenberg and Heisenberg (2016), p. 204.
[109] Heisenberg and Heisenberg (2016), p. 205.
[110] Heisenberg and Heisenberg (2016), p. 211.

It subdued me a lot that, even during the [Allied] invasion [of Normandy on June 6], the air raids do not stop.…In the long run, I cannot really withstand this nervous strain, especially when the day is also fully busy with work.[111]

By the end of 1944, everything had been moved to the south. The reactor experiments were moved to a wine cave in the nearby town of Haigerloch (Fig. 4.7),[112] run by Gerlach, who, like Heisenberg, was a German patriot but not a Nazi. Heisenberg worked mostly in Hechingen and commuted the ten miles to Haigerloch by bike. The fighting was now getting closer, and on December 2 he wrote Elisabeth from Hechingen:

**Fig. 4.7** Haigerloch, showing the cave into which the reactor experiments were moved. (Courtesy of the AIP Emilio Segrè Visual Archives, Goudsmit Collection. Photograph by Samuel Goudsmit)

---

[111] Heisenberg and Heisenberg (2016), pp. 212–213.

[112] Bernstein (2001), p. xxix and p. 99.

> Overall, I am doing so well here that one simply is left with a bad con-science …This is, after all, just a seventy-kilometer distance to the front! The work at the institute is also progressing nicely, right now, as is my own work.[113]

Even as the Allies got closer and the war was clearly coming to an end, Heisenberg made a trip to Zurich on December 18 to lecture on his S-matrix theory.[114] An interesting side story of this visit involves Heisenberg's near assassination. The American "baseball catcher turned spy," Morris Berg, traveled to Zurich and posed as a physicist to hear Heisenberg's lecture while carrying a pistol in his pocket to assassinate Heisenberg if it seemed that he was close to completing a bomb. Berg also went to dinner with Heisenberg a few days later. When the physicist Gregor Wentzel pointed out that the war was lost, Heisenberg replied "Yes, but it would have been so beautiful if we had won."[115] Berg decided that Heisenberg wasn't close to achieving an atomic bomb, and walked him back to his hotel safely.

Heisenberg and Gerlach's last experiments were carried out at the end of March and the beginning of April, 1945.[116] By this time Heisenberg had accepted the superiority of Diebner's uranium cubes over his plates. Comprising 1.5 tons of heavy water and a similar mass of uranium, surrounded by a graphite reflector to contain the neutrons, the reactor showed a sevenfold neutron increase, the best results ever, but it never reached criticality. Additional uranium cubes probably would have sufficed, but this was no longer possible in April 1945. The remaining heavy water and additional quantities of uranium blocks had been allotted to the Reich Research Council in Stadtilm, directed by Diebner. Remarkably, the Germans still couldn't work together and the infighting continued up to the very end. On February 1, Heisenberg wrote Elisabeth:

> In our nuclear physics group, the internal battle (Diebner vs. KWI) has broken out anew, probably as a result of the new wave of conscriptions and the threatening danger in the east.[117]

As the Allied forces approached from the west, food became scarcer in Hechingen, and even further east in Urfeld. On March 16, Heisenberg wrote:

> I am far from getting enough to eat.…I am becoming thinner and thinner, and less physically able. On your end it will, in the long run, be much worse, though.[118]

---

[113] Heisenberg and Heisenberg (2016), p. 210.
[114] Cornwell (2003), p. 328.
[115] Cornwell (2003), p. 328.
[116] Heisenberg, (1947).
[117] Heisenberg and Heisenberg (2016), p. 239.
[118] Heisenberg and Heisenberg (2016), p. 248.

And on April 6, Elisabeth replied:

> In the morning the children eat flour soup: simply stir salt and dry flour into boiling water. There will be clumps, but those will be especially desirable.[119]

On April 15, 1945, all the work shut down, and Heisenberg fled to his family in Urfeld. The journey from Hechingen to Urfeld is an adventure best described by Heisenberg in his diary:

> April 15 Frenzied departure activities in Hechingen and Haigerloch.
> April 19. Leaves 3:30 a.m. by bike (RR tracks destroyed). 4:15 gets on train at Jungingen. Gets off at Gammertingen (fear of planes in daylight). Bicycle to Kleintissen around 10 a.m. 7 p.m. depart by bicycle to Aulendorf, arrive 8:30. Train trouble, gets 8 km, continues by bicycle around 6 a.m.
> April 20. 8:15, arrives Grosse Mühle. Breakfasts and leaves 9:30. Watches bombers destroy Memmingen. Dinner in Krugzell, leaves 5 p.m. 8 p.m. arrives Kaufbeuren. 10 p.m. train for Shongau, arrives 10 p.m. 5 a.m. train departs for Weilheim.
> April 21. Arrives Weilheim 6:30 a.m. 9 a.m. train arrives Ohlstadt. Bicycle to Kochel. 11:30 clears pass on bicycle, soon arrives home.[120]

During the war years Heisenberg was considered by the Allies as the top theoretical physicist in Germany and the likely leader of any attempt to construct a German bomb. To determine if there was an active atomic bomb project in Germany and how far the scientists might have gotten with the project, the Allies established a "special ops" team, the Alsos mission, headed by Colonel Boris Pash. They started their reconnaissance missions as early as 1943, following Allied troops into Italy and France, interrogating scientists, and confiscating documents and reports along the way. On April 22, 1945, they went behind enemy lines into Germany, scouring the countryside from Heidelberg to Haigerloch. On April 24, the group found a textile mill that had been converted into a laboratory that the Germans used for their nuclear research, complete with a test reactor. They found Heisenberg's office, but not Heisenberg. The team did find 25 scientists, and from their interrogation discovered that many of the German research files were hidden in a watertight drum that they had sunk into a cesspool. After destroying the reactor and the laboratory, Pash began the hunt for the remaining German scientists. By questioning the Haigerloch scientists, Pash discovered that Heisenberg had

---

[119] Heisenberg and Heisenberg (2016), pp. 248–249.
[120] Heisenberg and Heisenberg (2016), pp. 253–255.

fled to the mountains of Bavaria, to the town of Urfeld, to join his family. When they got there, they found Heisenberg, suitcase packed, ready to go with the Allies. As related by Heisenberg's daughter: "On May 2, an American advance troop under Colonel Pash arrives to arrest Werner. The next day…Werner is transported to Heidelberg, then to several intermediary stops"[121] It's noteworthy that the Allies took special care of Heisenberg's family, providing food and a protection order for the house. Heisenberg writes in his diary: "the talk with Pash continues: he is under orders to arrest me, but will continue to take care of the family and me in every possible way."[122] Several other members of the Uranium Club were rounded up in the area, including Hahn, Wirtz, Bagge, and Korschung, along with von Laue, who was not actually a member of the Uranium Club. In May, Gerlach and Diebner were located in Munich. The ten scientists were taken first to Heidelberg, then eventually through Belgium, and finally to England where they were interned in extreme secrecy at a country manor called Farm Hall, near Cambridge, with no communication with the outside world (Fig. 4.8).

**Fig. 4.8** Farm Hall in Cambridge, England, where Heisenberg and others were detained. (National Archives and Records Administration, courtesy of AIP Emilio Segrè Visual Archives)

---

[121] Heisenberg and Heisenberg (2016), pp. 232–233.
[122] Heisenberg and Heisenberg (2016), p. 262.

Although the Germans never achieved a self-sustaining chain reaction, when he was captured by the Allies Heisenberg told his captor Goudsmit, who had actually been a colleague at an earlier time, that "If American colleagues wish to learn about the uranium problem I shall be glad to show them the results of our researches if they come to my laboratory."[123] He was unaware that the Americans had achieved a self-sustaining chain reaction more than two years earlier and would successfully test a bomb in two months. Goudsmit called this "pathetic." Part of the reason that the German nuclear program was so unsuccessful was complacency. They seem to have been aware of an Allied effort as early as December, 1941, and reported this to Schumann. But in their arrogance, the Germans believed that if they couldn't do it, then nobody else could. In 1943, in a letter to a Dr. Görnnert in Hermann Göring's office, Rudolf Mentzel (head of the Reich Research Council) wrote that

> though the work will not lead in a short time towards the production of practical useful engines or explosives, it gives on the other hand the certainty that in this field the enemy powers cannot have any surprises in store for us.[124]

Three months after the capitulation of Germany, on August 7, 1945, the USA dropped a uranium bomb on Hiroshima, and they dropped a plutonium bomb on Nagasaki a few days later. By then, Heisenberg and most of Hitler's "Uranium Club" were interned in Farm Hall, England, where their conversations were recorded. When they got the news of the atomic bombs from the BBC radio broadcast, they were shocked, unbelieving. The conversation went like this:

> Heisenberg: "Did they use the word uranium in connection with this atomic bomb?"
> All: "No."
> Heisenberg: "Then it's got nothing to do with atoms, but the equivalent of 20,000 tons of high explosive is terrific."
> Weizsäcker: "It corresponds exactly to the factor $10^4$."
> Gerlach: "Would it be possible that they have got an engine [nuclear reactor] running fairly well, that they have had it long enough to separate "93" [meaning plutonium]?"
> Hahn: "I don't believe it."

---

[123] Cornwell (2003), p. 337.
[124] Bernstein (2001), p. 45.

Heisenberg: "All I can suggest is that some dilletante in America who knows very little about it has bluffed them into saying: 'If you drop this it has the equivalent of 20,0000 tons of high explosive' and in reality doesn't work at all."[125]

## The Later Years

Following his capture, Heisenberg was asked at several points whether he wanted to go to America or stay in Germany. Goudsmit asked him during his interrogation in Heidelberg on the way to Farm Hall, and Patrick Blackett reiterated the question when he reached England. As he explains in a letter to Elisabeth in late January, 1946, after his release:

It is completely clear to me that in the next decades America will be the center of scientific life, and that working conditions for me in Germany will be much worse than over there. Exactly because of this, on the other hand, I am not needed there as much: many excellent, competent physicists are there. Here, however, it matters a great deal that an intellectual life should again become viable. Since 1933 it has been clear to me that here a terrible tragedy for Germany was in progress, only I could not have imagined the extent and the ending; and I stayed here at the time so that I might also be here afterward and help. This was exactly what I also told my American friends in the summer of 1939, and the best among them could understand it; this intention remains firm and will not be betrayed. Goudsmit understood it and thought it also quite correct.[126]

The ten German scientists interned at Farm Hall were repatriated on January 3, 1946. They had lived in relatively lavish conditions during their internment, and it was not until communication was restored that Heisenberg learned that his family had suffered significantly. Colonel Pash had been able to provide some groceries for the family, but this ended in the summer of 1945. After the "fat days of internment,"[127] some ten long months would pass before the Heisenberg family was reunited in Göttingen. As Heisenberg wrote to Elisabeth in 1946 in his first letter following his internment:

I may finally be permitted to tell you where I have been and where I am now. We got around quite a bit in Europe. First from Heidelberg to Versailles, then

---

[125] Bernstein (2001), p. 116.
[126] Heisenberg (2016), p. 286.
[127] Heisenberg and Heisenberg (2016), p. 129.

we stayed for a few weeks in a small suburb of Paris, even saw the city occasionally; in early June we arrived at a manor in Belgium near Liege, and finally, in early July—by plane—to England, where we stayed in a little village near Cambridge at a small country estate. That is where we remained these last six months, completely closed off from the world outside. We were treated decently…cared for excellently in material terms.[128]

The USA was not enthusiastic about allowing the German scientists to reestablish their research. On the other hand, the British were anxious to help rebuild German science and felt strongly that rebuilding the previous Kaiser Wilhelm Society (KWS) and its institutes with oversight by the Allies would eventually strengthen Germany's reinsertion into Europe. Therefore, following six months of internment in Britain, the group was released and taken to Alswede, a small village in the British Zone. Once free, the German group (principally Hahn and Heisenberg) emphasized that they needed a university town to restart their research. Göttingen (in the British zone) was chosen since it had suffered relatively little damage from the bombings. Coincidentally, Göttingen was where Heisenberg had studied under Born after Munich, and earned his *Habilitation*. Heisenberg and Hahn moved to Göttingen in 1947, occupying the Aerodynamic Testing Institute buildings, which had been spared bombing. The first year in Göttingen wasn't easy for Heisenberg, and at times things looked hopeless. Initially, the French refused to return any of the instrumentation that they had confiscated in Hechingen. In addition, Heisenberg had only three coworkers, with no laboratory materials and no association with the University of Göttingen. In a letter to Elisabeth in 1947 he wrote: "I'm totally exhausted; it is not just the hunger. I am no longer equal to this continuous organizing with all its disappointments."[129] In spite of everything, he continued to reject offers for work in America, where discouraged colleagues and acquaintances hoped to immigrate. It took almost a year for Heisenberg's research to begin to thrive. By 1947, he had located a suitable residence for his family just outside Göttingen, and they were able to rejoin him after many months of separation.

It was not initially clear if Hahn and Heisenberg would be allowed to rebuild their institutes or would simply continue what research they could on their own. In 1946, in agreement with the USA, the British resurrected the Kaiser Wilhelm Society first with Planck and then with Hahn as the President. Within two years, the society and its member institutes were renamed the

---

[128] Heisenberg and Heisenberg (2016), p. 274.
[129] Heisenberg and Heisenberg (2016), p. 130.

Max Planck Society. However, the research that the scientists could conduct at this point was controlled by the Allied Control Law Number 25 that banned any activity that touched on military areas of interest, including nuclear reactor research, isotope studies, cyclotron construction, and experimental high-energy physics. These projects were exactly what Heisenberg had hoped to restart in the postwar years. He was convinced that nuclear reactor technology would be the revival of German physical science. He, together with Hahn, Gerlach, Weizsäcker, and others argued vigorously with the Allies to allow reactor research. When the Control Law was finally relaxed in 1955, Heisenberg and others jumped immediately back into nuclear research. Within a decade Germany had become a leader in nuclear technology.

To accomplish Heisenberg's goal of becoming a leader who would rebuild German science, it was absolutely critical to regain the respect and trust of his European and American colleagues. Unfortunately, Heisenberg had lost much of this respect in part because of his science, which had not made any breakthroughs since his discovery of matrix mechanics, but also because of his pro-Nazi activities during the war. Scientifically, no one believed his theories of cosmic showers or his other attempts to extend quantum mechanics to higher energy, and on trips to the USA before the war and to Denmark, Netherlands, and Switzerland during the war, he had made remarks so naïve and offensive that he alienated all his former colleagues there. It's not clear that he ever understood the depth of the loss of respect and trust he suffered in both his personal and scientific life.

To begin his "rebirth" as a leader in the physics community and to regain the lost friendships, he began to present publicly his side of the story about German wartime research and the role that he and others in the Uranium Club had played in this research. Heisenberg desperately wanted the German scientists to be viewed not only as "morally untainted," but also as highly competent. One of the first papers that Heisenberg published after the war was an explanation of the role of the German nuclear scientists in nuclear research, which he published in English in the prestigious journal *Nature* so that scientists outside Germany could read it. According to this narrative, 1942 was the turning point. Although he and his group had initially made the case that building an atomic bomb was theoretically possible with sufficient materials and funding, by 1942 the war situation had worsened significantly for the Germans, and the Army Ordnance Office decided that it was likely that a usable atomic weapon would not be available before the war was over. After that, the Germans had made no attempt to produce an atomic bomb "because the project could not have succeeded under German war conditions."[130] He continued: "...any major project which did not promise quick returns was specifically forbidden." And finally, "In the upshot [the German physicists] were spared the decision as to whether or not they should aim at producing atomic bombs. The circumstances shaping policy in the critical year of 1942 guided their work...towards the problem of the utilization of nuclear energy in prime movers."[131]

Such self-exonerating statements were intended to avoid the eventual dispute that arose between American and German scientists about the actual role that the Germans had played in war research, and more specifically their reasons for not succeeding in building a bomb, or even a nuclear reactor.[132] Goudsmit was the loudest in his opposition to the story Heisenberg was trying to sell. He felt that the German scientists had grown complacent, convinced in their arrogance that if they couldn't produce a bomb, neither could the Allies. Goudsmit further made the case that the Germans had failed because the Nazis had removed many of Germany's top Jewish scientists, some of whom ended up working on bomb projects with the British or Americans. As Goudsmit wrote: "in actuality German scientists had only the vaguest notions of how a uranium bomb or even a reactor actually works."[133] These arguments continued through 1947, with Heisenberg continuing to defend Germany's scientific success in spite of the Nazi regime, while emphasizing their moral dilemma. On the other side, Goudsmit continued his belief that the German scientists had failed while trying to compromise within the Hitler regime. These arguments, which reflected the strained relationships between these two former colleagues, were representative of the view of many non-German scientists concerning Heisenberg.

Unfortunately, Heisenberg became isolated from former colleagues and increasingly marginalized in the global science community. He never became the central figure in rebuilding German physics as he had hoped. Once settled in Göttingen, he focused his efforts on rebuilding his Institute to continue his scientific research, and to focus on science policy related to nuclear research. As he became more involved in politics in postwar Germany, his influence in the physics world waned. Restarting his research program was a challenge. Aside from some secret scientific reports for the Uranium Club, he had published no original articles between 1938 and 1944, and many of the 30-plus articles he published up to 1970 were single author reviews or theory papers.

---

[130] Heisenberg (1947).

[131] Heisenberg (1947).

[132] Lucas, A.A. (2007). Revisiting Farm Hall. *Europhysics*, 38(4), 25–29.

[133] Cassidy (2010), p. 382.

He had three areas that he tried to expand during this time: superconductivity, turbulence, and mesons. He published only a few original articles in any of these areas, with titles such as *"The electrodynamic behavior of superconductors;" "On the statistical theory of turbulence;" "On the formation of mesons in multiple processes."* None of these papers had any major impact in the physics community.

In 1949, Heisenberg was appointed President of the German Research Council (GRC). He hoped to bring together under the umbrella of the Max Planck Society (MPS) the various West German academies of science. Heisenberg's view was that this GRC would be the "sole representative of all German science," and would promote German science in the international arena as well as influence the funding of research through the federal government. The GRC encountered opposition from the established Emergency Association for German Science, but through Heisenberg's influence the two merged in 1951, providing a more stable federal support agency for the MPS. In the 1950s, through Heisenberg's influence, German scientists were able to mobilize public opinion against acceptance by the German government of a NATO plan to allow the West German army to be provided with tactical nuclear weapons. To this end, Heisenberg and several of his colleagues (the "Göttingen 18") issued a manifesto in 1957 protesting the plan. The scientists were successful: Germany would remain nonnuclear, with a promise that the Americans would provide nuclear support in the event of a Soviet invasion. In the last chapter of his life, Heisenberg accepted the position of Director of the Max Planck Institute in Munich in 1958 and was appointed *ordentlicher Professor* (ordinary professor) at the University of Munich. This came 22 years after his initial failed attempt to succeed his mentor Arnold Sommerfeld at Munich. Now 56 years old, his focus turned to more travel, withdrawal from science policy affairs, and increased involvement in reshaping the Institute. His intellectual interests turned more toward philosophy, as his search for a "unified field theory," a theory to unite all the forces of nature, except, perhaps, gravity, didn't bear fruit.

By this time Heisenberg was not a well man (Fig. 4.9). He had begun to suffer from a liver condition that caused weakness, dizziness, and depression. He enjoyed spending time at his cabin in Urfeld, among the mountains he loved. Finally, surgery revealed advanced cancer of the kidneys and gall bladder. He died at his home in Munich on February 1, 1976.

Goudsmit wrote an obituary of the man that he had "so admired, and yet so reviled," a statement that expressed both the frustration and the pity that many other colleagues must have felt about this great scientist. In his closing paragraph, Goudsmit wrote:

**Fig. 4.9** Heisenberg while visiting Harvard in 1973. (Photograph by John H. Martin)

Heisenberg was a very great physicist, a deep thinker, a fine human being and also a courageous person. He defended his science under dangerous circumstances. He chose to stay in his beloved country even though he was surrounded by hostility. Many of us hoped he would have been more outspoken in condemning the Nazi regime. He was one of the greatest physicists of our time, but he suffered severely under unwarranted attacks of fanatical colleagues. In my opinion he must be considered to have been in some respects a victim of the Nazi regime.[134]

## Who Was Werner Heisenberg, and Why Did He Fail?

In the end, we come back to the questions we introduced at the beginning of this chapter: Who was Werner Heisenberg, really, and why did he fail to build an atomic bomb?

Three reasons for Germany's failure to develop the bomb have been advanced. First, Heisenberg did not have the means or resources to build the bomb in wartime Germany. This has been refuted by Manfred Popp.[135]

---

[134] Niels Bohr Library and Archives, Box 11, Folder 99 (n.d.) Werner Heisenberg obituaries, 1976–1977. https://repository.aip.org/islandora/object/nbla%3A238173#page/1/mode/2up.

[135] Popp, M. (2016). Misinterpreted documents and ignored physical facts: The history of 'Hitler's Atomic Bomb' needs to be corrected. *Berichte zur Wissenschafts-Geschichte* 39(3), 265–282 (2016).

Second, Heisenberg was incompetent, either as a scientist or a leader; this point of view has been advanced by Goudsmit[136] and Bernstein[137] but refuted by Robert Jungk.[138] And third, Heisenberg didn't want to build the bomb and delayed or sabotaged the effort; this has been argued by Jungk and Thomas Powers[139] but is refuted by evidence that appeared later. We examine these possibilities in order.

First, could Heisenberg, or more broadly the *Uranverein*, have built the bomb in wartime Germany under the Nazis? Arguably, yes. Funding was secure until the end of 1941 when the *Wehrmacht* was stopped outside Moscow, and Allied bombing became serious only in 1943, when daylight bombing became possible. Starting from the September 26, 1939, kick-off meeting, this gave the *Uranverein* nearly four years to develop a working nuclear reactor. By the end of this time, they had sufficient heavy water for one successful reactor, but not for three. In fact, there was enough uranium available for Harteck's dry ice reactor as well, since it didn't require heavy water, although processing sufficient uranium was sometimes a problem. But Heisenberg and Diebner didn't respect each other and couldn't work together, so they ended up with three unsuccessful efforts and left Harteck out completely. Had the Germans collaborated in their thinking and concentrated their resources on any one of these experiments they might have achieved a self-sustaining reactor by 1942. With a working reactor they could probably have gotten vast funding from Speer and Milch at the meeting in June, 1942. It was unthinkable to consider enriching enough uranium for a bomb in wartime Germany: most of the 120,000 workers and over half the $2B cost of the Allied program went to uranium enrichment, and the enrichment plant in Oak Ridge consumed one seventh of all the electricity produced in the United States, equivalent to nearly half as much electricity as was being produced in all of Germany at that time. Such an undertaking would not have been possible in wartime Germany. But excluding the cost of uranium enrichment, the cost of the rest of the Allied atomic weapon program was comparable to the cost of the V-2 rocket program in Germany. By 1940, the German team knew that plutonium was a fissile material that would be produced in a working reactor, and a reactor could have been concealed and placed in a bomb-proof bunker as, in fact, it was at the end of the war. With this they might have produced enough plutonium for a bomb. Of course, the Germans were

---

[136] Goudsmit, S. (1947), *Alsos* (Washington, DC: American Institute of Physics).

[137] Bernstein (2001).

[138] Jungk, R. (1958). Brighter than a thousand suns. Houghton Mifflin.

[139] Powers (1993), p. 305.

unaware of the difficulties facing the assembly of a critical mass of plutonium. Because it is so radioactive, plutonium must be assembled rapidly by explosively compressing a solid mass of the material. Otherwise, it will pre-detonate and fizzle. This was the most challenging problem that the Allies faced, but the German team would have become aware of this problem once they had a small amount of plutonium, and the Germans had substantial expertise in the necessary explosive techniques, which they used in other weapons. At least in principle, had they been better organized, the German team had the resources to build the bomb.

Second, was Heisenberg competent for this undertaking? Clearly, Heisenberg understood the theoretical considerations for a nuclear reactor, although he insisted on his inefficient design using plates of uranium and resisted Diebner's better design using small uranium cubes. And, despite the fact that the theory of the critical mass for a bomb was known and had been openly published in England, it is not clear to this day whether Heisenberg ever calculated the critical mass of uranium or plutonium needed to make a bomb until he did it after the war at Farm Hall. There is evidence that someone might have done this calculation since there are references to roughly the correct value of the critical mass in one of the secret reports of the *Uranverein*. And in the June, 1942, meeting with government officials Heisenberg himself said a critical mass would be about the size of a pineapple, which is about right. In fact, the entire theory of an atomic bomb, including the self-disassembly of the exploding bomb, was described in the Frisch-Peierls memorandum in March, 1940, which the Germans should have been equally able to produce. But building a reactor or a bomb is an experimental program, and Heisenberg was incompetent as an experimenter. After the war, Harteck complained:

> But how can you be a leader in such technological matters when you have never run an experiment in your whole life? That's ridiculous. That's no excuse whatsoever.
>
> While Heisenberg is one of the best theoreticians of our age and Weizsäcker, in addition to being a very good physicist and philosopher, could also expound his views very well, nevertheless both had never been involved in a large experimental venture before. How could they think they could lead the development of a new technology? That was poor judgement; it is almost unbelievable.[140]

---

[140] Bernstein (2001), p 40.

Though he was not as good a physicist as Heisenberg, Diebner's reactors worked better than Heisenberg's. How much farther might they have gotten if Heisenberg and Diebner had respected one another and collaborated, instead of competing?

Heisenberg was never trained as a manager of a large and complex program. Neither was Robert Oppenheimer, the scientific leader of the American effort, but Oppenheimer had Army Gen. Leslie Groves to organize and drive the American program. The demands he made and the risks he took make an amazing story. Groves ran the American nuclear program like a military project. For example, the graphite reactors at Hanford used for breeding plutonium were built quickly on a daring gamble. Based on Fermi's first reactor with little data to go on, they had the usual "unexpected" problems, such as "poisons" coming from the nuclear reactions themselves that stopped the chain reaction periodically, but they solved the problems one by one. And in another urgent move, when there wasn't enough copper for the magnet windings in the electromagnetic enrichment process, Groves made the magnets from silver that he got from the Treasury.

The high-risk, go-for-broke attitude of the Americans was based on their respect for the German scientists and their fear that Hitler might get the bomb first. Word of a German effort came from at least two sources. The first was a message from Fritz Houtermanns, who was working in a private laboratory near Berlin belonging to Manfred von Ardenne. Shocked in 1941 when he discovered that plutonium might be a quick route to a bomb, he sent a message through a refugee who was leaving for America that the Allied effort (if any existed) should be speeded up. The other source was Niels Bohr, whom Heisenberg visited in September, 1941, and to whom Heisenberg admitted that he had been working on nuclear energy. Bohr, in turn was in contact with Allied scientists, and eventually, after he fled Denmark a year later, he assisted in the Allied nuclear program. In contrast to this, Heisenberg and the others, in their arrogance, were never concerned that the Americans could build a bomb if they couldn't.

Third, did Heisenberg intentionally block the bomb? There is no evidence that Heisenberg would not have built a bomb for Germany, even under Hitler, if he could have. In one draft of the letter that he wrote but never actually sent to Heisenberg concerning their conversation in Copenhagen in 1941, Bohr writes

I am greatly amazed to see how much your memory has deceived you. [You gave] me the firm impression that, under your leadership, everything was being done in Germany to develop atomic weapons and that you…had spent the last two years working more or less exclusively on such preparations.[141]

While Heisenberg and the others were detained at Farm Hall after the war, they found out that the Americans dropped an atomic bomb on Japan. After they had gotten over their shock that the Americans had accomplished what they could not, von Weizsäcker introduced the excuse:

> I believe the reason we didn't do it was because all the physicists didn't want to do it, on principle. If we had all wanted Germany to win the war we would have succeeded.[142]

Hahn disagreed strongly, replying "I don't believe that but I am thankful we didn't succeed."[143] The evidence supports Hahn. Later, von Weizsäcker began to create the "*Lesart*" (story) often repeated in later years:

> History will record that the Americans and the English made a bomb, and that at the same time the Germans, under the Hitler regime, produced a workable engine. In other words, the peaceful development of the uranium engine was made under the Hitler regime, whereas the Americans and the English developed this ghastly weapon of war.[144]

Of course, the Germans never did develop a "workable engine," but there is no evidence that this was deliberate. Throughout the war, Heisenberg remained a dedicated German nationalist. He seems also to have feared the Russians more than he feared Hitler. As he told Bohr when he visited him in occupied Copenhagen, Hitler was needed to defend the West against Russia. Although he doesn't seem to have been a Nazi, Heisenberg defended the Nazi regime in discussions he had with colleagues in the USA in 1939, in Denmark in 1941, in the Netherlands, and Poland in 1943, and even in Switzerland in the last months of the war, and he expressed his belief that in time the Nazis would become better.

Evidently, the German scientists in the *Uranverein* lacked the courage to undertake a high-risk, go-for-broke program. Thus, early in the war Heisenberg held out the possibility of a bomb, but never asked for enough money to build

---

[141] Winters, J. (2003). Evidence emerges: Heisenberg attempted an A-bomb for Hitler. Discover, 24(1), p. 70.

[142] Bernstein (2001), p. 122.

[143] Bernstein (2001), p. 122.

[144] Bernstein (2001), p. 138.

it, just enough to build a reactor and keep himself and his scientists from being conscripted into the *Wehrmacht* and sent to the front. As Heisenberg says in a conversation recorded at Farm Hall, after the war "We wouldn't have had the moral courage to recommend to the government in the spring of 1942 that they should employ 120,000 men just for building the thing up."[145] The figure of 120,000 men was about the size of the Manhattan Project that built the Allied bombs, as Heisenberg had just learned from the BBC announcement of the dropping of the first atomic bomb on Hiroshima. But most of this investment was for isotope separation, which would have been unnecessary for a plutonium bomb. In fact, they had been warned at various times by Esau and Speer that failure of such a large project might place their lives in danger. The fear of being sent to fight at the front, balanced against the fear of promising too much and failing, might have affected them more than the moral considerations.

Summing up: had it been easy, Werner Heisenberg would have built an atomic bomb for Germany, even for Hitler. While he was never a Nazi and had been persecuted as a "white Jew," he remained a patriotic German to the end. But it wasn't easy to build an atomic bomb, and Heisenberg, the brilliant theorist, was incompetent both technically and as a program leader to do it.

## Bibliography


Bernstein, J. (2013). *Hitler's Uranium Club: The Secret Recordings at Farm Hall.* Springer.

Cassidy, D. C. (2010). *Beyond Uncertainty: Heisenberg, Quantum Physics, and the Bomb.* Bellevue Literary Press.

Cornwell, J. (2003). *Hitler's Scientists: Science, War, and the Devil's Pact.* Penguin Books.

Heisenberg, W., & Heisenberg, E. (2016). *My Dear Li: Correspondence 1937–1946.* Yale University Press.

Walker, M. (1995). *Nazi Science.* Perseus Publishing.


---

[145] Bernstein (2001), p.122.

# 5

## The Magic Crystal

### Shockley, Bardeen, and Brattain—1956 Nobel Prize in Physics *for their researches on semiconductors and their discovery of the transistor effect*

*I don't know the future. I didn't come here to tell you how this is going to end.*
*I came here to tell you how it's going to begin.*
*Spoken by Neo, in the movie The Matrix.*[1]

In 1947, a small team of scientists at Bell Telephone Laboratories made the critical modification that changed the crystal in a "crystal radio" from a device that detects radio signals into a device that amplifies the signals. They called it a transistor. It is now the fundamental component at the heart of all the digital devices that we use today. The newest smart phones have more than ten billion transistors in them, and larger computers have quadrillions. By comparison, the human brain, viewed as a computer, contains about 86 billion neurons. Programmed with artificial intelligence (AI), digital computers can now perform many of the functions of a human brain, and do it faster. The concept of artificial intelligence goes back at least to Alan Turing, the brilliant mathematician who also conceived the stored-program computer, and built the first computers to break the Nazi codes in World War II. In 1950, he proposed what is now called the "Turing test" to determine if a computer is intelligent. Roughly speaking, if you're talking to someone on the phone and you can't tell if it is a person or a computer, then it is intelligent. By this standard, it seems like computers have now achieved intelligence.

Of course, brains have an entirely different architecture than digital computers. They process information through the connections they make between neurons, called synapses, which form neural networks. Although digital

---

[1] The Matrix, a 1999 movie by Lana Wachowski and Lilly Wachowski.

© The Author(s), under exclusive license to Springer Nature Switzerland AG 2025
V. Shepherd, C. Brau, *Beyond the Genius*, Copernicus Books,
https://doi.org/10.1007/978-3-032-02735-1_5

computers will probably always win at "number crunching," like computing $\pi$ to 100 trillion decimal places, neural networks are more efficient than digital computers and better suited to cognitive processes like pattern recognition. For example, honeybees have only about a million neurons and a billion synapses, but they can process visual information, control flight, identify food, navigate back to the hive, communicate with other bees, recognize threats, and organize the colony, all with tiny, flyable, user-friendly "computers." Moreover, synapses are not "on–off" devices like the transistor switches used in computers. As our brains store information, some synapses get faster, others slower, to remember the information, but nothing turns on or off. Digital computers, functioning with "on–off" switches, have to be programmed to mimic this behavior. Inspired by this, scientists have now developed "synaptic transistors"[2] made from exotic new materials like graphene, the single layer of carbon atoms that won Andre Geim and Konstantin Novoselov the Nobel Prize in Physics in 2010. As explained by Amanda Morris, a synaptic transistor[3] can "recognize patterns and demonstrate associative learning, a form of higher-level cognition, even with imperfect input," functioning in a fashion more nearly like a brain.

Going beyond artificial intelligence, will a computer ever become conscious, like a human? What will be the "Turing test for consciousness?" Will it ever feel sadness, joy, curiosity or remorse, or be aware of its own existence? Do we need to be concerned about the future?

And now, this breaking news: "Elon Musk's Neuralink Has Implanted Its First Chip in a Human Brain. What's Next?"[4] The Borg would tell us: "Resistance is Futile."[5]

This chapter discusses what may be the most important invention in history, one that is completely changing the arc of human existence. The invention of the transistor is the work of many people, but especially three men working together, each with his own strengths and idiosyncrasies: Walter Brattain was a gifted experimenter who could turn ideas into hardware; John Bardeen was a brilliant theorist who would leave Bell Telephone Laboratories for the University of Illinois, discover the theory of superconductivity, and win a second Nobel Prize in Physics; and William Shockley, the group leader, was a visionary who would leave Bell Telephone Laboratories to create "Silicon Valley"…and promote racist eugenics. It's the story of a remarkable idea born

---

[2] Dume, I. (2024). Moiré material makes a synaptic transistor for neuromorphic computing. *Physics World*. Yan, X. Moiré synaptic transistor with room-temperature neuromorphic functionality. *Nature* 624, 21/28.

[3] Morris, A. (2023) Synaptic transistor mirrors human brain function. *Neuroscience*, December 29. https://neurosciencenews.com/synaptic-transistor-ai-25402/.

[4] Guarino, B. (2024). Elon Musk's neuralink has implanted its first chip in a human brain. What's next? *Scientific American*, January 30. https://www.scientificamerican.com/article/elon-musks-neuralink-has-implanted-its-first-chip-in-a-human-brain-whats-next/

[5] Communicated by The Borg. https://en.wikipedia.org/wiki/Borg.

in the research laboratory of a comfortable monopoly with no serious competition. How did all this happen?

## The Beginning

The transistor was born in the laboratories of the Bell Telephone Company, a company that was formed to help people speak to one another from a long distance. Faced in the 1930s with growth beyond its technological limitations, it had the resources to address the problem, and the visionary leadership to concieve a transformational technology. But births are never easy, and the transistor was no exception.

On March 10, 1876, Alexander Graham Bell spoke the first words on a telephone to his assistant in the next room: "Mr. Watson, come here; I want to see you."[6] Bell had gotten a patent on his invention of an "apparatus for transmitting vocal or other sounds telegraphically" just three days before. On the basis of this patent, the Bell Telephone Company was organized in Boston, Massachusetts, on July 9, 1877. The first telephones were rented in pairs: you could call home from the office or you could call the office from home. That was it. But customers wanted to call lots of people, and within a year the first switchboard began operating in Boston. Operators would manually connect a wire from the caller to the destination on a big board covered with an array of plugs, thus the name "switchboard." Eventually, there were hundreds of thousands of operators working for the Bell Telephone Company at huge boards, like the one shown in Fig. 5.1. Sometimes the operators working at the largest boards wore roller skates. In small towns, you might know the operator by name and chat for a bit, but wherever you lived or worked, you would ask her to connect you to your friend Fanny Browne, explaining, for example, "that's Browne with an e."

Then, in 1888, a disgruntled undertaker in Kansas City, Missouri, had a brilliant idea. Adam Strowager had heard rumors that an operator was diverting his calls to other businesses. He wanted to put an end to that, so he invented the first automatic telephone switching system. His automatic system was based on a rotary switch, like the "low, medium, high" switch on your slow cooker, but his mechanical switch was rotated electrically in steps following the electrical pulses from a rotary telephone. Arguably, this marked the beginning of the digital age: "Fanny Browne, that's Browne with an e," now became just "2-4601." And, you couldn't chat with the operator. Interestingly, the need for this "personal touch" and the power of the labor

---

[6] Bruce, R.V. (1973) *Alexander Graham Bell and the conquest of solitude*, Boston, Little, Brown, p. 181.

**Fig. 5.1**   Telephone switchboard ca. 1914

movement delayed the introduction of automatic switching systems at Bell Telephone for a couple of decades. It wasn't until 1918 that automatic switchboards had replaced operators in many cities. But humans still had to complete calls between cities, and the average time to connect a long-distance call was about 15 min. By the 1930s, fully automated systems, called crossbars, occupied huge rooms with large cabinets, each filled with thousands of Strowager's electromechanical switches.

Anticipating competition from regional telephone systems, Bell Telephone Company gave strong support to research. They wanted to offer the best technology, and above all to defend against a competitor getting a patent that they needed. They also focused on wide-area and long-distance service. Since long-distance calls had to be "boosted" every couple of hundred miles, they needed amplifiers, and the only technology for this was the "Audion" tube. They were forced to buy the patent for the Audion tube from Lee de Forest. They never wanted that to happen again, so in 1925 they formed Bell Telephone Laboratory, colloquially called "Bell Labs," or just "the Labs," from the engineering department of Western Electric, the manufacturing division of the Bell Telephone Company. Bell Labs' mission was to address fundamental scientific problems that might prove useful for the Bell Telephone system. "There was more academic freedom at Bell Labs' research department than at most universities. You could do what you wanted to. There was very little pressure to do anything practical," recalled Charles Kittel, a physicist who worked at the Labs from 1947 to 1951.[7] The discoveries made at Bell Labs have been nothing short of astonishing, and make an epic story by themselves. The Labs got their first Nobel Prize in 1937 for Clinton Davisson's and Lester Germer's 1927 discovery of the wave nature of electrons scattered from a nickel surface; Philip Anderson was awarded the 1977 Nobel prize for his theoretical work on metals and insulators; and Arno Penzias and Robert Wilson received the 1978 Nobel Prize for their 1964 discovery of the cosmic microwave background radiation, the remnant of the "big bang" origin of the universe. And while there is no Nobel Prize for mathematics, Claude Shannon founded the mathematical theory of communication in 1948, a massive contribution with far-reaching consequences. He showed that the rate of transmission of information is limited by "bandwidth" and "noise," which is why we always want more bandwidth and less noise. Was all this fundamental research really useful to the Bell Telephone system? As we've seen before, you never know where fundamental research might lead, but in the course of their research on semiconductors, Bardeen, Brattain and Shockley discovered the transistor in 1948, for which they shared the 1956 Nobel Prize.

In 1936, Mervin Kelly, the research director at Bell Labs, realized that a new technological path was needed. The Bell Telephone System, which stretched from coast to coast, was already the largest machine in the world, and it was difficult to imagine how much larger it could get. Stowager's electromechanical switches were too bulky and too slow. Vacuum tubes wouldn't work either. They were hot, they burned out, and they were too big. But before vacuum-tube radios became popular, the old-fashioned "crystal radios" worked with just a tiny piece of a special crystal. If a solution was to be found, it might lie in the little lumps of material in crystal radios. With no moving parts, no heat, and smaller than a grain of rice, the crystals would be ideal for Bell's switching problems.

But how could they make a crystal into a switch? At that time, scientists barely understood how the crystals in those radios worked. Electrically, they were poor conductors and poor insulators, so they called them "semiconductors." But they had some strange properties. For instance, they would conduct electricity in one direction, but not the other. Maybe these strange properties could be exploited. So, in 1936, Kelly formed a small group at Bell Labs to begin fundamental research on these semiconductors. The first member of the group, Walter Brattain was already at Bell Labs doing experiments on semiconductors for other uses, but Kelly knew Brattain needed theory to back him up. His first hire was William Shockley.

---

[7] Kessler, R. (1997, April 6) Absent at the creation; How one scientist made off with the biggest invention since the light bulb. *The Washington Post Magazine*, p. 16.

Then World War II interrupted. Both Brattain and Shockley were given leaves of absence to work on military projects. Much of the rest of Bell Labs was shifted to military research, but some of it was related to semiconductors because crystal radios made good receivers for radar. In fact, they were the only thing that would work with the new high-frequency "microwave" radars then being developed.[8] At the end of the war, Kelly got Brattain and Shockley back to Bell Labs to work on semiconductors. The first glimpse of the future came at Christmas in 1947, when Walter Brattain and a new hire, John Bardeen, invented the point-contact transistor, followed soon after by Shockley's junction transistor. After that there was a cascade of new technologies, starting with a simple pocket radio and leading to an entire computer on a single chip. It's a fascinating story.

## The Crystal

The heart of a crystal radio was a tiny bit of material, a crystal, with strange properties. It conducted electricity, but not very well, and in only one direction. It was called a semiconductor. No one understood how it worked until the 1920s, when the first theoretical explanation appeared. Understanding how it worked provided the springboard for new devices, including the transistor.

The simplest device to use a semiconductor is called a rectifier, it changes alternating current (AC) to direct current (DC). We use it to charge our batteries from ordinary household power. Rectification was discovered by the German physicist Ferdinand Braun in 1874 while he was trying to measure the electrical conductivity of galena (a crystalline mineral). In his experiments, Braun attached two wires to his galena crystal to measure the conductivity, but he was getting strange results that he didn't understand. He finally traced them to the fact that the current through the crystal depended on the polarity of his connections: current would flow in one direction but not the other. If he applied AC, which flows back and forth, half of each cycle was blocked and he was left with DC, which flows in only one direction. This is called rectification. Around this time, radio was in its infancy, and in 1898, Braun used rectification to "detect" radio waves. Radio waves alternate at very high frequency, too high for our ears, or even headphones or loudspeakers, to respond.

---

[8] Microwaves and the "magnetron" used to create them are now the heart of your microwave oven. The first commercial microwave oven was called the "Radarange." A six-foot tall commercial version appeared in 1947, and a countertop version in 1967. It sold for $495, equivalent to more than $4000, today.

But by rectifying the signal, the high-frequency AC is converted to DC, and headphones can respond to that. If we turn the radio signal on and off in specific patterns, we get Morse code, and if the signal is modulated at an audio frequency, we hear a voice. The conversion of a high-frequency radio signal to a useful on–off or audio signal is called "detection."

In the early days of radio, crystal receivers based on Braun's discovery were the only option available. Since only certain points on the surface of the crystal (galena, for example) would act as a rectifier, these points had to be found by touching a metallic wire called a "cat's whisker" to the crystal. "Cat's whisker" crystal radios were used by soldiers in the trenches in World War I. In fact, since they used no electricity and got all their power from the radio waves themselves, they were undetectable by the enemy and could be used by soldiers in the foxholes of World War II. And a fancy crystal wasn't necessary; even the copper-oxide coating on an oxidized copper penny had sufficient rectifying properties to be used as a detector. For this and other contributions to radio communication, including tuned circuits (which we use to select a specific radio station) and cathode-ray tubes (which eventually became the old-fashioned TV screens), Braun shared the 1909 Nobel Prize in Physics with Guglielmo Marconi.

In the 1920s, vacuum tubes replaced crystals in radios because vacuum tubes could both detect the signal and amplify it. In fact, beyond just modulating the current flowing through the tube, a strong input signal could completely shut it off like a switch. But vacuum tubes have their own problems. In the first place, they operate only at low frequencies, suitable for AM radio and, later, FM radio. But they don't work at the much higher "microwave" frequencies used for radar, so crystal detectors were critical for radar in World War II. In addition, vacuum tubes, with their glass envelopes, hot filaments, and other electrodes, are fragile, they burn out, they consume power and they give off heat. This makes them unsuitable for large applications demanding extreme reliability…such as telephone switching systems. Something as small, simple, and reliable as the crystals in the old "crystal" radios was needed.

But how does the crystal in a radio work and how could a rectifier be made into a device to amplify voices or switch electric current on and off? What strange properties of Braun's crystals made Mervin Kelly think it could be done? The first few things they tried failed, and then serendipity stepped in.

Before the discovery of quantum mechanics, there was no way to explain why some materials, like metals, conduct electricity, and other materials, like salt, don't. Then in 1913, Niels Bohr discovered that the electrons in atoms can't have just any energy, but only the energy levels of certain discrete orbits around the nucleus. Since there can be at most two electrons in each orbit, the electrons settle into the lowest available energy levels, filling them up until the last electrons occupy what are called the valence levels of the atom, as shown in Fig. 5.2. The valence electrons are in the outermost orbits of the atom and are responsible for the chemical properties (valence) of the atom. When billions of identical atoms are assembled into a crystal, the same rules apply…sort of. Except for the valence electrons, nothing much changes, but for the valence electrons on the outside of the atoms, the individual atomic orbits disappear and the valence electrons travel in orbits that spread out to fill the entire crystal. When this happens, the discrete energy levels of the atoms become bands of very closely spaced energy levels, as shown in Fig. 5.2. Not surprisingly, the valence electrons are responsible for conducting electricity in metals, but a funny thing happens in insulators. If the valence electrons completely fill all the levels of the valence band in the crystal, we get an insulator. If this seems counterintuitive, we can imagine a pool table with no side pockets. When we have just a few balls on the table and we tilt the table, the balls run downhill to the other end of the table, like the electrons in a conductor when we apply an electric field. But if we completely fill the table with balls, the balls can't move, even when we tilt the table. When the balls are electrons in a crystal, they can't move and we have an insulator, like salt. But if the next higher energy band lies only slightly higher than the filled valence band of an insulator, a few of the electrons will be thermally excited from the valence band into the empty higher band, even at room temperature. Once they are in the empty band, they can carry electrical current. This makes the crystal a weak conductor that we call a semiconductor.

Finally, if the valence band is nearly filled, but not quite, the conduction of electricity is basically the same, but we view it differently. Imagine, again, the

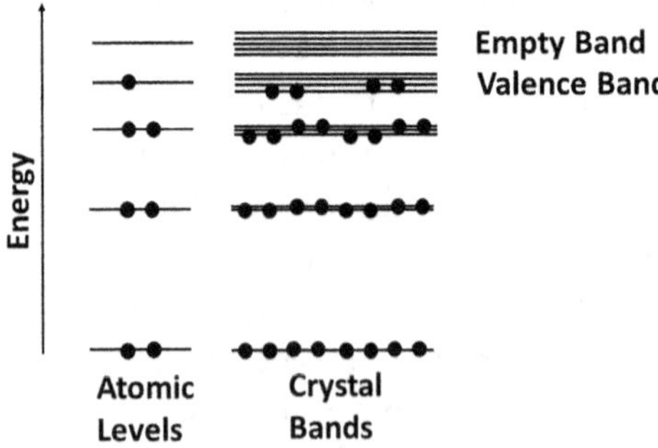


**Fig. 5.2** Energy bands in an atom and in a crystal

pool table almost filled with balls with just a few "holes" in the pattern where a ball is missing. If we tilt the table now, the ball just above the "hole" will roll downhill into the empty space and the hole will appear to move in the opposite direction, up the table to the spot left by the ball. In a crystal, a "hole" is a missing electron, and it leaves behind a positive charge, due to the left-over charge of the atom when the electron is absent. When an electric field is applied to the crystal, the electrons will move like the balls when the table is tilted, and the hole will move in the opposite direction. This is called hole conductivity, since the holes appear to move long distances in successive steps while the electrons hardly move at all. As we shall see, the concept of hole conductivity is important for understanding rectifiers and transistors. Remember that since the electrons are negatively charged, they travel in the direction opposite the direction the electrical current is flowing, but since the holes appear as positive charges, they travel in the direction of the electrical current. In semiconductors, the mobility of holes is comparable to that of electrons.

There's just one more thing. It has to do with "doping:" If we add just a tiny bit of some impurity to the crystal, we can change its behavior. If the impurity atoms donate electrons to the empty band, as shown in Fig. 5.3, they will conduct electricity in that partially filled band. This makes the crystal an n-type semiconductor, so-called because the electrons are negatively charged, and the nearly empty band is now called the conduction band. If, on the other hand, the impurity atoms absorb electrons from the valence band of the crystal, as shown in Fig. 5.3, they leave behind holes, which can conduct electricity. Now we have a p-type semiconductor so-called because the holes are positively charged, and the nearly filled valence band is called the conduction

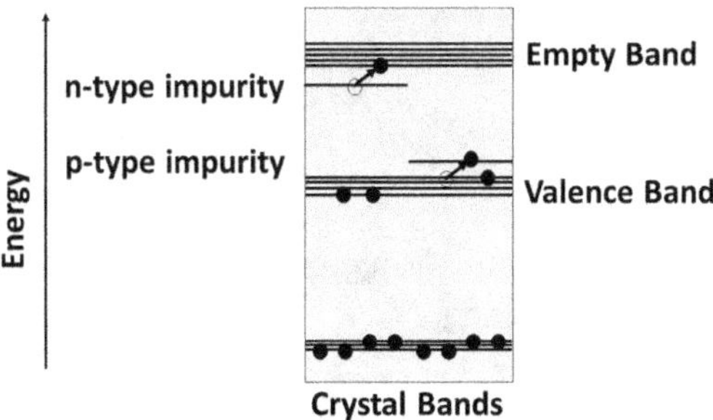


**Fig. 5.3**   Doping in a crystal

band. With no doping, the conductivity is very small at room temperature. With n-type and p-type semiconductors, one type of conductivity, either electron conductivity or hole conductivity, dominates and is much larger. With doping, we can make interesting devices.

Of course, natural crystals, and even synthetic crystals, contain small impurities. Since the impurities aren't uniformly distributed in the crystal, this creates regions of n-type and p-type semiconductors. In the old days of crystal radios, a "cat's whisker" was used to find a "good" spot on the crystal surface where rectification and detection occurred. The "cat's whisker" was a thin metal wire, and where it made contact with an n-type spot in the crystal, the electrons in the conduction band of the crystal were at a higher energy than the electrons in the metal "whisker," as shown in Fig. 5.4. If we put a negative voltage on the crystal, called forward polarity, the electrons can "spill" across the junction into the metal, like a waterfall, and electricity flows. But when the electron leaves the crystal, it leaves behind a hole, which moves in the opposite direction. So, instead of an electron falling into the metal, we may equivalently say that the metal has injected a hole into the crystal, which carries the current. On the other hand, if we put a positive voltage on the crystal, called reverse polarity, the electrons in the metal don't have enough energy to jump up into the conduction band of the crystal, sort of like salmon trying to climb a waterfall that is too high, so no current flows. Voilà, we have rectification. The arguments are similar, but with the voltages and currents reversed, if the wire contacts a p-type region of the crystal. This explains how a so-called point-contact rectifier works. Two of these rectifiers make a point-contact transistor. We'll come back to that later.

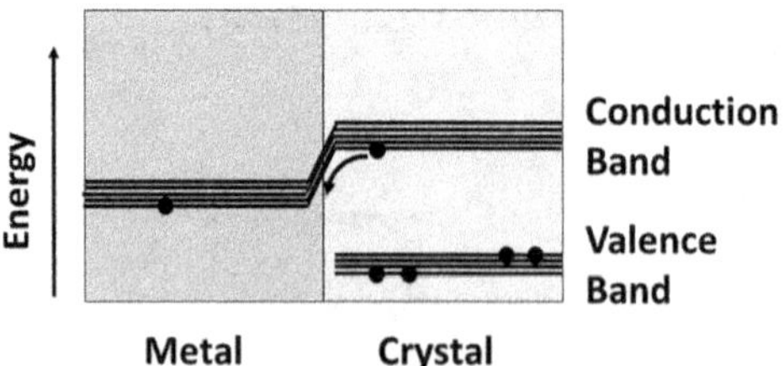


**Fig. 5.4** Rectifier with a forward voltage on the crystal

## The Labs

A scientific laboratory is more than a room full of test tubes and voltmeters. It's really a place where ideas are born in the minds of bright people with visionary management and the resources to support them. Bell Telephone Laboratories was fortunate to have people like Walter Brattain, William Shockley, and John Bardeen with the resources of a huge company behind them.

No one understood semiconducting crystals when Bell Labs was started in 1925, but semiconductor rectifiers, especially copper-oxide rectifiers, were becoming important for many applications, from battery chargers to power supplies. Consisting of a copper-oxide coating on a copper surface, they worked like the cat's-whisker detectors used for crystal radios, but were much bigger so they could carry more current. This is where Bell Labs' semiconductor research began in the 1930s. Unfortunately, the world-wide depression of the 1930s limited what could be accomplished, even at Bell Labs, which was supported by a powerful monopoly. The coming war would redefine the focus of research at the Labs.

Of the three scientists who invented the transistor, the first to arrive at Bell Labs was Walter Brattain (Fig. 5.5). A small-town country boy, Brattain was born on February 10, 1902, in Amoy, China, to American schoolteachers working

**Fig. 5.5**  Walter Brattain. (Courtesy of the Nobel Foundation)

there. Walter and his mother returned to the USA in 1903 and moved to Spokane, WA, followed the next year by his father. In 1911, they moved to Tonasket, WA, a rural area in the Columbia River watershed near the Canadian border, where they lived and worked a cattle ranch. Walter rode to elementary school each day on horseback, and he became an expert horseman and marksman. Growing up on a rural farm, he learned to do things himself, including how to tear down and reassemble an automobile engine. He also learned self-reliance by taking their 300 head of cattle up into the Forest Service lands in the summer, spending a week at a time alone with just the cattle and a few books. After schoolwork and chores, Walter had little time left over, but his mother helped with his education, especially in math, and he skipped seventh grade. To help him get into Whitman College, his parents (with help from his Aunt Bertha) sent him to Moran School, a private high school on Puget Sound across from Seattle, where he learned his first physics and graduated with honors in 1920.

He matriculated at Whitman College that fall. A small liberal arts college, Whitman had one physics instructor and one math instructor who prepared him for postgraduate work. He graduated from Whitman in 1924 with a double major in physics and mathematics and went on to get a Master of Arts from the University of Oregon in 1926, and a Ph.D. from the University of Minnesota in 1929. At Minnesota, Brattain was introduced to quantum mechanics in a course taught by John Van Vleck, one of the first to bring the new ideas of quantum theory from Europe to the USA. For his thesis, working under John Tate, Brattain did an experiment in which he bombarded mercury atoms with electrons to excite them to the higher-energy orbits predicted by Bohr. By the time he got his Ph.D., the formal education he received in school combined with the "hands-on" skills and self-reliance that he developed on the farm constituted an excellent preparation for a career in physics research, especially experimental research.

After graduating from Wisconsin, Brattain joined the National Bureau of Standards in Washington, DC. At the Bureau he worked on comparing frequency standards from around the world, but before long he became restless and realized that he really wanted to do research. In January, 1929, the American Physical Society had its annual meeting in Washington, and Brattain decided to take the opportunity to look for another job. There, he met Joseph Becker, who was working at Bell Labs and was looking for someone to hire. The next evening, they had dinner with Karl Darrow, who was just then bringing back the latest ideas of quantum mechanics from Europe, and Clinton Davisson, who had recently discovered electron waves scattering from a crystal surface. This was impressive company for a young Ph.D.! Brattain was hired, and he joined the lab the following August, 1929.

At that time, vacuum triodes offered the only means for amplifying signals in the repeaters that made transcontinental telephone conversations possible. When he got to Bell Labs, Brattain began working with Becker on surface modifications of the hot filaments in the vacuum tubes to allow them to emit electrons (called thermionic emission) at lower temperatures, hoping to reduce their power consumption and extend their lifetime. But as Brattain explained, "the theoretical explanation of thermionic emission was very much a question up in the air."[9] Then, in 1931, Brattain had the opportunity to attend a summer school at the University of Michigan, where he heard lectures on the new quantum mechanics from the giants of the field, including Wolfgang Pauli and Arnold Sommerfeld. In his lectures, Sommerfeld explained how electrons arrange themselves within the energy bands of metals. With these new ideas, Brattain was finally able to understand and calculate the thermionic emission he was measuring from his filaments. He and Becker published their results, and in the fall, Brattain began lecturing on quantum mechanics at the Labs.

In 1931, Brattain shifted his attention from thermionic emission to copper-oxide rectifiers. Working like the point-contact "cat's-whisker" detectors discussed above, only bigger, they were already being used both as simple rectifiers in power supplies and as detectors when several voices were multiplexed on a single wire. But as important as they were technologically, copper-oxide rectifiers couldn't be understood theoretically. As Brattain reflected, "There was … the intuitive idea that somehow work-function differences were involved, or the ability of electron charges to flow out of one material and not out of the other."[10]

But this was the early 1930s, and the world was in turmoil. To avoid layoffs during the Great Depression, Bell Labs cut the work week from six days to four. In Europe, Hitler came to power in 1933. The good news, if it could be called that, was that many physicists fled Germany and came to the USA. Scientifically, quantum mechanics was beginning to explain some of the mysteries scientists had been facing as they tried to understand atomic and solid-state physics. The band theory of electrons in crystals began to pull back the curtain on some of these mysteries. Solid-state physics—the physics of solid materials—was born.

Mervin Kelly (Fig. 5.6), an irascible but visionary Ph.D. physicist at Bell Labs, was the head of the vacuum-tube department until 1936, when he replaced Oliver Buckley as research director of the Labs. Recognizing what he

---

[9] Riordan, M. & Hoddeson, L. (1997). *Crystal Fire*. Norton, New York, p. 62.

[10] Riordan and Hoddeson (1997), p. 64.

**Fig. 5.6**   Mervin Kelly. (Courtesy of Nokia Corporation and AT&T Archives)

considered the fatal limitations of vacuum tubes, Kelly saw them as a dead end. Together, Kelly and Buckley began to think about solid-state (crystalline) devices to replace the vacuum tubes and mechanical relays in the switching systems the Bell system needed to connect the vast and growing number of telephone calls being made. But the shortened work week forced on the Labs by the Depression made room for the Lab employees, including Brattain, to learn more about solid-state physics. At the same time, the experienced physicists driven out by Hitler began to build up the universities and labs in the USA. When the Depression-era hiring freeze was lifted in 1936, Kelly and Buckley were ready to make their move to recruit physicists who understood the new quantum theory of solids. Their first hire was William Shockley.

William Shockley (Fig. 5.7) was born on February 13, 1910, in London. His mother May described him a few days later as "a fine, well-developed, lusty-voiced infant and already knows what he wants when he wants it."[11] May was 27 at the time; William's father was 51. A mining engineer who stimulated Billie's early interest in science, he traveled on business much of the time. William was a burden May wasn't prepared to face mostly alone; he was frequently sick, demanding, and had serious tantrums. May decided that William would be her only child. The couple returned from London to the USA in April, 1913, settling in Palo Alto, near where May had lived while she was attending Stanford.

---

[11] Riordan and Hoddeson (1997), p. 22.

**Fig. 5.7** William Shockley. (Courtesy of the Nobel Foundation)

Growing up, Billy had few friends; he always liked to take charge, and exhibited sudden fits of uncontrollable rage. "Anger is about the only emotion he displays, with a little love at times," his father wrote. "He is not spanked because he has a 'friend'—a slender bamboo sprout from the yard who sometimes speaks to Billy forcibly."[12] Billy was clearly different, and they schooled him at home until he was eight. His father nurtured his interest in science, and May taught him arithmetic. In 1918, they enrolled him in Mrs. Gurmell's Private School, but two years later, he transferred to Palo Alto Military Academy; Billy needed stricter discipline. Billy also got an informal introduction to physics from the Stanford physicist Perley Ross, whose daughters were two of young Billy's friends. "He tried to explain wave motion to me, and I had a great deal of difficulty getting any grasp on that at all," Shockley later recalled.[13]

After Billy graduated from Palo Alto Military Academy in 1922, the Shockleys spent a year traveling on the East Coast. By this time, Billy's father's health was beginning to deteriorate. When they returned to California, they moved to a house in Hollywood where Billy started at Hollywood High School. His interest in science continued, and he took a course in physics at the Los Angeles Coaching School. But his father's health continued to decline, and in early May 1925, he was brought home unconscious from "apoplexy" (stroke). He died on May 26, and teenage Billy became the leading male in

---

[12] Riordan and Hoddeson (1997), p. 24.

[13] Riordan and Hoddeson (1997), p. 25.

the family. Billy's father had left a modest estate, and a few months later, on his sixteenth birthday, Billy and his mother went out and bought a new car.

Billy did well at Hollywood High School, especially in science and math. Sometimes he outpaced his teachers, and he didn't hesitate to correct and challenge them. He graduated in 1927, an outstanding student … but not necessarily liked by the faculty. By this time, however, he knew that he liked physics and realized that he was rather good at it. But immersed in the milieu of sons and daughters of Hollywood's movers and shakers, Billy also developed the swagger and chutzpah that would characterize him later, especially in college. As his friend and colleague Frederick Seitz later remarked, Shockley was "strongly influenced by the Hollywood culture of the day, fancying himself a cross between Douglas Fairbanks, Sr. and Bulldog Drummond, with perhaps a dash of Ronald Colman."[14]

After finishing high school, Shockley entered the University of California-Southern Branch (now UCLA), which was within walking distance of his home. But in 1928 he transferred to the California Institute of Technology in Pasadena. At that time, Cal Tech was just starting up, with money enough ($4 million) to lure the well-known physicist Robert Millikan, who had won the Nobel Prize in Physics in 1923 for measuring the charge on an electron. With his fame and donors' money, Millikan was able to attract superstars from established schools like Harvard and MIT. Some, like Albert Einstein, Edwin Hubble, who discovered the expanding universe, and Albert Michelson, who measured the speed of light, stayed as visitors for shorter or longer periods, making the faculty club an exciting place to be. Some, like Linus Pauling, who won a Nobel Prize for chemical bonding, and later a second Nobel Prize for peace, stayed as faculty. Linus Pauling was a particularly important mentor to Shockley and suggested that he learn quantum mechanics by reading P. A. M. Dirac's book on relativistic quantum mechanics. Dirac's abstract formalism was beautiful in its compactness and led to the mathematical discovery of antimatter, but it was really hard to understand. Reading one of Dirac's papers, even Einstein once wrote, "I am toiling over Dirac. This balancing on the dizzying path between genius and madness is awful." But Dirac's elegant approach, beautiful in the abstract, was less useful for calculating anything practical, like the wave functions of electrons in crystals, which was the kind of stuff that Shockley would do later.

Shockley earned his Bachelor of Science degree from Caltech in 1932, but along the way, other aspects of his personality began to show themselves. Throughout his life, Shockley's hobbies included sleight-of-hand and rock

---

[14] Seitz, F. (1994). *On the frontier: My life in science.* New York: American Institute of Physics, p.67.

climbing, and he had an abiding interest in training ant colonies; he would spend hours training ants to work through mazes with teeter-totters and other obstacles.[15] He also was a fitness buff and good-looking enough to be a model in advertisements for strength training equipment. He also was fond of practical jokes, and Caltech was just the place for this. One semester, he and some compatriots (allegedly including Robert Oppenheimer) enrolled a completely fictitious student named Helvar Scavi in the class of Fritz Zwicky, a particularly disliked professor. Having previously taken the course, they turned in tests with answers to some of the questions before scribbling, "Well, I'm too damn drunk to write anymore." But the answers were so brilliant that Zwicki gave Scavi an *A* in the course, and, grading on a curve, the rest of the class a *C-*.[16]

In September, 1932, as the Depression deepened, Shockley headed east to MIT, riding in his roadster with fellow physics student Frederick Seitz, who had just graduated from Stanford and was headed to Princeton. Along the way they rode past overloaded jalopies heading west to escape the dustbowl that was beginning in Oklahoma, hoping to find better times in California. On the ride, Seitz got a sense of Shockley's Hollywood persona when he learned that Shockley kept a loaded pistol in the glove compartment. As they drove through Texas, camping at night, Shockley went out into the desert one night and fired the pistol to scare off some coyotes. Some locals reported "desperados" to the police, but it wasn't until the pair got to New Jersey that they got into trouble; there the police stopped some "suspicious" persons and found the gun.

While Shockley was on his way to MIT, Seitz was traveling to Princeton to study with Eugene Wigner, who had been driven out of Germany because he was a Jew. Wigner is interesting because he was a brilliant mathematician who, together with Hermann Weyl, laid the foundation for the abstract theory of symmetries in quantum mechanics. But Wigner could also do calculations, and when war came, he played an important role in the Manhattan Project to develop the atomic bomb. Up to this point, research in America was largely experimental and very practical, and emphasized the physics of materials, which we call solid-state physics. The elegant theoretical research was centered in Europe where Bohr contemplated the philosophical relationship between the observer and the subject (think of Schrödinger's cat[17]), Paul Dirac discovered antimatter in his equations, and Heisenberg sought (in vain)

---

[15] Moll, J.L. (1955) *William Bradley Shockley, 1910–1989, A biographical memoir*, National Academies Press, Washington, DC, 1955, p. 310.

[16] Shurkin, J. N. (2006) *Broken genius*, Macmillan, New York, p. 20.

[17] To try to reconcile Heisenberg's uncertainty principle for the microscopic world with the certainty expected in the macroscopic world, Bohr proposed that the state of a system is determined only when it

to discover the "next" quantum mechanics. Pauli, for one, famously remarked, "*Festkörperphysik ist eine Schmutzphysik*"—solid-state physics is the physics of dirt.[18] For most European physicists, it was "beneath them" to work on practical problems. When Hitler came to power, many of the theorists, including Felix Bloch, Rudolf Peierls, Hans Bethe, and others, fled Germany, never to return. The center of semiconductor research shifted to America, where European theory combined with American pragmatism to enrich both.

To properly describe the behavior of electrons in a crystal, it's necessary to treat them as waves, as described by Erwin Schrödinger. The first to describe the electron waves in a crystal was Felix Bloch, which he did for his doctoral thesis in 1928.[19] This pointed the way, but Block never did any calculations and, as they say, the devil is in the details. In the Wigner-Seitz method, which they developed for Seitz' Ph.D. thesis, the crystal is divided up into an infinite number of identical cells each centered around an atom in the crystal, and the waves are calculated numerically, by hand with a mechanical calculator, repeatedly refining a previous calculation. Needless to say, this was a tedious process since there were no electronic computers in the 1930s. Shockley would use the Wigner-Seitz method in his own thesis.

After dropping Seitz off at Princeton, Shockley continued on to MIT. Upon arriving in Cambridge, he found an apartment he could afford on his $77 per month appointment as a teaching assistant. Located near MIT in an industrial section of Cambridge, Shockley described it as being on "an Irish street...surrounded by rather impossible streets, negro and ramshackle," revealing, perhaps, some of the racist feelings that he would exhibit later in life.[20]

In August 1933, Shockley abruptly married Jean Bailey in California. They drove back to Boston and found an apartment in Cambridge in September. Jean worked part time, and Shockley's mother chipped in to supplement Shockley's stipend from MIT. The young couple fit easily into the graduate student social scene. Their first child, Alison, was born in March. Around this same time, Shockley took and passed the exams that were required before he could begin to work on his thesis.

---

is observed. As a challenge, Schrödinger proposed the question: A cat is in a box with a vial of poison that is opened by a random device. Before the box is opened, is the cat alive or dead?

[18] Commentary (2018) Condensed matter's image problem. *Physics Today*. AIP Publishing. 19 December https://doi.org/10.1063/pt.6.3.20181219a ISSN 1945-0699.

[19] Felix Bloch was Werner Heisenberg's first graduate student. Being Jewish, he fled to the USA when Hitler came to power, and worked briefly in the atomic bomb program at Los Alamos before switching to problems in radar at Harvard. In 1962, he shared the Nobel Prize in Physics with Edward Purcell for his work on nuclear magnetic resonance which led, later, to medical MRI, magnetic resonance imaging.

[20] Shockley, W. letter to May Shockley (1932), September 24. *Stanford Archives*.

He also continued his penchant for mischief. At MIT he got to be friends with James Fisk and even wrote a scientific paper with him. They would work together again when Fisk joined Bell Labs a few years after Shockley. But one night, they crept into the administration building and rewired the elevator. When Karl Compton, the president of MIT, arrived that morning and tried to go to his office, every button he pushed took him to a different floor. Finally, he had to walk to his office.

At MIT Shockley completed his thesis under John Slater. Slater had studied in Europe but came back to America to work on real-world stuff. He became a powerful force in American physics and politics. Shockley's assignment was to calculate the electron waves in sodium chloride, common table salt, which contains two different kinds of atoms, sodium and chlorine, using the methods developed by Wigner and Seitz at Princeton. As Shockley later admitted, "the main essence of that thesis was really the discipline of sitting down and running calculating machines for a very long period of time."[21] In the end, he was able to say that "I drew the first realistic pictures of energy bands in—actually calculated complex energy bands for—a real crystal."

Shockley received his Ph.D. from MIT in 1936, just as the hiring freeze was lifted at Bell Labs. Kelly was now the research director, and he was determined to strengthen fundamental research at the Labs with people who really knew theoretical solid-state physics. He visited MIT, which was becoming a strong presence in this field, and hired Shockley on the spot, his first hire after the freeze. Shockley joined the Labs in June, at 463 West Street, in New York, where they were located at that time. Initially, he was assigned to a sequence of groups that gave him a broad view of research at the Labs. As he said later, "I was put into a pretty rigorous sort of indoctrination period by being sent around to several different places within the laboratory, all having to do with electronics."[22] In these various settings, Shockley demonstrated his unusual ability to deal with both theory and experiment. Most physicists deal effectively with either theory, like Heisenberg, or experiment, like Brattain, but not both. In this regard Shockley was favorably compared to Fermi,[23] but Fermi was truly extraordinary: half the particles in the universe are called Fermions because of his fundamental theoretical research, but Fermi's Nobel Prize was actually awarded for his experimental research on the transmutation of elements by neutron bombardment, and during World War II he constructed the world's first working nuclear reactor. Shockley never made

---

[21] Shockley, W. interview by Hoddeson, L. (1974). American Institute of Physics, p. 7.

[22] Riordan and Hoddeson (1997), p. 82.

[23] Moll (1955), p. 308.

fundamental discoveries comparable to Fermi's, but he was constantly inventive. His first patent, in 1938, was for an improvement in so-called photomultiplier tubes.[24] Modern embodiments of this device are at the heart of night-vision devices. Besides his inventiveness, Shockley's great strength was the way he could break a problem into its parts to find a solution, and he could do real-world calculations to explain what was being observed experimentally. He was always a fountain of new ideas, more than enough to keep the others in his group busy working on his own (Shockley's) ideas. By the end of his career, Shockley accumulated more than 90 patents. But Fermi could go into the lab and do an experiment himself; Shockley couldn't. His early attempts to make a field-effect transistor failed, and sometimes he actually discouraged younger members in his group from pursuing experiments of their own that later led to breakthroughs.

Shockley was also flamboyant: he often drew bouquets seemingly out of the air at the end of lectures and climbed the rock walls of his lab clinging to grooves with his fingertips. But he seemed unable to feel empathy for others or to understand human relationships. As we shall see, this led to his failure to advance at Bell Labs and to the collapse of his ventures in Silicon Valley.

While Shockley worked at the Labs in New York, he and Jean lived in an apartment that they found in Greenwich Village, where most of their friends lived. It was an "artsy" area populated by writers, artists and musicians. Shockley could walk to the Lab, dropping Alison off at nursery school on his way. Jean found a job at the YMCA to supplement their income. Shockley exercised to stay in peak physical condition, and they spent their summer vacations camping on an island in Lake George, in upstate New York, where he introduced Jean to rock climbing. Alison remembered those vacations as the family's happiest times.

In 1937, Bell Labs began construction of a new facility in Murray Hill, New Jersey, that was finished in 1941. In the meantime, the Labs moved to temporary quarters in Whippany, New Jersey. In anticipation of the move, the Shockleys moved to a house in the village of Gillette. The house was surrounded by woods, where Shockley took Alison on walks and collected snakes. He also built a firing range in the cellar. After working out in the morning, Shockley took the train to New York, until the Whippany facility opened. After that, he drove himself.

In 1938, Kelly formed a group specifically to look into the possibility of solid-state electronic switches to replace the mechanical relays then in use. Since Shockley had previously organized an informal "journal club" that met

---

[24] Riordan and Hoddeson (1997), p. 82.

weekly to discuss recent developments in quantum mechanics, the group was placed under Shockley and included Foster Nix, a metallurgist, and Dean Woolridge, a physicist. They were given enormous independence to do "fundamental research" to discover new materials and methods that might eventually be useful for the Bell Telephone System, following their own ideas wherever they might lead. In December of that year, Otto Hahn discovered nuclear fission. Shockley, Brattain, and Fisk, from Bell Labs, were in the audience when Bohr described their discovery in a talk at Columbia University. The following spring, Kelly asked Shockley and Fisk if it would be possible to use nuclear fission to produce useful power. It didn't take long for Shockley to recognize that by putting uranium chunks in a moderator to slow the neutrons, a nuclear reactor could be constructed, and he and Fisk set about to calculate the optimum dimensions for the lumps and for the whole assembly. They also considered the possibility of an explosive. Two months later, they reported their results to Kelly. The results were so promising that their work was immediately shut down and kept quiet. Independently and at the same time, Fermi had conceived the same ideas and had begun experiments at Columbia that led to the world's first nuclear reactor.

The new group at Bell Labs soon became interested in the work being done on copper-oxide semiconductors by Brattain and Becker. In the winter of 1939–40, Shockley, working somewhat secretively by himself, conceived the idea of a solid-state amplifier based on an electrode to control the flow through a copper-oxide junction. When his crude experiments didn't work, he finally asked for help from Brattain and Becker, who were better experimenters. They promptly informed him that they had tried that idea a couple of years earlier and it hadn't worked, but Shockley prevailed upon them to try it again. Brattain recalled, "I was quite sure it wouldn't work. But I said, 'Bill, it's so damned important that if you'll tell me how you want it made, and if it's possible, we'll make it that way. We'll try it." It still didn't work. "The result was nil. I mean, there was no evidence of anything."[25]

Then, in the spring of 1940, two events changed the course of research at Bell Labs. On March 6, Becker and Brattain were called to Kelly's office to watch an amazing demonstration by Russel Ohl, a scientist from the radio laboratory in Holmdel, NJ. When Ohl shined a flashlight on a piece of silicon with two wires attached to a voltmeter, the response was ten times anything that they had observed in other photocells. Ohl had discovered this while he was working with silicon for detection of microwave radio signals, but he couldn't understand what was happening. He also noticed some other bizarre

---

[25] Hoddeson, L. & Daich, V. (2002). *True genius: The life and science of John Bardeen: The only winner of two Nobel Prizes in physics.* Joseph Henry Press, p. 126.

behavior that he attributed to a crack in the silicon rod. Brattain suggested that a barrier in the silicon rod caused the photovoltaic effect and other peculiarities that Ohl had observed. Maybe silicon could be manipulated for other semiconductor applications.

Following up on Brattain's suggestion, Ohl went back to his lab to look for the putative barrier that was causing the bizarre behavior in the silicon rod. By sliding a contact along the rod, he found that there was indeed a boundary in the silicon rod; half the rod had impurities that made it n-type silicon, and the other half p-type. They had serendipitously created the world's first p-n junction, the beginning of a whole new class of semiconductor devices: the junction itself formed a rectifier within the silicon rod. With positive voltage on the n-type end of the rod and negative on the p-type end (reverse polarity), the electrons and holes were drawn away from the junction and no current flowed. With the opposite voltage (forward polarity) the electrons and holes flowed toward the junction, allowing current to flow across the junction, as shown in Fig. 5.8. With a p-n junction, rectification and detection took place entirely within the silicon crystal itself. Moreover, when light was applied to the crystal, it excited electrons in the n-type material into the conduction band. When these electrons migrated to the junction, they "fell" across the junction into the holes in the p-type material, generating a voltage. The large photovoltaic effect that Ohl had discovered was the progenitor of today's solar cells.

Then, on April 9, 1940, Hitler invaded Norway and on May 10 he followed with the full-scale invasion of Luxembourg, the Netherlands, Belgium, and France. Paris was occupied on June 14, and France surrendered on June 22. In the skies over England, radar played a critical role in the ability of the RAF (the British Royal Air Force) to defeat the *Luftwaffe* (the German air

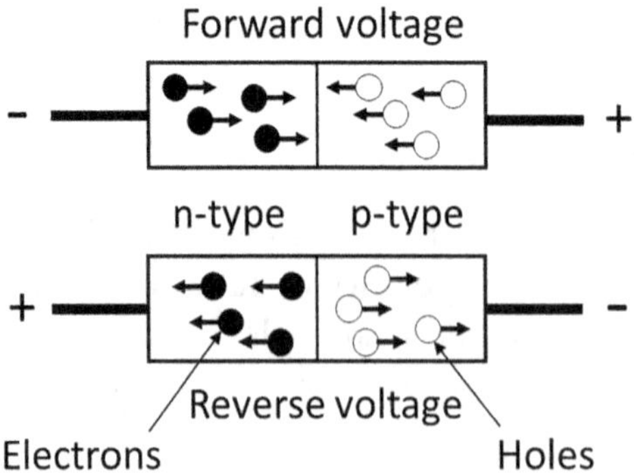


**Fig. 5.8**  p-n junction rectifier

force) in the Battle of Britain, and radar helped the British Navy to defeat the German submarines in the Battle of the Atlantic. Crucial to all this, John Randall and Harry Boot at the University of Birmingham had just developed what they called a "cavity magnetron." Nowadays that's the little tube in your microwave oven that heats your coffee in the morning and your supper in the evening, but it was arguably the most important invention of World War II. The atomic bomb ended the war, but the magnetron won it. The cavity magnetron was small enough to put in a plane and could create large amounts of microwave radiation, and the microwave wavelengths it produced were short enough (around 10 centimeters, or four inches) to locate objects as small as a submarine periscope. But detection of the short-wavelength radiation was problematic. Vacuum tubes wouldn't work; it took longer for the electrons to cross a vacuum tube than for the oscillating microwaves to reverse their fields. But cat's whisker detectors would work. Ohl and his silicon crystals might be just the ticket. There was much to learn. Despite Ohl's efforts at purification, silicon (like other semiconductors), was plagued with impurities and exhibited erratic behavior.

Faced with critical wartime needs, Kelly urged Becker and Brattain to focus their work on silicon detectors. As Brattain later recalled, "I guess we kind of quit the copper oxide work."[26] Positioned as they were in the semiconductor business, the Labs immediately began a crash program to develop detectors for radar. In fact, the Labs had already been secretly working on radar for the Navy using longer wavelengths, where vacuum tubes could work. The first detectors for the new microwaves were based on the old cat's-whisker technology, but cat's whiskers were inevitably noisy and fragile. This could be improved with higher-purity silicon, and DuPont, working with Seitz, who was by then at the University of Pennsylvania, produced silicon that was 99.99% pure, and eventually 99.999% pure, just ten parts in one million. Ohl and his group went on to identify the impurities, the dopants, needed to produce p-type and n-type behavior. With these materials, reliable point-contact detectors could be made, and this information was shared with other labs in the USA and Great Britain. When they figured out how to make them, p-n junctions would be quieter and more reliable, but work on the p-n junction was kept secret within Bell Labs.

Then, on December 7, 1941, the Japanese bombed Pearl Harbor. Much of Bell Labs was converted to wartime research. Brattain was the first to leave the semiconductor work. Harvey Fletcher from Bell Labs and John Tate, Brattain's thesis advisor at Minnesota, had formed a group headquartered at Columbia

---

[26] Riordan and Hoddeson (1997), p. 98.

University to work on the magnetic detection of submarines. In January 1942, Brattain was recruited to join this effort and spent two years at the Quonset Naval Air Station, in Rhode Island, and Cold Spring Harbor, on Long Island.[27] Both places were within a few hours' drive from the Labs, so he could go home from time to time. With the USA now in the war, German submarines started sinking ships along the Atlantic coast. Often the ships were sunk within sight of land, even using the lights of New York City to illuminate targets. The USA was unprepared, and coastal shipping offered many targets. The German submariners called this "the second happy time," referring back to the "happy time" a few years earlier when the British were unprepared for the Battle of the Atlantic. To carry out their experiments, Brattain sometimes had to fly out over the Atlantic to search for German submarines lurking there.

In May 1942, Shockley took a leave of absence from Bell Labs to become research director of the Anti-Submarine Warfare Operations Research Group, which had been set up at Columbia by Phillip Morse, who had taught Shockley quantum mechanics at MIT. A branch of applied mathematics, operations research uses analytical and statistical methods to break an organizational problem down into its essential elements and find an optimal approach.[28] This was a perfect opportunity for Shockley: his ability to break a problem down and get to its essential elements was remarked on by others, and his natural penchant for telling people how they should go about their business had been apparent since birth. He was good at it and important military and government leaders sought him out. The problems that he analyzed included anti-submarine warfare, how to organize a convoy and arrange depth-charge patterns, for example, and strategic bombing. Despite the difficulty of getting data from the Navy, he was able to get enough actual data to show that slower planes flying farther from the coast would sink more submarines. With statistical analysis of weather and other factors, he was able to show that the Germans didn't use radar to find the Allied convoys, so they could hide in storms. He used statistics to show that strategic bombing was economically cost effective in Europe but not in the Pacific. He even played a major role in training bomber pilots in the use of the new radar bomb sights. Near the end of the war Shockley authored an important report projecting the military and civilian casualties that might be expected from an invasion of Japan, on the order of a half million Allied deaths and several millions of

---

[27] Riordan and Hoddeson (1997), p. 104.

[28] Operations research was pioneered by Charles Babbage (1791–1871), a British mathematician who in 1834 also conceived the first computer. A mechanical device, his "Analytical Engine" had stored memory and conditional branching. He was never able to complete its construction.

Japanese deaths.[29] Although Shockley knew little of the Manhattan Project to develop nuclear weapons that was underway, the report supported the decision to drop the atomic bombs. After atomic weapons were dropped on Hiroshima and Nagasaki, he analyzed the costs and results and concluded that nuclear weapons were enormously cost effective, costing the Japanese economy about 300 times what they cost us. His work during the war was so important and consequential that in 1946 he was awarded the National Merit of Honor.

This work frequently took Shockley away on travel. In the early years, it was mostly to Washington, and he got a suite at the University Club, where he could swim for exercise. Later, when he was working on the bombsights and training pilots to use them, he took a three-month tour to England, Italy, Egypt, India, Australia, and the Marianas Islands in the Pacific to learn about the problems facing the bombers in actual combat, analyze them, and pass the analysis on to the pilots and bombardiers. Often, he flew at night in planes blackened to avoid detection.

But his long absences from home and the importance and excitement of his war work—compared to changing diapers when he was home—led to stresses in his marriage. Back in Madison, New Jersey, Jean, Alison and little Billy, born while Shockley was away in August, 1942, were alone much of the time, and since he was constantly on the move, it was difficult to communicate with him when he was away. Often, he couldn't disclose his whereabouts, and even when he was home, Shockley wasn't totally there. Inevitably, he brought his work home with him, and since it was secret, he couldn't talk to Jean about it. At one point, the tension got so bad that Shockley contemplated divorce and suicide. Playing Russian roulette, the gun didn't fire, and he put it away. It isn't known that he ever tried suicide again.

By May 1943, "Black May" as it was known to the German submariners, the Allies had won the Battle of the Atlantic due to the convergence of several technologies including operations research to improve antisubmarine strategies and microwave radar to find the German submarines. With no more need for magnetic submarine detectors, Brattain returned to Bell Labs in December 1943. By this time, the Labs were heavily engaged in war-related research, and Brattain joined Becker in the development of infrared detectors that might be used in bombsights. Brattain had worked with Becker since he first arrived at Bell Labs, first on thermionic emission in vacuum tubes and then on

---

[29] Compton, K.T. (1946) If the atomic bomb had not been used. The *Atlantic Monthly*, CLXXVIII, 54–56. Giangreco, D.M. (1997). Casualty projections for the U.S. invasions of Japan, 1945–1946: Planning and policy implications. *Journal of Military History*. 61 (3): 521–581.

copper-copper oxide rectifiers, and he looked forward to working with him again. But Kelly was already thinking of more important work for Brattain when the war ended.

## The Group

By 1943, the initial shock of the Nazi *Blitzkrieg* and the bombing of Pearl Harbor were wearing off. The Americans had by then captured Guadalcanal and were moving north across the Pacific, and the Russians had thrown the German army back 250 km from Moscow. Confidence had grown to the point that leaders in government and industry were already looking beyond the war to their next projects. In May of that year, Kelly wrote a lengthy memorandum that focused on his concern that vacuum tubes were a dead end and semiconductor switches and amplifiers were needed. Recognizing the enormous strides that science had made under conditions of wartime urgency, he understood that tremendous changes and competition lay ahead. Bell Labs needed a program of fundamental research in solid-state physics to meet this challenge. He envisioned a reorganization of the Labs with a new section to work on fundamental research in solid-state physics. With his knowledge of quantum mechanics and his demonstrated inventiveness, William Shockley was just the man to lead this effort.

Toward the end of the war, Shockley was splitting his time between war work in Washington and working with Kelly to plan the future at Bell Labs. At the Labs he also had some time to think about solid-state devices. Stimulated by visits to Ohl, who was working with the newly available doped high-purity silicon at Holmdel, Shockley focused his attention on this substance. On the weekend of April 13–15, 1945, he drew up a diagram, supported by calculations, for a new type of switch or amplifier to replace a vacuum tube, and taped it into his research notebook. In this idea, an electrode placed close to a thin silicon film would create an electric field large enough to draw electrons or holes to the surface to increase the conductivity and control the current. In May 1945, just as Germany surrendered, Shockley tried some experiments at the new lab facilities in Murray Hill. His calculations of the "field effect" predicted a large effect, but the experiments didn't work. There was no response to as much as a thousand volts on the control electrode. "Quite mysterious," Shockley recalled.[30]

---

[30] Riorden and Hoddeson (1997), p. 113.

In July, Kelly announced his reorganization of the Labs. Three new groups were formed in the physics department: one, dedicated to solid-state physics, would be led by Shockley and the chemist Stanley Morgan. In his memorandum authorizing this shakeup, Kelly wrote:

> Employing the new theoretical methods of solid state quantum physics and the corresponding advances in experimental techniques, a unified approach to all of our solid state problems offers great promise. Hence all of the research activity in the area of solids is now being consolidated in order to achieve the unified approach to the theoretical and experimental work of the solid state area.[31]

Shockley refused to be burdened by administration, so most of this was picked up by Morgan. Brattain was relieved to be included in this group. Together with Walter Pearson, a physicist with whom Brattain had worked in the past, and two technicians, Shockley formed a sub-group specifically focused on semiconductors for possible application to the Bell Telephone system.

In August 1945, the Shockley family took a vacation in the Adirondack Mountains. They didn't hear about the atomic bombs until they got back. The war was over, but it proved difficult for Shockley to extract himself from commitments to the War Department. He was still in demand, and constantly on travel. It was immediately clear that Shockley's small group needed more people, especially in theoretical solid-state quantum physics, and Shockley set about to find them. His first hire was John Bardeen.

John Bardeen (Fig. 5.9) was born on May 23, 1908, in Madison, Wisconsin. His father, Charles Bardeen, was the first Dean of the University of Wisconsin Medical School. His mother, Althea, had studied and briefly taught art. John was the second of four children, two years younger than his brother William, and was followed by a sister and another brother. It soon became apparent that John was different. "John is the concentrated essence of brain," wrote his mother.[32] John became his father's favorite, but John and William were boyhood friends, and remained fond of each other throughout their lives. Living near Lake Mendota, not far from the University, they enjoyed swimming and boating in the summer and ice skating in the winter, just like other kids. John started elementary school in 1914, when he was six, but after the third grade they enrolled him in University High. Elementary school was too easy. By skipping three grades, John became the youngest in his class, sometimes taking the same classes as his brother William. But even so young, John stood

---

[31] Riordan and Hoddeson (1997), p. 116.
[32] Riordan and Hoddeson (1997), p. 16.

**Fig. 5.9**   John Bardeen. (Creative Commons Attribution-Share Alike license)

out, especially in math. "John has undoubtedly a genius for mathematics," his mother wrote.[33] Unfortunately, in 1918, she was diagnosed with breast cancer. Following a radical mastectomy, she seemed to improve, but more cancer appeared. She traveled to Milwaukee and Chicago for treatment, returning in March 1920, when it appeared hopeless, but the children were not told how serious it was. "I remember stopping in to see her [in the hospital] on the day before she died," John recalled. "I thought she looked well that day, and cheerful, and I was shocked to hear the next day that she had passed away."[34] His father remarried that fall, but John was heartbroken and his studies suffered.

He finished University High in 1922, at the age of 14, but feeling not old enough to go off to college, he stayed for another year at Madison Central High School, taking more physics and math courses. He graduated in 1923, still only 15 years old. That fall, he entered the University of Wisconsin, where he could still live at home.

At Wisconsin, Bardeen majored in electrical engineering. Radio was just taking off around this time, and like many other kids, John had built himself a crystal radio set. With a tuning coil wrapped around a Quaker Oats box, a galena crystal, and a cat's whisker he could pick up signals from as far away as Chicago. He also liked electrical engineering for its good job prospects and its

---

[33] Riordan and Hoddeson (1997), p. 17.

[34] Riordan and Hoddeson (1997), p. 18.

use of mathematics, but finding it easy, he took additional courses in math and physics. He joined Zeta Psi fraternity and lettered in swimming and water polo. Always on the quiet side, this seemed to bring out another side of his personality. A summer job at Western Electric, that he extended into the fall, delayed his graduation to 1928, but after that he stayed at Wisconsin to get a master's degree in electrical engineering in 1930. Along the way he took more courses in physics, learning quantum mechanics from the likes of Van Vleck and Dirac.

He graduated in the depths of the Great Depression and took a job as a geophysicist with Gulf Oil company at their new research laboratory in Pittsburgh. There he worked with his old master's thesis advisor, Leo Peters, analyzing the earth's magnetic field to find new oil fields. But after three years, he got bored and decided to go back to school to get his doctorate. The only place he applied was Princeton, which readily accepted him. "I picked Princeton because there was an outstanding mathematics department there, as well as the Institute for Advanced Study. They had some of the leading mathematicians in the world and some of the leaders in theoretical physics," including Albert Einstein and John von Neumann.

At Princeton, Bardeen joined the group working under Wigner, which included Shockley's friend Seitz, with whom he became fast friends. He also played bridge with Walter Brattain's brother Robert, who introduced him to Walter (the same people keep popping up; the world of physics was small back then). Bardeen was always soft-spoken and modest, and he was comfortable with the Brattains, whose rural upbringing showed in their unassuming manner.[35]

For his thesis, Wigner had Bardeen calculate the work function of sodium. The "work function" is the energy needed to extract an electron from the metal, which had played a central role in Einstein's theory of the photoelectric effect in 1905. Einstein never calculated the work function since there was no quantum mechanics back then, but his theory offered the first proof that quanta of light existed. But to actually calculate the work function using quantum mechanics required not only the laborious computations that had absorbed so much time and energy from Seitz and from Shockley, but had added complications associated with the behavior of the electrons near the surface. The negatively-charged electrons tend to spread outside the surface of the sodium, leaving a net positive charge just inside the surface. These layers of positive and negative charge at the surface complicate things. Later, when

---

[35] Hoddeson & Daich (2002).

he got to Bell Labs, his understanding of electrons near surfaces, which he had gained from so much labor at Princeton, would help him understand why Shockley's first ideas for a transistor didn't work.

In 1935, when he completed these calculations, he was offered a Junior Fellowship at Harvard University. This is an enormously prestigious opportunity. "Junior" doesn't mean lesser, it merely distinguishes the students chosen by senior members of the Harvard Society of Fellows. The Junior Fellows are given financial support for three years while they conduct independent research at Harvard in any discipline, with no formal requirements or grades. Other Junior Fellows have included well-known and important figures like the behaviorist B. F. Skinner; economist Paul Samuelson; historian Arthur M. Schlesinger; presidential advisor McGeorge Bundy; linguist and activist Noam Chomsky; biologist E. O. Wilson; economist and whistleblower Daniel Ellsberg; and James Fisk, with whom Bardeen would work at Bell Labs. They comprised a remarkable crowd. Bardeen was on an upward trajectory.

While he was at Harvard, Bardeen got the opportunity to get to know Shockley, Shockley's advisor Slater, and others at MIT, which is just down Massachusetts Avenue from Harvard. As a Junior Fellow, he worked with the theorist Van Vleck and the experimenter Percy Bridgeman, both Nobel Prize winners. By the time he got to Bell Labs, his experience as an engineer and his experience with Van Vleck and Bridgeman gave Bardeen, who was himself a theorist, the ability to work with experimenters. This would prove crucial. He also had the chance to see Jane Maxwell, a biologist he had met and dated while he was working at Gulf Oil in Pittsburgh, who had taken a teaching job in Wellesley, just west of Cambridge. "I saw a great deal of her during my last year as a Junior Fellow," Bardeen recalled.[36] They were married in July, 1938, just as Bardeen finished his Fellowship, and the couple moved to Minnesota.

At Minnesota, Bardeen took a position as Assistant Professor of Physics, but before long the clouds of war began to appear. Following the fall of France in the spring of 1940, the USA began seriously to prepare for war. Bardeen was called to head the group working on magnetic mines and torpedoes at the Naval Ordnance Laboratory near Washington, DC, from 1941 to 1944. While he was there, Jane gave birth to a son and a daughter.

By the end of the war, Shockley had formed a small but growing group to do fundamental research in solid-state physics at Bell Labs. Brattain, Pearson, and a chemist Robert Gibney were already aboard, but they needed another good theorist. Having met Bardeen while he (Shockley) was at MIT and Bardeen was at Harvard, Shockley knew he had his man, and Kelly made him

---

[36] Riordan and Hoddeson (1997), p. 80.

an offer at double the money he would get back at Minnesota.[37] With a family to support, he took the offer and reported to Bell Labs on Monday, October 15, 1945.

A week later, Shockley gave him his first assignment: find out why the "field effect" that Shockley had predicted didn't work to control the current in the silicon. It was characteristic of Shockley to assign people to work on his (Shockley's) own ideas. When he had conceived the idea six months earlier, Shockley had calculated a large effect, but experiments hadn't worked. In the next couple of weeks, Bardeen used his own methods to compute the size of the effect he would expect from an experiment, and came up with results similar to Shockley's: there should be a large effect, but there wasn't. Then he remembered the surface effects that he had calculated for his Ph.D. thesis on the work function of sodium. Perfect! Perhaps enough electrons were getting trapped in surface states to cancel the applied fields.

## The Christmas Present

At this point, the group switched from testing semiconductor devices to understanding fundamental solid-state physics. They all collaborated closely: Bardeen worked on the theory while Brattain did experiments on surface effects suggested by Bardeen's theory, contributing his own insights as well. At the same time, Pearson investigated the bulk properties of silicon, and another material, germanium. Sitting just below silicon in the periodic table, germanium had shown promise as point-contact detectors in experiments at Purdue during the war. Shockley, the group leader, poked his head in once in a while, but largely pursued his own interests. Rapport, sharing, and excitement characterized the group, and by February, 1947, they had made enough progress for Bardeen to publish his results in *Physical Review*, a top journal of physics in the U.S. After traveling to Europe with Bardeen that June to visit labs there, Shockley was able to brag "we are quite far ahead on [the] theory of rectification."[38] In August, Brattain was able to publish his experimental results on surface states in silicon and germanium in the same journal. But with the demise of his idea for a "field-effect" device, Shockley's interest in semiconductors declined, and following the trip to Europe he became more interested in the phenomena that make metals soft while other crystals are

---

[37] Riordan and Hoddeson (1997), p. 118.
[38] Riordan and Hoddeson (1997), p. 128.

rigid. Meanwhile, Bardeen and Brattain continued to work on the properties of semiconductors.

Then, in November, in a series of experiments, Brattain discovered that a film of liquid electrolyte, even water, could be used to control or even cancel the electrons trapped in the surface states by varying the voltage on the liquid. A couple of days later, Bardeen suggested immersing a point-contact rectifier in a drop of water, using wax to insulate the point contact from the water, and varying the voltage applied to the liquid to see if this would modulate the current in the point contact.[39] Brattain immediately fashioned a simple experiment. When they tried it that afternoon, it worked! It was the world's first solid-state amplifier or, at least, the first non-vacuum-tube amplifier. The electrolyte was, after all, a liquid, not a solid. On his way home, Brattain told his car-pool that "I'd taken part in the most important experiment that I'd ever do in my life."[40] During the succeeding weeks, the group frantically tried different configurations using both silicon and germanium with various liquid electrolytes. But the effects were always small and limited to very low frequencies. Then, on December 9, 1947, they tried a crystal of what they called "high-back-voltage" germanium and got a power gain of 6000! But the frequency response was still slow, and the sign of the voltage on the electrolyte was wrong. Later, they figured out that this was due to holes replacing the electrons normally responsible for conductivity in their germanium. In other experiments using a solid electrolyte, they got only a small effect, but good frequency response.

Then, Bardeen suggested that they might be able to replace the electrolyte by putting a second point contact next to the first contact. But it had to be less than two-thousandths of an inch away, less than the thickness of a piece of paper, a distance even smaller than the point contacts themselves. To try it, Brattain came up with the clever idea of wrapping a strip of gold foil around the tip of a plastic wedge and slitting it with a razor blade to form two closely spaced electrodes. Pressed against the surface of a germanium crystal, it would form the two contacts that Bardeen needed. Brattain's crude apparatus is shown in Fig. 5.10. Connected individually, each gold foil contact behaved exactly as a point-contact rectifier should. The next day, December 16, 1947, they applied −10 V to one contact, shown here on the right, called the output circuit. Since this provided reverse polarity for the n-type germanium, no current flowed. But when they applied just +1 Volt forward polarity to the other contact on the left, current started to flow in the input circuit, and it appeared

---

[39] Shurkin (2006), p. 105.
[40] Riordan and Hoddeson (1997), p. 131.

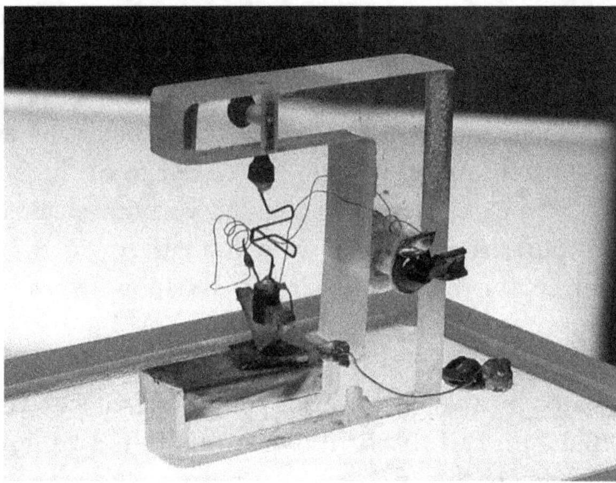

**Fig. 5.10** Photograph of Bardeen and Brattain's first experiment set in a clear plastic display frame. A squiggly wire forms a spring to press the tip of a small triangular plastic wedge against a germanium crystal on the bottom of the frame. (Creative Commons Attribution-Share Alike 3.0 Unported license. Original photo converted to black and white)

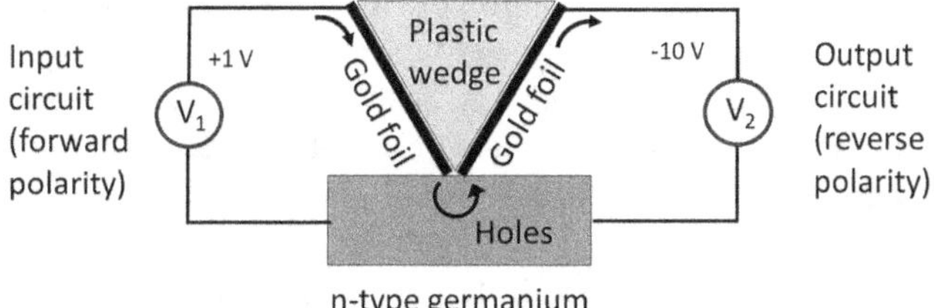


**Fig. 5.11** Schematic diagram of the first point-contact transistor. When the current flows in the input circuit (left side) the holes flow across to the other contact and current flows in the output circuit (right side)

in the output circuit. They now had a way of controlling the current in the high-voltage circuit from the low-voltage circuit! This was the solid-state amplifier, or switch, that they were looking for. When Bardeen got home that evening, he mentioned to Jane, "We discovered something important today." "That's great," she replied, and that was the end of the discussion for the evening.[41] John rarely discussed his work at home.

How did this first transistor work? A schematic diagram of the device is shown in Fig. 5.11. As shown there, there are two circuits, one on the left and one on the right. The output circuit on the right forms a point-contact

---

[41] Riordan and Hoddeson (1997), p. 138.

rectifier with the polarity reversed, so no current flows even with as much as −10 V. The input circuit on the left forms another point-contact rectifier, but this time with forward polarity so current flows in this circuit even with just 1 V. When current flows in the input circuit, holes are injected into the germanium from the gold foil. Then, due to the high voltage of the other gold foil, it sucks up the holes like a vacuum cleaner and current flows in the output circuit. In this way, the current in the input circuit controls the current in the output circuit. But now the voltage has been increased, in this case by a factor of ten, from 1 V input to 10 V output. Depending on the conditions, the increase can be much larger. If the input signal is modulated by a voice, for example, the signal (the voice) is amplified. But the transistor can also turn the current in the output circuit completely off, or on. In this way, the transistor works as an electrically controlled switch, just what Mervin Kelly was looking for to connect massive numbers of telephone calls.

The next day they called Shockley to tell him what they had found. Brattain thought he was excited by the news, but Shockley admitted later that:

> Frankly, Bardeen and Brattain's point-contact transistor provoked conflicting emotions in me. My elation with the group's success was balanced by the frustration of not being one of the inventors. I experienced some frustration that my personal efforts, started more than eight years before, had not resulted in a significant inventive contribution of my own.[42]

It was not enough for Shockley to be proud that his group had done it. He wanted not just to share the credit and the self-esteem; he wanted all of it. This was the start of problems for the group. Whereas before this the group had been open and cooperative, Shockley now became guarded and secretive. Things continued to get worse in the coming weeks, and since he was the group leader, he was in a position to cause trouble.

To demonstrate the use of their new discovery, Bardeen and Brattain connected a microphone to the input circuit and headphones to the output, and could hear their voices amplified. But cautious of announcing a discovery of such potential importance, a solid-state amplifier, they did more tests before showing it to Kelly. A week later, December 23, Bardeen, Brattain, Shockley and some others in the group gave a demonstration to Harvey Fletcher and R. Benjamin Brown, who were next up in the hierarchy at the Labs. All were amazed, but Brown suggested a final test: connect the output to the input to

---

[42] Shurkin (2006), p. 107.

form an oscillator, the same way a PA system squeals when the volume is turned up too high. With no other input, this would prove that the device really acted as an amplifier. They tried it the next day, which was Christmas eve, 1947, and it worked, an astonishing Christmas present to the Labs. For the next few weeks, they did some more tests before showing it to Kelly. He immediately put a lid of secrecy on the invention, and the patent attorneys got to work. Patents were important to Bell Labs not because they wanted to sell transistors; their product was communications, telephones and wires. But they didn't want anyone else to patent something they needed or use it to compete with their telephone monopoly. Today, multiple cell-phone and cable companies compete, using technologies developed in many different places. But it all started with Bell Labs' transistor.

The patent group soon found that the idea of a solid-state amplifier had been conceived and patented in Canada in 1925 by the Austrian-Hungarian physicist Julius Lilienfeld,[43] and in 1934 by German physicist Oskar Heil.[44] But these devices were different enough not to interfere with the present invention—later they would be called field-effect transistors—and neither inventor seems to have done any experiments. But who would be called the inventors of the point-contact transistor? Since Bardeen and Brattain had happened upon the point-contact transistor somewhat serendipitously while Shockley was working on other things, it was decided that Bardeen and Brattain's names would be on the patent. Moreover, Shockley's ideas on field effects were too close to those covered by the earlier patents of Lilienfeld and Heil, so he was left off. Shockley was furious. After all, he was the group leader and wanted the credit for himself. Management tried to smooth things over, Bardeen recalled. To stoke Shockley's ego, "Orders came down the line, presumably because he had contact with Kelly, that no pictures be taken of Bardeen and me without [Shockley] being present,"[45] as shown, for example, in Fig. 5.12. Shockley was, after all, Kelly's man, his first hire to fulfill his vision of a future without vacuum tubes, and a constant source of ideas. He would be the last to leave the physics group, and he would take many ideas with him. The patent application was finally filed in June, 1948, and the first public announcement of the invention was made the same month. Not surprisingly, this began to tear the group apart.

---

[43] United Stated Patent US1745175 Lilienfeld, J. Method and apparatus for controlling electric current; first filed in Canada on October 22, 1925.

[44] GB 439457 Heil, O. Improvements in or relating to electrical amplifiers and other control arrangements and devices; first filed in Germany March 2, 1934.

[45] Riordan and Hoddeson (1997), p. 156.

**Fig. 5.12**   Bardeen (left), Brattain (right), and Shockley (seated)

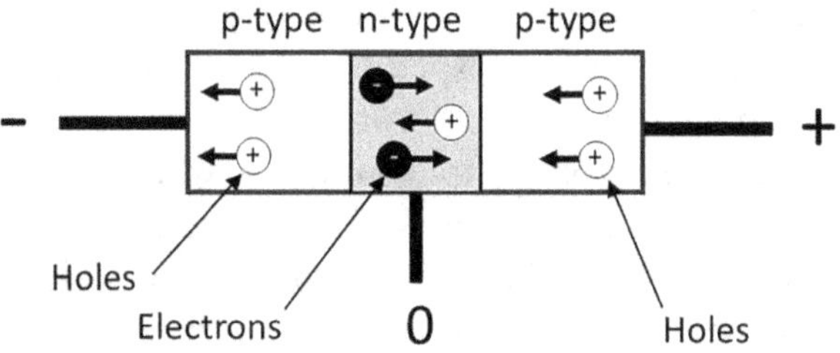


**Fig. 5.13**   p-n-p bipolar-junction transistor

Work continued on point-contact transistors, but they were noisy and fragile. Their high-frequency capability, like that of their point-contact rectifier precedents, made them useful, but something better was needed. Angered at being left off the patent, and recognizing the problems of the point-contact transistor, Shockley immediately went off by himself and began in secret to fill his notebook with other ideas. He wanted his name on a patent, and maybe at least a share of a Nobel Prize. Some of these ideas were related to the field-effect transistor, Shockley's original idea that he had thought of in 1945, but on January 23, 1948, he conceived the idea of a "bipolar-junction" transistor. Continuing his secrecy, he kept it from Bardeen and Brattain. Whereas the point-contact transistor looks like two point-contact rectifiers arranged side-by-side, as shown in Fig. 5.11, the bipolar-junction transistor looks like two p-n-junction rectifiers arranged back-to-back, comprising two p-type regions with a thin n-type region sandwiched in between, as shown in Fig. 5.13.

Accordingly, the junction transistor is sometimes called a p-n-p, or with the doping reversed, an n-p-n transistor. Its operation is similar to that of its predecessor, the point-contact transistor. The one junction has reverse polarity and allows no current. But when the other junction has forward polarity, current flows and holes are injected into the central region. If this region is thin enough, the holes are sucked up by the high voltage and reach the other side so current flows in the output circuit, just as it does in the point-contact transistor. If it could be made, and if it worked, the advantages expected of the junction transistor might be twofold: first, it should be more rugged than a point-contact transistor, and second, it should have less noise. But how could anyone fabricate such a device?

Remarkably, at the same time that Shockley arrived at this concept, John Shive, who by this time was working with Bardeen and Brattain, tried a new configuration for the point-contact transistor. Finding it difficult to place the two point-contacts close enough together on the surface of the germanium, he ground and etched a sliver of germanium to a thickness of about two thousandths of an inch and placed the point contacts on opposite sides of the germanium, forming the point-contact equivalent of Shockley's bipolar junction. When he tested it, he observed strong amplification. At Becker's urging, on February 18 he showed his results to a meeting of the group, including Shockley. Shockley was "startled when Shive presented his findings."[46] Realizing that Bardeen and Brattain were getting close to his concept of a junction transistor and afraid of getting scooped, Shockley quickly stood up and presented his own ideas and analysis to the group. It was immediately apparent to everyone that Shockley had been working on this for some time in secret, sharing nothing with the others in the group. The openness and rapport that had characterized the group earlier was shattered. It would never be restored.

Shive's results also showed that the holes could travel right through the bulk of the n-doped germanium. Until then it had been thought that the current traveled through a surface layer. This new understanding of the transistor proved that holes were emitted by the first point contact and collected by the second, as described above. Shive's result also showed that Shockley's concept might work. The challenge was to find a way to fabricate such a structure.

The semiconductor physics group now split into two camps, one with Bardeen and Brattain, who continued to work on the point-contact transistor and its basic physics, the other with Shockley and those who went with him, including Gerald Pearson. Administratively, the entire group fell under

---

[46] Riordan and Hoddeson (1997), p. 154.

Shockley, but since Shockley eschewed the administrative responsibilities, these continued to fall to Morgan, who kept the two groups apart. After Bardeen and Brattain's patent was filed on June 17, 1948, Shockley filed patents on his own concepts on June 26, nine days later.

The point-contact transistor was announced to the world in a press conference on June 30, 1948, held in the Bell Labs headquarters in New York. In preparation for the event, it was decided that they needed a new name for the device. "Semiconductor triode" and other names they'd been using in-house weren't "catchy" enough for something this important. After a search for names and much anguish, they finally decided on the name "transistor." All the wives attended the event, and there were official photographs and elegant dinners. Shockley gave the technical presentation and answered questions. He was probably the most comfortable speaker and best suited to give the talk, but like the photographs, this presented him to the world as the father of the invention, when at best he was the uncle.[47] Shockley, of course, wasn't about to do anything to dispel that notion. Shortly afterward, they published three scientific papers in the July 15 issue of *Physical Review*, one by Bardeen and Brattain on the point-contact transistor, one by Brattain and Bardeen on the physics of point contacts, and one by Shockley and Pearson on their field-effect experiments. But nobody outside Bell recognized the enormous importance of the invention, and the press reports were muted. The *New York Times* gave it only a few inches on page 46 of the July 1 edition.[48]

To keep the scientists focused on fundamental research, Kelly put development of an industrial version of the point-contact transistor in another group under Jack Morton. Kelly still believed in the importance of fundamental research and didn't want to lose that. Morton was an engineer who had successfully directed a project to develop tiny vacuum tubes for microwave transmission, and the group he formed included Becker and Shive. But initially, the transistors that the group developed showed large variations in their performance, sometimes as much as 50 percent. Brattain kibbitzed with them and developed some procedures that seemed like "witchcraft," but gradually the performance improved. By mid-1949, Morton's experimental production line had produced 3700 transistors for testing. They gave some 2700 of them to other companies for their tests, hoping to learn as much as possible, as soon as possible, about their problems and their potential uses. But point-contact

---

[47] Kessler (1997).

[48] *New York Times,* July 1.1948, p. 46.

transistors weren't destined to become a huge market. Fragile and noisy, they would be overtaken by Shockley's junction transistors.

While all this was going on, the split in the group grew wider. When new physics experiments by Dick Hanes proved that the charge carriers in n-type germanium were actually positive holes, the results were communicated to the rest of the solid-state physics group, but not to Bardeen and Brattain. Shockley shut them out of the new work. And when they all moved to new quarters in Murray Hill, Bardeen and Brattain's offices were on the floor below the others.

The first problem Shockley's group faced was how to fabricate a junction transistor. A p-n-p transistor required two segments of p-type germanium separated by an exquisitely thin section of n-type germanium, just two thousandths of an inch thick, the thickness of a piece of paper. That March, Morgan Sparks in Shockley's group literally used a saw to split a drop of p-type germanium on an n-type slab to form a junction device that looked much like Brattain's first point-contact transistor but with p-type germanium replacing Brattain's gold foil. It was just a proof-of-principle experiment, but it worked. The road to a real device went through some work Gordon Teal was doing on growing single-crystal germanium filaments. By slowly pulling a seed out of melted germanium, they could form a filament of high-purity germanium and by putting a tiny bit, less than 100 µg, of dopant in the melt they could produce p-type or n-type germanium. They then found that by adding p-type dopant to an n-type melt they could get p-type behavior, and by adding more n-type dopant they could get back the n-type behavior. By early 1950, they had learned how to make an n-p-n junction transistor with this double-doping technique, and with extremely slow pulling (10 s for the thin p-type layer) they could get p-type layers about as thick as a few sheets of paper. By etching away some of the n-type material, they were able to attach a lead to the p-type region, and on April 12, they tested it. It worked. The frequency response was poor, limited to audio frequencies, but thinner p-regions would improve that. By January, 1951, with even slower pulling they were able to make p-type regions only 1–2 thousandths of an inch thick. This expanded the frequency range to AM radio frequencies. Point-contact transistors could do much better in this regard, but the junction transistors had thousands of times less noise.

Shockley announced the existence of the junction transistor in talks he gave late in 1950, and on May 24, 1951, Shockley, Sparks and Teal submitted a technical paper to *Physical Review* that was published in July:

Used as "n-p-n transistors" these units will operate on currents as low as 10 microamperes and voltages as low as 0.1 volt…Their current-voltage characteristics are in good agreement with the diffusion theory.[49]

The official public announcement was made on July 4, 1951, in the same West Street auditorium that had been used to announce the point-contact transistor three years earlier. The talks touted the low noise and high efficiency, a million times more efficient than vacuum tubes. This was important because heat removal was one of the problems that made vacuum tubes unworkable in large switching systems. None of the talks mentioned how they fabricated the transistor; for a while, this remained a Bell Labs secret.

By this same time, point-contact transistors had been improved to the point that they were satisfactory for use in the Bell Telephone system. New techniques and high-purity materials had reduced the noise and variability to acceptable levels. Fabricated by Western Electric, the manufacturing branch of Bell Telephone, they entered service in 1952. By the mid-1950s, military needs for transistors became so great that the military funded half the development costs at Bell Labs and underwrote construction of a Western Electric plant in Pennsylvania. This, in turn, gave the government rights to the ideas, and they supported the development of production capabilities at General Electric, RCA, Raytheon, and Sylvania. Now there was a whole new industry to fabricate solid-state components and systems to use them. The first application of the new transistors was in hearing aids in 1952. Pocket radios would soon follow.

## The Breakup

Brilliant scientists with inventive minds are not always cut out to be managers. Sometimes they can understand semiconductor physics, but not people. Shockley was one of these, and people soon realized that they couldn't work with him. Unfortunately, it took a while for Bell Labs management to understand the problem, and by the time they did, it was too late.

Throughout the development of the junction transistor, Bardeen and Brattain continued to be shut out by Shockley and they were limited to research on the old point-contact transistors. Moreover, the fundamental research for which Kelly had formed the group in the beginning had now become device-oriented development under Shockley. This wasn't Bardeen's

---

[49] *Shockley, W., Sparks, M., & Teal, G.K. (1951).* p-n junction transistors. *Physical Review, 83, 151.*

kind of work. Taking up the mandate to do fundamental research on anything that could be potentially useful to Bell Labs, Bardeen shifted his interest to superconductivity.

Superconductivity had been discovered by the Dutch physicist Heike Kamerlingh Onnes in 1911. In 1908, Kammerling Onnes had been the first to liquify helium at the near-absolute-zero temperature of 4.2 °K, and had begun to use it to explore the physics of matter at this temperature. As he lowered the temperature of mercury, he measured its resistance and saw it decrease as expected as the temperature got lower. Then, suddenly, at a temperature of 4.2 K, the resistance vanished. It didn't get smaller, it completely vanished! Electrical currents just kept going forever, without any electric field. But nobody knew why; superconductivity had come as a surprise, and it remained a complete mystery. It was a great challenge for Bardeen (and others), and a superconductor might someday be important for Bell Labs. But since superconductors require extremely low temperatures, superconductivity seemed less interesting and immediate than semiconductors had been.[50] Although Bell Labs, and Kelly in particular, was willing to let him pursue this work, there was no one else interested in it to talk to. In 1950, Bardeen began to think about leaving.

In October of that year, he ran into Seitz at a conference, and told him how he felt. Did Seitz, who by now was a professor at the University of Illinois, have any ideas? Seitz went back to Urbana and asked around. For a physicist of Bardeen's stature, they had to find a way to hire him. In March 1951, Illinois offered Bardeen a joint appointment in Physics and Electrical Engineering, but at a salary lower than what he was getting at Bell Labs. Bardeen was concerned about the teaching load, but didn't care about the salary.

Back at Murray Hill, management began to realize that they might lose a brilliant theorist. In fact, it wasn't just Bardeen. Dissatisfaction with Shockley's dictatorial style of micromanagement was widespread, but the Bell Labs management always backed Shockley. Finally, Bardeen and Brattain complained to Fisk, who was by then Shockley's supervisor. Fisk split the group so that they no longer had to report to Shockley. But even this wasn't enough, and despite a counter-offer of more salary and his own group, three years after he and Brattain discovered the point-contact transistor, Bardeen accepted the offer from Illinois at the end of April. In a memorandum to Kelly dated July

---

[50] Recent research has led to superconductors that work at higher temperature, above the temperature of liquid nitrogen, but nothing has yet been discovered that is superconducting at room temperature and ordinary pressure.

24, 1951, Bardeen wrote that Shockley involved them "only as he thought of problems of his own that he wanted investigated experimentally…In short, he used the group largely to exploit his own ideas."[51] At Illinois, Bardeen would be able to carry out his research among colleagues and students who valued his work. This move couldn't have worked out better. In 1972, Bardeen shared a second Nobel Prize in Physics with two of his students for the theory of superconductivity.[52] He remains the only person ever to get two Nobel Prizes in Physics.

Now that he was no longer under Shockley's thumb, Brattain stayed on at Bell Labs. He finally retired in 1967, aged 65, and became an adjunct professor at his old alma mater, Whitman College, where he stayed until 1972. He died in 1987 of Alzheimer's disease.

At home, Shockley's family situation became even more difficult. Problems had begun in World War II when Shockley was off on long trips for his work in operations research, but they only got worse over time. Then, in February 1953, Jean was diagnosed with uterine cancer.[53] After radiation therapy and surgery, Jean was released on April 1st, but Shockley wasn't satisfied. He used his influence to reach the best oncologists in America, and threw himself into research on cancer treatments. It seemed to him like another intellectual challenge. Oncology wasn't an exact science, and opinions varied. Shockley decided that massive radiation after surgery was the best idea, and that summer Jean was treated in New York. Since she survived, this may have been the right decision, but while she was undergoing treatment, Shockley visited her in the hospital and informed her that he was going to leave her.

While all this was going on, in addition to the work on junction transistors, Shockley vigorously pushed other ideas, many related to his 1945 idea for a field-effect transistor. Despite being elected to the National Academy of Sciences in 1951, one of the youngest persons ever to receive this honor, Shockley was jealous of being left off the transistor patent and pressed the patent department to file new patent applications. But a report that had been authored in part by Bardeen showed that the device patented by Lilienfeld would work and was similar to Shockley's, so the patent department abandoned Shockley's field-effect transistor. This didn't help relations between Shockley and Bardeen.

---

[51] Kessler (1997), p. 16.

[52] Bardeen, J., Cooper, L.N., & Schrieffer, J.R. (1957). Microscopic theory of superconductivity. *Physical Review* 106 (1): 162–164.

[53] Shurkin (2006), p. 134.

Bardeen and Brattain weren't the only ones who were unhappy working under Shockley. Next to leave was Gordon Teal, who with Morgan Sparks, had fashioned the first successful junction transistor from filaments of high-purity germanium. When Teal began his research on high-purity filaments, Shockley had refused to support it since he "considered this unnecessary for our research."[54] Teal was forced to do the research at night, working into the early morning hours. According to Teal, Shockley "was pretty pig-headed on this."[55] Responding to an ad in the *New York Times* in 1952, Teal was invited to Texas Instruments for a visit. At that time, TI was a small company that was making germanium point-contact transistors under license from Western Electric and was beginning to make junction transistors as well. But silicon has some advantages over germanium. Although its higher melting temperature makes silicon more difficult to work with than germanium, it is less sensitive to temperature. Moreover, in the "off" state, silicon transistors wouldn't "leak" current the way germanium transistors did. For most switching applications, when a switch is "off," it needs to be completely off, not just mostly off. With this in mind, Teal had already begun research on silicon, had fabricated p-n junctions in silicon, and had published his results. He was just the guy for TI.

At TI, Teal got the support and encouragement he needed. That was good, because it took more than a year after he got there to make the first silicon junction transistor, and the fear of being beaten by a competitor such as Raytheon was palpable. As soon as he succeeded, TI ramped up production. Finally, on May 10, 1954, Teal brought three of his transistors to the National Conference on Airborne Electronics. He was the next-to-the-last speaker that afternoon, and as he recalled: "One after the other" the preceding speakers had "remarked about how hopeless it was to expect the development of a silicon transistor in less than several years."[56] When Teal took three of his production silicon transistors out of his pocket and demonstrated their ability to operate in hot oil, the audience was shocked. They rushed to the back of the auditorium to get copies of his paper. The final speaker addressed an empty room.

Before long, TI developed the first transistor radio, the Regency TR1 pocket radio, shown in Fig. 5.14. At a list price of $49.95, they sold only 1500 the first year, 1954, but over 100,000 in the following year. It was sold out everywhere, but the price wasn't high enough for TI to make a profit. The

---

[54] Riordan and Hoddeson (1997), p. 174.
[55] Riordan and Hoddeson (1997), p. 174.
[56] Riordan and Hoddeson (1997), p. 208.

**Fig. 5.14** Regency TR1 transistor radio. (Creative Commons Attribution-Share Alike 2.0 Generic license. Original figure converted to black and white)

big break came when Thomas J. Watson, the president of IBM, gave the radios to his top executives, admonishing them that if a little company like TI could do that, IBM could, too. Two years later, TI supplied the transistors for IBM's first fully transistorized computer. First demonstrated in October 1954, the IBM 604 used over 2200 transistors, compared with 1250 tubes in the vacuum-tube version, the IBM 603. But the 604 consumed only 5% as much power.

Meanwhile, in postwar Japan, Masaru Ibuka, Akio Morita and their little company Totsuko, applied for a license from Western Electric to manufacture transistors. At $25,000, it was a big investment for a little company that made audio recorders for a niche market. At first, both Western Electric and the Japanese bureaucrats in the Ministry of International Trade and Industry blocked them because of their small size (120 employees), but they finally got the license. In early 1954, they made their first junction transistor and designed a pocket radio that used four of them. With this little radio, the struggling company found a market much larger than their audio recorders. It seems to have been a success. A few years later, they changed the company name from Totsuko to Sony.

But junction transistors at that time were still limited in their high-frequency performance, unable to reach the 100-MHz range of FM radio, which was becoming popular by then. Fortunately, new processes were being developed at Bell Labs to make thinner gaps between the emitter and collector regions. The most successful method used diffusion: beginning with an extremely pure semiconductor crystal, a vapor containing the dopant was allowed to diffuse into the surface of the crystal. Besides making it possible to fabricate layers thinner than a human hair, this technique could be used to make large area junctions, in contrast to the filaments that could be drawn from a melt. The problem is that diffusion in solids requires high temperatures. In silicon, the required temperature is about 1300 °C, hot enough to glow orange. But by early 1954, Calvin Fuller and Gerald Pearson could make a p-n junction the size of a postage stamp. With its large area, it functioned as a solar cell. On the front page of the *New York Times* for April 26, 1954, the headline hyped "Vast Power of the Sun is Trapped by Battery Using Sand Ingredient."

## Dreams of Riches

The only person who didn't recognize William Shockley's limitations as a manager was William Shockley. Unable to get the recognition and rewards that he felt he deserved, Shockley decided that Bell Labs wasn't big enough for him, and he looked for alternatives. Maybe he could even become a billionaire. In the end, he created many billionaires, but he wasn't one of them.

By the mid-1950s, William Shockley was stuck at Bell Labs and having a mid-life crisis. At the Labs, he was a brilliant theorist and a visionary inventor, but Bell Labs had finally come to recognize that he was a terrible manager. While others, including Fisk and Brown, were getting promoted above him, Shockley remained locked in middle management. In the spring of 1951, when he left for the University of Illinois, Bardeen sent Kelly a scathing memo about Shockley's management style. A year later in 1952, when Teal left for TI, Bell Labs management finally woke up and reorganized, leaving Shockley in place leading research in a middle-management position. He complained to his mother, "Recognition of outstanding individual contribution is frequently lacking when organizational actions are taken."[57] Feeling left out and somewhat at sea, in 1954, Shockley took a leave of absence from Bell Labs. The first thing he tried was a temporary teaching position at Cal Tech, but he didn't find that satisfying either.

Then the Pentagon called, pulling him back into operations research, this time as Research Director of a new Weapons Systems Evaluation Group. While he was in Washington, DC, in March, 1954, to discuss the new position, he

---

[57] Riordan and Hoddeson (1997), p. 225.

met Emmy Lanning, who taught nursing at a psychiatric facility in nearby Rockland, Maryland. She had the temerity to challenge him, and she understood people. They soon became soulmates, giving Shockley another reason to spend time in Washington. His life now resembled the time he had spent in Washington during the war when he only went home to see Jean and the kids a couple of weekends each month. But in January, 1955, Emmy left Washington to take a position teaching nursing at the Ohio State University, and Shockley began to lose interest in his operations research. Meanwhile, Shockley's deteriorating marriage became more fraught: two years after her first cancer, Jean's cancer returned, and the prognosis was not good. And while he was away from Bell Labs, exciting developments were happening without him.

In early 1955, Morris Tanenbaum, in Shockley's group, succeeded in making the first good junction transistor using the diffusion technique. Returning quickly from Europe, Morton canceled development of all other fabrication techniques and focused completely on diffusion. With their thin layers, the diffused-base transistors would operate at high frequencies, beyond 100 MHz. Bell Labs went public with these developments in June 1955, at the IRE Semiconductor Device Conference in Philadelphia. TI and Sony were both there. Unwilling to work for someone else, Shockley decided to strike out on his own.

Unable to find a university position that would interest him on his terms, he began to think about starting his own company. On June 1, 1955, he wrote to Emmy: "Think I shall try to raise some capital and start on my own." He continued, "After all, it is obvious I am smarter, more energetic and understand people better than most of these other folks."[58] This seems a rather remarkable self-assessment, considering what had gone on, and what was yet to come. To his mother, he wrote: "I am having a fine time, now, what with [my role at] the Pentagon coming to an end and lots of people willing to back me up in a new venture to the tune of $500,000 plus."[59] But it wouldn't be that easy; Shockley was turned down in many places, including Bell Labs, before he found Arnold Beckman.

Meanwhile, as she waited for a divorce in Reno that August 1955, Jean wrote to Shockley's mother May, "I had planned the summer quite otherwise, but Bill changed his job and felt it desirable to change his family status at the same time."[60]

---

[58] Riordan and Hoddeson (1997), p. 232.

[59] Riordan and Hoddeson (1997), p. 233.

[60] Riordan and Hoddeson (1997), p. 233.

Beckman (Fig. 5.15) was a successful scientist turned businessman whom Shockley had met in the 1930s when Beckman was a chemistry professor at Cal Tech. By this time, Beckman Instruments was grossing more than $20 million/year. When Shockley called in August 1955, Beckman invited him out to California. After a week of discussions, they had an agreement: the initial project would be to produce diffused-base transistors using Tannenbaum's process, and Shockley would get "maximum personal satisfaction…and financial reward commensurate with performance…."[61] Shockley flew back to Washington to get his belongings, hopped into his Jaguar, and picked up Emmy in Columbus. They were married on November 23 in a simple ceremony with no friends present. Shockley then continued on to California; Emmy finished up in Columbus and flew to join Shockley at Christmas time. They remained faithful and supportive to one another until the end of their lives.

Beckman wanted the new venture to be located in the Los Angeles area, where Beckman's other industries were located, but Shockley preferred San Francisco, his home grounds. When Frederick Terman, the Stanford provost and Dean of Engineering and fellow member of the National Academy of Sciences, offered Shockley a location near the Stanford Campus, that settled it. Among other things, Terman agreed with Shockley's philosophy to hire only top people at top salaries, rather than many lower-level people: "Better to have one seven-foot jumper on your team than any number of six-foot

**Fig. 5.15**  Arnold Beckman. (Courtesy of the Arnold and Mabel Beckman Foundation; used with permission)

---

[61] Riordan and Hoddeson (1997), p. 234.

jumpers."[62] Shockley immediately began recruiting, and by February, 1956, when the venture was formally announced, he had four scientists and engineers. But they would learn too late that even on a sports team, top players who don't get playing time soon leave.

At first, Shockley tried to raid Bell Labs for their top people, but to Shockley's surprise, they all turned him down. The families didn't want to leave New Jersey, they told him. After that, he tried other top companies and universities. In February 1956, he recruited Robert Noyce, then 29 years old, a Ph.D. from MIT, who at Philco had developed the highest-frequency transistors available at the time. "It was like picking up the phone and talking to God," Noyce recalled[63]; Shockley was that big a draw. But before he would hire him, Shockley had one unusual, quirky requirement: Noyce would have to take a series of psychological tests, IQ tests, ink-blot tests, and other tests to see if he would fit in with the group that Shockley envisioned. Noyce passed the tests and was hired. Next on Shockley's list was Gordon Moore, 27 at the time, a physical chemist from Cal Tech working at Johns Hopkins. When Shockley interviewed Moore, he grilled him with a series of questions and timed his answers with a stopwatch. Remarkably, these youngsters were so excited to work with Shockley that they accepted these bizarre encounters. Noyce and Moore arrived at Shockley Semiconductor Laboratories in April, 1956. By June there were more than 20 people. All but one were under 30. Shockley was very exciting to young people, according to Termin, "but hard as hell to work for." Shockley was demeaning and insensitive. R. V. Jones found it so insulting that he resigned after only a couple of weeks. Others threatened to follow him.

By this time, the Justice Department had forced Bell Labs to give up its rights to the basic patents for the point-contact and junction transistors, since the government had supported the development, but Bell retained the process patents for making diffused-base transistors. Beckman licensed these rights, but even with Shockley's inside connections to Bell Labs, they were unable to fabricate anything. The requirements for the diffused junctions were just too demanding. Shockley's management style didn't help. "Working with Shockley turned out to be really interesting," said Moore, "But he had some peculiar ideas about motivating people."[64] When the young staff members expressed a need to publish their work in order to further their careers, Shockley gave them his own ideas to flesh out and publish. Sometimes he took them off

---

[62] Riordan and Hoddeson (1997), p. 235.

[63] Riordan and Hoddeson (1997), p. 237.

[64] Riordan and Hoddesonm (1997), p. 241.

whatever they were engaged in to work on his latest pet idea. He even put his top scientists on the production line just for the experience.

That fall, on November 1, 1956, the Swedish Academy announced that Shockley, Bardeen, and Brattain had been awarded the Nobel Prize in Physics for inventing the transistor. Shockley first heard about it from a reporter who called. Bardeen heard about it on the radio as he was scrambling eggs, and when Brattain heard about it, he was skeptical and went to the Labs, where he got confirmation from his excited colleagues. The *New York Times* put a brief note on the front page and the Wall Street Journal didn't even mention it, but there was jubilation in Palo Alto, Urbana, and Murray Hill.

Bardeen and Brattain flew together to Sweden with their wives and Bardeen's son. Stopping in Copenhagen on their way, they visited Niels Bohr, and arrived in Stockholm on the 6th of December for the ceremony four days later. Before the ceremony there were elegant dinners and receptions, concerts, visits to institutes and schools and almost anything the Swedes could offer. Shockley flew with Emmy and his mother, May, and arrived just in time to make it to the ceremonies. The Nobel ceremonies themselves are very formal and rich in tradition, as one might expect, and the awards were handed to the Laureates by King Gustav VI Adolf of Sweden himself. After the awards ceremony, back at the hotel, Bardeen and Brattain invited Shockley and Emmy to join them and their wives, and they celebrated together for the first time in a long while, and for the last time ever.

Back in Palo Alto, Shockley had to get down to business. It was almost a year since the company started, and they hadn't yet produced a transistor they could sell. Already the grumbling had started among the employees. Shockley reacted with resentment and paranoia. Moore describes an incident that happened that March, when a secretary gashed her hand on a sharp object stuck in her door. Shockley suspected sabotage and began to investigate a pair of technicians that he accused of disloyalty, subjecting them, among other things, to polygraph tests. When they were cleared, he accused Sheldon Roberts, one of his first hires. Roberts responded by examining the sharp object under a microscope: it was the tip of a push-pin. Moore recalled that "Someone had pinned something up there with a bad pin, and the head had fallen off."[65]

By this time, Beckman was becoming concerned that Shockley Semiconductor Lab was costing money, producing nothing, and Shockley was the problem. But when Beckman visited the Lab in late May and proposed some changes, Shockley fired back "I can take this group and get

---

[65] Riordan and Hoddeson (1997), p. 247.

support any place else!" and left the room in a huff.[66] Back at the Lab, eight of the top scientists, including Moore, got together and resolved to contact Beckman directly. Responding to a call from Moore, Beckman came up to the Bay Area to meet with them. They informed him that Shockley had to go or they would, and they wouldn't go anywhere with Shockley. Not only was it difficult to produce diffused-base transistors, this was no longer Shockley's principal interest. He was totally obsessed with his four-layer, n-p-n-p "Shockley diode."

Known nowadays as a thyristor, the Shockley diode can exist in either of two states, on or off, and remains in that state until it is triggered. It still finds use in high-power applications, but for microelectronics the same function can be satisfied by two tiny transistors in a so-called "flip-flop" circuit, which is much simpler and smaller. By taking the place of several components, the Shockley diode is like a small "integrated circuit," but it completely missed the coming revolution. In any case, it was Shockley's idea, conceived while he was still at Bell Labs, and to him that was of overriding importance. It also reflects another important point: Shockley was, by this time, 46 years old. Scientists, and especially theorists, are most productive and innovative in their early years. Newton was only 24 when he invented calculus, Einstein was 26 when he discovered the theory of relativity, and Heisenberg was 24 when he conceived quantum mechanics. Shockley in 1956 was past his most creative years. The future would have been better left to the younger scientists in his lab. On top of this, Shockley was a lousy business manager, and contrary to his own self-image, he did not understand people. He was cruising on his reputation and his Nobel Prize.

Following the meeting with the eight dissidents, Beckman called Shockley with a new plan. On June 10, he met with Shockley and the eight to discuss it: the lab would split into two groups. Moore and the other disgruntled employees would continue to work on developing diffused-base transistors, while Shockley and a few loyalists would move to a nearby facility and work on the Shockley diode. Administratively, a committee would sit under Shockley until a business manager could be found, but Shockley would have the final say. He would remain "the boss." Out of loyalty, Beckman couldn't deprive him of that. But the two groups continued to diverge; it was just an uneasy truce and wouldn't last. It was a decision that Beckman would come to regret.

By midsummer, Noyce, Moore and the others were serious about leaving. Since they worked so well together, they looked at first for a firm that would

---

[66] Shurkin (2006), p. 176.

pick them all up and keep them together, but they couldn't find one. Then someone suggested that they should form their own company. Through his family in New York, Eugene Kleiner, an engineer in the dissident group, was put in touch with the investment firm of Hayden, Stone & Co, who took an interest. Since at that time most of the money and technology was on the East Coast and they wanted to stay in Mountain View, it was difficult to find an investor, but Sherman Fairchild, founder of Fairchild Aircraft and Fairchild Camera, stepped up with a loan of $1.38 million to found Fairchild Semiconductor. On September 18, 1957, eight of the brilliant scientists and engineers that Shockley had recruited all resigned, including Noyce, Moore, and Roberts. Except for a small group working on the Shockley diode, there was nothing left. Shockley was stunned, and called them the "traitorous eight." In his notebook, he merely noted "Wed 18 Sept—Group resigns."[67]

Shockley continued to work on the Shockley diode, but with four layers it was difficult to fabricate. By 1958, Beckman had poured another million dollars into Shockley Transistor Corporation, as it had become known, but while Shockley could produce hundreds of the devices each day, their performance was variable and unreliable. There was no market for them. And an effort to make his field-effect transistors was getting nowhere. In April, 1960, Beckman sold Shockley Transistor to Clevite Transistor, in Waltham, Massachusetts. As the losses continued to mount, in 1968, Clevite sold the division to the telecommunications company ITT, where it disappeared. As Seitz recalled of Shockley,

> I called him the Moses of Silicon Valley. He led his people there, but never made it himself into the promised land.[68]

Meanwhile, Fairchild Semiconductor was making millions of dollars with the people that Shockley had driven off. With bright, young minds, and better management, they were soon producing diffused-base transistors. By December 1958, they had sales of over half a million dollars and were making a profit. And they were ready for new ideas. Across the country, in June 1958, at a press gathering in honor of the tenth anniversary of the invention of the transistor, Jack Morton remarked, "Perhaps the safest prediction one can make is that transistor electronics has a great future—that it will go in new

---

[67] Riordan and Hoddeson (1997), p. 251.

[68] Hiltzik, M.A. (2001, December 2). The twisted legacy of William Shockley. *The Los Angeles Times*.

directions we cannot foresee today at all."[69] The next revolution was about to happen.

## The Tyranny of Numbers

As the number of transistors in a device increased from dozens to thousands and millions, problems arose that even junction transistors couldn't deal with. New ideas were needed. They came in the form of integrated circuits and field-effect transistors, and they came from innovative, agile organizations, spin-offs populated with people who could "think outside the box." Soon, devices that once filled a room could fit in your pocket, and the end wasn't yet in sight.

By avoiding the heat, size, and fragility of vacuum tubes, transistors made it possible to think about systems of increasing complexity. The first transistorized computer was built in November, 1953, by Tom Kilburn, Richard Grimsdale and Douglas Webb at Sussex University.[70] It used 92 point-contact transistors and 550 diodes. By 1958, IBM introduced the 7070. It had about 30,000 germanium junction transistors and 22,000 germanium diodes, individually mounted and wired together by hand on 14,000 circuit boards. It was as tall as a man and weighed 10 tons. A large part of the size and weight owed to the cooling that was required even by junction transistors. It was almost impossible to imagine a system much larger than this. This was called the "tyranny of numbers," and new technology would be needed. It came in two forms: the size was addressed by the invention of integrated circuits, and the heat was addressed by the development of metal-oxide-semiconductor field-effect transistors, called MOSFETs.

As the number of components in a computer or other device increased, it became necessary to break the circuit down into modules, which were placed on circuit boards, but even these had their limits. Working alone one July, while the rest of TI was on vacation, Jack Kilby had an inspiration: "Extreme miniaturization of many electrical circuits could be achieved by making resistors, capacitors, and transistors & diodes on a single slice of silicon."[71] By the time his boss, Willis Adcock, returned from vacation, Kilby had figured out how to make resistors from the natural resistance of the silicon and capacitors

---

[69] Riordan and Hoddeson (1997), p. 255.

[70] Kilburn, T., Grimsdale, R.L. &. Webb, D.C. (1956). A transistor digital computer with a magnetic-drum store. *Proceedings of the IEEE Convention on Digital Computer Techniques* 103 (Part B).

[71] Riordan and Hoddeson (1997), p. 258.

from a p-n junction. Adcock suggested he try it, using germanium instead of silicon, since by then TI didn't yet have diffused-base technology in silicon. Kilby put together the simple circuit shown in Fig. 5.16, an oscillator with one transistor, connecting the components together with gold wire, and demonstrated it to Adcock and a few others on September 12, 1958. It oscillated at 1 MHz. A week later, he had a flip-flop, using two transistors and resistors on a single chip. Soon, others at TI were using silicon instead of germanium. On February 6, 1959, TI filed a patent for "Miniaturized Electronic Circuits," claiming

> In accordance with the principles of the invention, the ultimate in circuit miniaturization is attained using only one material for all circuit elements and a limited number of compatible process steps for the production thereof.[72]

This was a step toward the new technology needed to address the "tyranny of numbers."

Meanwhile, at Fairchild Semiconductor, Noyce was thinking about using an insulating layer of silicon dioxide to protect their devices, when it occurred to him that he could etch through the silicon dioxide to the components underneath and connect them with wires printed right on the insulating layer. Now, with the precision available from photolithography, hundreds of components could be placed on a single chip and wired together automatically.

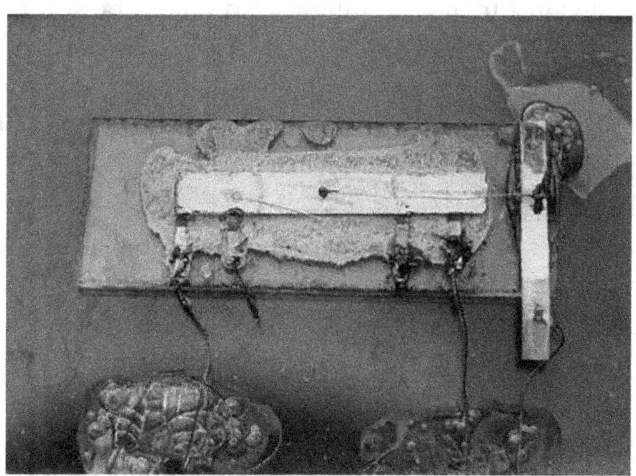

**Fig. 5.16** Jack Kilby's original hybrid integrated circuit from 1958. (Courtesy of the National Museum of American History. Original figure converted to black and white)

---

[72] Kilby, J.S. United Stated Patent US53138743A (1959). Miniatured electronic circuits.

On July 30, 1959, Fairchild filed a patent for "Semiconductor-and-Lead Structure," claiming "to facilitate the inclusion of numerous semiconductor devices within a single body of material."[73] The integrated circuit was born. For his concept of an integrated circuit, Kilby was awarded the Nobel Prize in Physics for 2000. In 1998, Noyce, together with Moore and the others, left Fairchild, founded Intel and became billionaires.

The final obstacle in the "tyranny of numbers" was the heat given off by the transistors. Even junction transistors, while more efficient than point-contact devices, consumed too much power and released too much heat. The solution came in 1960 when Mohammed "John" Atalla and Dawon Kahng developed the first MOSFET. This is the same basic concept conceived in 1925 by Lilienfeld, independently by Hail in 1934, and by Shockley in 1945. Of the three inventors, only Shockley had tried some experiments and his experiments failed. After repeating Shockley's calculations, Bardeen concluded that the field couldn't penetrate into the silicon or germanium due to electrons trapped in surface states. They mostly gave up on the idea, although Shockley kept it on a back burner. The solution appeared in 1959, when Atalla and Kahng, then at Bell Labs, discovered that by forming a silicon dioxide layer on a silicon surface they prevented the formation of the troublesome surface states. When a voltage was applied to a metallic electrode placed on the silicon dioxide layer, called a "gate" electrode, the electric field could now penetrate into the silicon below. This field either attracted electrons or repelled them, depending on the sign of the voltage on the gate, and controlled the flow of current in the silicon surface, forming a field-effect transistor. This had two important advantages over the junction transistor: it was simpler to fabricate, and since it didn't draw any current, it required less power to operate, and produced less heat. And it was tiny. In 1960, Atalla and Kahng fabricated the first MOSFET with a gate only 20 microns (millionths of a meter) wide. Oddly, Bell Labs overlooked this development, possibly because Atalla and Kahng were engineers, newly hired at Bell, rather than Ph.D. physicists and chemists with a track record. Ignored by management, Atalla left Bell Labs in 1962 and moved to Hewlett Packard.

Now the floodgates were open. With the development of integrated circuits and MOSFETs, the way was clear to ever greater complexity. In 1965, with only a few data points, Gordon Moore observed that the number of components on a silicon chip had doubled every two years since 1959, and predicted that this would continue. As shown in Fig. 5.17, "Moore's Law" has proved

---

[73] United States Patent US2981877A – Semiconductor device-and-lead structure, Robert Noyce, 1959.

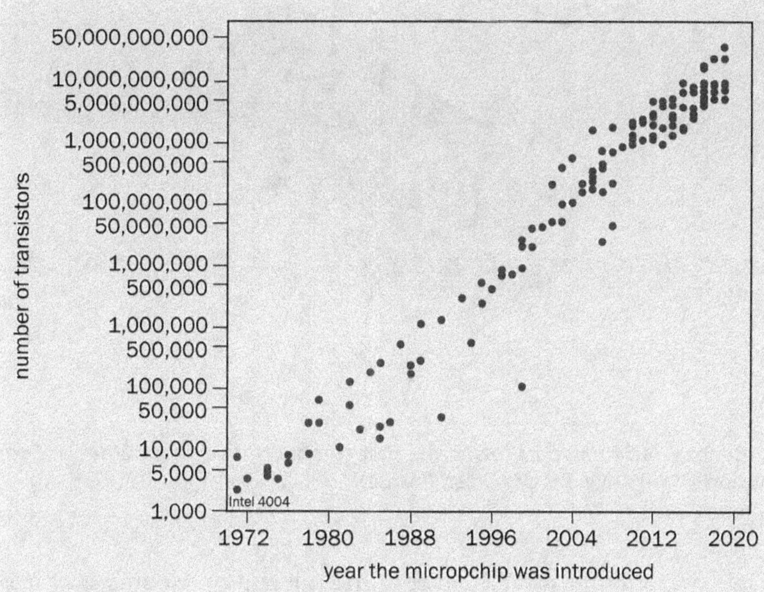


**Fig. 5.17** Moore's law, showing that the number of transistors on a chip doubles every two years. (© IOP Publishing. Reused with permission. All rights reserved. Original created by Hannah Ritchie and Max Roser, converted to black and white)

true from then until now, six decades later, rising from one transistor on Kilby's first chip to 50 billion in 2020.[74]

In 1968, Intel hired Ted Hoff, a postdoc at Stanford, as their twelfth employee. By 1969 Hoff had developed the first microprocessor, with about 2000 transistors. A microprocessor is a complete working computer on a single chip.[75] Two years later, in 1971, Intel introduced the first commercially available microprocessor, the Intel 4004, as shown in Fig. 5.18. A complete computer on a chip, it sold for $60.

How long can Moore's law continue to hold? What are the physical limits? The smallest transistors are still a long way from the size of a single atom, but gate dimensions are now on the order of a hundred atoms. Will even a single atom be the limit? What about quantum computing? Biological microprocessors are said to be smaller and more efficient than silicon devices, and since the processing occurs by means of the connections, called synapses, between neurons rather than the neurons individually, what does this say about the future of silicon microprocessors? Is there another architecture in the future using

---

[74] McKenzie, J. (2023). Quantum advantage. *Physics World*, June 20, p. 19.
[75] Perry, T.S. (1994). How Ted Hoff invented the first microprocessor. *IEEE Spectrum*.

**Fig. 5.18** Intel 4004 microprocessor, the first computer on a chip. (Creative Commons Attribution-Share Alike 3.0 Unported license)

the new "synaptic transistors?" Every time the end of Moore's Law has been predicted, it has been exceeded.

In a certain sense, it might be said that the modern computer owes nothing to Bardeen, Brattain, and Shockley. Point-contact transistors (Bardeen and Brattain) and junction transistors (Shockley) aren't found in a modern computer. Of course, this isn't really fair; they developed the first solid-state switches from which modern computers are descended. Perhaps more credit should be given to Mervin Kelly, Research Director at Bell Labs, who as early as 1936 already recognized the "tyranny of numbers" facing the telephone system. With hardly anything to go on at the time, he decided that solid-state components would be the answer and started a program of fundamental research. That decision might represent the most penetrating vision and the most daring leap of faith in this entire story.

## Darker Thoughts

Unable to admit, or to understand, that it was his own fault that he had missed the revolution in technology that he had created, Shockley became increasingly inward-looking and cross with the world. In the end, he had only Emmy, and his dark thoughts.

On a Sunday evening in July, 1961, Shockley and Emmy were driving with their son Dick down the Coast Highway when an oncoming car swerved into their lane. The collision left Emmy and Shockley in the hospital. He had head injuries and a broken pelvis, and she had a fractured leg. Neither had any memory of the accident. Shockley remained in the hospital for a month, where he continued to work in bed. Emmy remained in the hospital longer. Although Shockley recovered and went home in September, Seitz felt that his head injuries and his time in the hospital had changed Shockley in subtle ways.[76] His racist views and paranoia became more pronounced.

Two years later, when Clevite got tired of Shockley's management style and removed him as director of Shockley Transistor Corporation, his old friend Frederick Terman, the Stanford provost who had offered Shockley a location near the campus for his laboratory back in 1955, offered him an endowed chair as the Alexander Poniatoff Professor of Engineering and Applied Science. Shockley would hold this post until he retired as professor emeritus in 1975. While he was at Stanford, Shockley's interests moved away from solid-state physics and devices. Among other things, he taught popular courses on problem solving to both graduate students and undergraduates. As noted before, Shockley was very good at breaking a problem down into its elements and solving them individually, as well as using statistics to draw inferences from complex data. This was clear in the work he did on operations research in World War II, and the students loved his courses. During this same time, Shockley also renewed his connections to Bell Labs as an "executive consultant" beginning in 1965, and among other things that he did there, he taught his problem-solving methods to new staff members.

But he also became interested in darker subjects. At first, he was interested in the connection between intelligence and genetics, but increasingly he focused attention on racial differences in intelligence and the degeneration of the human race. In a November, 1965 article in U.S. News & World Report entitled "Is Quality of U.S. Population Declining?" Shockley argued that

> With improvements in technology…inferior strains have increased chances for survival and reproduction at the same time that birth control has tended to reduce family size among the superior elements…But the whole subject is being swept under the rug…If you look at the median Negro I.Q., it almost always turns out not to be as good as the median white I.Q…How much of this is genetic in origin? How much is environmental?[77]

To answer this question, in a paper published in the prestigious journal *Science* in 1966,[78] Shockley used his knowledge of statistics to propose a way to correlate intelligence with racial genetics; he called it his *H* factor. Basically,

---


[76] Seitz, F. (1989) In defense of Shockley. *Nature*, 342, 474.

[77] Hiltzik (2001).

[78] Shockley, W. (1966). Possible transfer of metallurgical and astronomical approaches to the problem of environment versus ethnic heredity. *Science*, 154, p. 438.

Shockley believed that Blacks were genetically inferior to Whites, and wanted to prove that the problem wasn't simply their environment in America. Two years later he read a paper to the National Academy of Science with the title "Proposed Research to Reduce Racial Aspects of the Environment-Heredity Uncertainty."[79] Needless to say, this touched off a furor, and his old friend Seitz, who by then was the president of the Academy, would thereafter have nothing to do with him. In a 1974 interview, Shockley concluded that

> My research leads me inescapably to the opinion that the major cause of the American Negro's intellectual and social deficits is hereditary and racially genetic in origin and, thus, not remediable to a major degree by practical improvements in the environment.[80]

Shockley defended himself against charges that he was a "White Supremacist" by arguing that Jews and Asians were more intelligent than Whites. He even argued that his children, despite getting degrees from Radcliffe and Stanford, represented "regression" since their mother, Jean, "had not as high an academic-achievement standing as I had."[81]

Beyond believing in the racial inferiority of Blacks, Shockley was seriously concerned with the decline of the human race as a whole due to dysgenics. This is the idea that intelligence is hereditary and that less intelligent people have more children, leading over time to degeneration of the human population. His obsession with dysgenics may have begun while he was in the hospital following his automobile accident when he had the chance to read about an individual with an IQ of 70 throwing acid in the face of a store clerk.[82] Shockley claimed that the young man was one of a dozen illegitimate children of a woman on welfare with an IQ of 55.[83] The remedy, of course, is eugenics, that is, sterilization of inferior people. In fact, he proposed that persons with below-average intelligence should be offered an incentive to get sterilized. He suggested $1000 for each IQ point below 100.[84] The similarity of Shockley's proposals to Nazi programs led the science writer Roger Witherspoon to write an article for the *Atlanta Constitution* in 1981 noting the comparison. Shockley

---

[79] Shockley, W. (1968). Proposed research to reduce racial aspects of the environment-heredity uncertainty. Proposal read by Shockley before the National Academy of Science on April 24, 1968.

[80] Buckley, W.F. (1974) Firing Line with William F. Buckley Jr.: Shockley's thesis (Episode S0145, Recorded on June 10, 1974).

[81] Boyer, E. J. (1989, August 14). "Controversial Nobel laureate Shockley dies". *The Los Angeles Times*.

[82] Moll (1955), p. 316.

[83] Kessler (1997).

[84] Kallikak, H. (1989). William Shockley (1910—1989). *Nature*, 341, 190.

sued, claiming defamation, and won the case three years later. He was awarded $1.

If sterilization and more drastic measures are regarded as negative eugenics, positive eugenics is the notion of selective breeding. This idea has been around for thousands of years and was also promoted by the Nazis, but most recently, it has led to selective sperm banks, such as the Repository for Germinal Choice, which operated from 1980 to 1999. This sperm bank claimed to have sperm from several Nobel Prize winners, but only Shockley acknowledged his donation. His notoriety may have discouraged other Nobel laureates from donating or admitting it.

Unsurprisingly, Shockley's extreme views led to serious protests. Students at Stanford and other places where he gave speeches protested loudly. A 1968 meeting of the scientific honor society Sigma Xi at Brooklyn Polytechnic was canceled because Shockley was scheduled to speak and they feared violence. On February 16, 1972, a group of students at Stanford burned him in effigy, and on December 8, 1973, students at Princeton followed suit when he debated his views there (Fig. 5.19).[85] Shockley never shied away from the controversy. His ego seemed to relish the attention, and his competitiveness seemed to make him more determined than ever to prove that he was right, telling one interviewer that "I have faith in the long-term values of open discussion."[86] But the faculty at Stanford were infuriated by his association with the University, and most of those who had been close to him pulled away.

Increasingly, in later years, Shockley exhibited increased signs of paranoia. He had recording devices installed on all his phones, both at home and in his offices, and he recorded all his conversations and interviews. Every piece of correspondence, every scrap of paper, was retained and filed. When his secretary quit, Emmy took over all responsibility for this, even to the point that when the tape recorder was full, she would listen in on calls and take dictation. Loving him as she did, Emmy completely submerged herself in Shockley. Friends and colleagues stopped calling.

In 1989 William Shockley died of prostate cancer. By this time, he had few friends or colleagues. His ashes were buried in Alta Mesa Memorial Park, with no memorial service. That was his choice, and anyway, who would have come? Emmy stayed with him to the end; they truly seemed to be soulmates. Without her, he would have died alone. She lived another 18 years, leaving the house exactly as it had been the day he died. Nearly deaf, she died in 2007, but the controversy didn't end there. After her death, the Auburn, California, park

---

[85] Pollack, A. (1973, December 8) *Daily Princetonian.*
[86] Is Quality of U.S. Population Declining? (1965, November 22). *US News & World Report*, pp.68–71.

**Fig. 5.19** Shockley is burned in effigy at Princeton. (© 2022 The Daily Princetonian Publishing Co./republished with permission)

district accepted her donation of an 11-Hectare wooded area to be called "Nobel Laureate William B. Shockley and His Wife Emmy L. Shockley Memorial Park."[87] Protests erupted and no such park exists.

At the time of his death, Shockley hadn't spoken to his kids for years. As he lay on his deathbed, he forbade Emmy to notify his three children that the end was near.[88] They only learned of his death from the newspapers.[89] Alison, the eldest child, was closest to him, and when she married in 1957, he gave her away, but after that they became estranged. William, the elder son, dropped out of college, but Dick, the younger son, followed in his father's footsteps. He graduated with distinction from Stanford and got a Ph.D. in physics from the University of Southern California. Like his father, he was also an avid rock climber. His climbing friends described him as brilliant: "[Dick] Shockley was so smart, so intelligent, but his childhood was pretty messed up."[90] The last of the children to die, Dick Shockley loathed his father.

---

[87] Holden, C. (Ed.) (2009, May 29). Shockley dilemma. *Science*, 234, p. 1123.

[88] Kesler (1997).

[89] https://pbs.org/transistor/album1/shockley/shockley3.html.

[90] Potter, S. https://www.climbing.com/people/a-climber-we-lost-richard-dick-shockley/.

## The Dream Team

It's difficult to imagine a more diverse team than the three physicists who invented the first transistor: Walter Brattain, the gifted experimenter; John Bardeen, the "essence of brain;" and William Shockley, the visionary inventor. It seems fair to say that the transistor could not have been invented by any two without the third. In the beginning, collaboration and sharing of ideas were strong and the atmosphere was exciting. While Brattain and Bardeen were doing most of the work that led to the point-contact transistor, Shockley also contributed ideas and, after all, he was the leader. But when he was left off the patent, Shockley's competitive nature split him from the other two, and he shut them out of the further research that led to junction transistor.

The importance of collaboration and competition in science could hardly be more apparent than they were at the beginning of this project. But after that first success, the star burst and the group ceased to exist. Bardeen and Shockley left Bell Labs and went off in different directions; only Brattain stayed. It's interesting to speculate what might have happened if they'd been able to stay together. Shockley's original concept, the field-effect transistor, is now the most important version of the transistor, but he couldn't make it work. Bardeen identified the problem of electrons in surface states, and Brattain's liquid electrolytes showed how to suppress the surface states. But it wasn't until after Bardeen and Shockley had left that Atalla and Kahng, newly at Bell Labs, used a solid metal oxide in place of Brattain's liquid electrolyte and invented the field-effect transistor that is now the heart of all our phones and computers. Would cooperation have led to this discovery sooner? Or are teams of superstars bound by their nature to explode at some point? After leaving Bell Labs, Shockley tried to create his own team of superstars. It didn't last, but the debris from the explosion created Silicon Valley. Is competition good or bad?

We see in this story that scientists, even geniuses, are real people, with all their flaws. Some have more flaws than others. Brattain was sometimes irascible, but steady, and returned to his *alma mater* after retiring from Bell Labs. Bardeen, "whispering John," was quiet until he couldn't take any more from Shockley and went on to win another Nobel Prize at the University of Illinois. Shockley was flamboyant and adventurous. He left Bell Labs to seek his fortune, and ended up an outcast with racist thoughts.

## Bibliography

Riordan, M., & Hoddeson, L. (1997). *Crystal Fire: The Birth of the Information Age*. Norton.

Shurkin, J. N. (2006). *Broken Genius: The Rise and Fall of William Shockley, Creator of the Electronic Age*. Macmillan.

# 6

## The Prion Diseases

**Daniel Carleton Gajdusek—1976 Nobel Prize in Physiology or Medicine (Shared with Baruch Blumberg)** *for their discoveries concerning new mechanisms for the origin and dissemination of infectious diseases*

**Stanley Prusiner—1992 Nobel Prize in Physiology or Medicine** *for his discovery of prions—a new biological principle of infection*

*The kuru and scrapie viruses may well prove to be the first infectious 'micro-organisms' without nucleic acids for their genetic information. If they are…replicating proteins or membranes without DNA or RNA, we will have opened up a very new chapter of microbiology.*[1]

C. Gajdusek, 1976.

This story begins in the wild country of the highlands of Papua New Guinea in the 1950s and ends in a research laboratory in California where prions were first discovered. Dr. Carleton Gajdusek (GHY-da-shek), the brilliant but deeply flawed pediatrician, intrigued by his years studying rabies, plague, scurvy, and arbovirus infections, wanted to find a rare disease that he could call his own. He found this disease—kuru—in Papua New Guinea. After years of studying victims of this disease, he finally declared in the 1970s that an unconventional slow virus caused the disease that was spread by cannibalism. He received the Nobel Prize in 1976 for his work on new mechanisms for the dissemination of infectious diseases. Dr. Stanley Prusiner, a neurologist at the University of California at San Francisco, emerged in the 1970s, with the goal of identifying the infectious agent which by this time included several diseases in addition to kuru now called transmissible spongiform

---

[1] Anderson, W. (2008). The Collectors of Lost Souls. In Johns Hopkins University Press eBooks. https://doi.org/10.56021/9781421433608.

encephalopathies (TSEs): Enkephalos from the Greek for brain and pathos meaning disease. His work led to the theory that the infectious agent was not a virus as proposed by Gajdusek but was a protein which he called a prion. The story of these two Nobelists is clouded by the personalities of each awardee: Gajdusek was described as a scientific genius with the emotional maturity of a child. He would eventually end up disgraced and in exile from the U.S. Prusiner was described by his colleagues as aggressive and egotistical, and as one colleague put it, he had never seen anybody covet the Nobel as openly as Prusiner.

The story of what is now known as transmissible spongiform encephalopathies (TSEs) began with the observation by a single pathologist in the early twentieth century. As he looked at the brain slices from a woman who had died from a mysterious disease, he was sure that it was linked to some form of brain damage. Serendipitous connections between this woman's brain, the strange disease that was taking the lives of women and children in a wild, uncivilized island that few white men had ever ventured into, and the disease known as scrapie where sheep madly scraped their sides against fences before collapsing and dying would lead to investments of millions of dollars into research and the awarding of two Nobel prizes. The TSEs are rare, but research continues in hopes of finding a treatment or cure or to be prepared in case of another outbreak, and the goal of a fundamental discovery that will expand our understanding of all neurodegenerative diseases.

## A Feast of Love

> What is the matter with us are we mad? Here is good food and we have neglected to eat it. In future we shall always eat the dead, men, women, and children. Why should we throw away good meat? It is not right![2]

New Guinea is the second largest island in the world: 1500 miles long and 400 miles wide (Fig. 6.1). The island lies in the Western Pacific below the equator and just north of Australia, with a mountainous interior behind impenetrable rain forests. The inhabitants are Melanesians, divided into hundreds of warring tribes. Islanders speak 700 separate languages, representing half of the languages on Earth. In 1950, it was considered the last wild place on earth, and its dark reputation had repelled explorers for hundreds of years.

---

[2] Rhodes, R. (1998). *Deadly feasts: the 'prion" controversy and the public's health.* Simon and Schuster, p. 25.

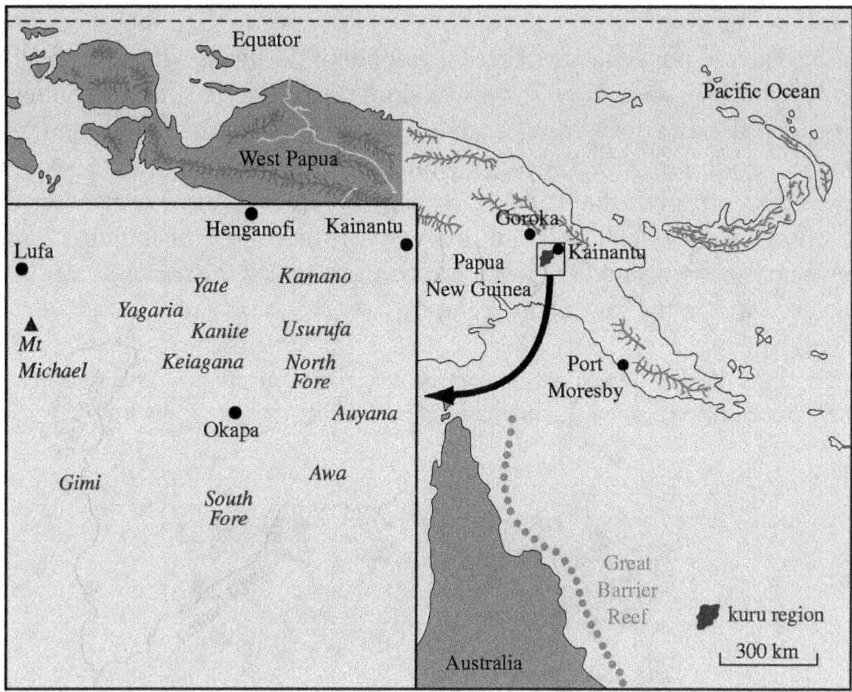


**Fig. 6.1**  Map of New Guinea

During World War II, the New Guinea campaign became one of the major military conflicts when Japan tried to capture Port Moresby toward the eastern end of the island. Over 240,000 Japanese, Australian, and U.S. servicemen died during the conflict, mostly from disease. Papua New Guinea is the Eastern half of the New Guinea island and was officially put under Australian control after World War II. Australian Administration patrols entered the remote Okapa region in 1948–1949, and by 1951 it had become clear to the colonial authorities that there were at least 14,000 members of the primitive Fore community living in small clearings and grasslands of the Eastern Highlands. The Fore grew vegetables such as sweet potatoes, yams, and corn, and the sole source of meat appeared to be pigs. But there was in fact another source of meat that led to the tragic appearance of the disease called kuru.

Officer Gerry Toogood was one of the first Australian officials to hike into the unmapped territory of Papua New Guinea, an area south of Kainantu in the Eastern Highlands occupied by the North and South Fore communities. As he and his patrol studied the culture of the Fore, they reported that a strange illness had appeared in their villages, afflicting as many as 200 persons per year, mostly women and children. In 1954, John McArthur was one of the first to identify this disease as kuru, meaning shivering or trembling. The word derives from the Fore word *kuria* or *guira*, meaning to shake, referring to the body tremors that are a classic symptom of the disease. It was also called *negi-nagi*, which meant foolish person, since the victims laughed at spontaneous intervals. The disease eventually caused frightening neurologic and psychiatric symptoms (Fig. 6.2). Victims became unsteady and wobbled uncontrollably. The whole body eventually began to shake. Patients lost the ability to swallow, talk, or think logically. "Patients eventually grow mute, become demented, and die."[3] John McArthur recorded the first description of kuru in 1953 in the south Fore region:

> Nearing one of the dwellings, I observed a small girl sitting down beside a fire. She was shivering violently and her head was jerking spasmodically from side to

**Fig. 6.2** Child with advanced kuru. (Courtesy of Creative Commons Attribution 3.0 Unported license)

---

[3] Klitzman, R.L (1998). *The trembling mountain: A personal account of kuru, cannibals, and mad cow disease.* Perseus Publishing, p. 111.

side. I was told that she was a victim of sorcery and would continue thus, shivering and unable to eat, until death claimed her within a few weeks.[4]

By 1950, kuru was claiming women in every Fore village. The villagers believed that sorcery, an active practice among the Fore peoples, caused the disease. Sorcerers would take something intimately connected with the victim—hair, feces, or discarded food—and wrap it in leaves and bury it in swampy ground. As the leaves decayed, the victim would worsen and die. The patrol officers could not convince the Fore people that sorcery was "mere superstition."[5] The belief in sorcery seemed to offer to the Fore an explanation for unexpected illnesses and death. Clinging to this belief possibly delayed scientific investigations and the ultimate elimination of kuru.[6]

From the late 1940s to the 1960s, anthropologists arrived in Papua New Guinea to study the culture of the Fore people. Graduate students Ronald and Catherine Berndt entered Papua New Guinea in the early 1950s. They went first to Maira on Kogu and realized to their dismay that the area was mountainous, heavily wooded, and unexpectedly cold. A day's walk south was still unpatrolled, and only a few outsiders had visited the region. In 1952, they moved on to the Fore villages where they found the conditions even harsher. They focused on culture, violence, and social control. The Berndts struggled "to comprehend the provisional and relational character of personhood or self among the highlanders."[7] They were in the midst of warring communities and tribes that routinely practiced cannibalism and firmly believed that sorcery was the cause of any and all diseases. Although the Berndts had experience with Aboriginal and European communities, they were completely unprepared for how to connect to these sometimes violent peoples. In looking back at the relationship of these anthropologists with the Fore, the Fore people adjusted more readily to the anthropologists than the Berndts ever adjusted to the Fore.[8] Police agents reported that the Fore were losing as many as 100 women and children per year. The anthropologists and the Australian government officers realized that kuru was moving rapidly through the Fore community, and neither the anthropologists nor the officers had the skills to study

---

[4] Lindenbaum, S. *Kuru sorcery.* (2013). Paradigm Publishers, p. 9.

[5] Radford, N.A., & Scragg, N.R. (2013). Discovery of kuru Revisited: How Anthropology hindered then enhanced kuru research. *Health and History, 15*(2), 29. https://doi.org/10.5401/healthhist.15.2.0029.

[6] Radford and Scragg (2013), p. 30.

[7] Anderson (2008), p. 20.

[8] Amderson (2008), p. 26.

a disease and develop treatments and/or containment. What was needed at this point were medical scientists.

Carleton Gajdusek was the first American scientist to arrive in the Okapa region of Papua New Guinea in 1959. Gajdusek was…well…unique. Perhaps the best description of him was stated in hindsight in his obituary: "he was the most outlandish and peripatetic of the microbe hunters."[9] Stories about scientists fascinated Gajdusek as a child, and he read Paul de Kruif's book, the *Microbe Hunters*, about the pioneers in infectious diseases. He scrawled the names of the scientists and doctors on the stairs leading to the childhood laboratory he had created in his attic: Walter Reed, Louis Pasteur, Robert Koch, and Anton von Leeuwenhoek. He was closest to his aunt, who was an entomologist, and as Gajdusek wrote: "She had me doing experiments just after I was a toddler."

Gajdusek was determined to become a famous scientist and got his undergraduate degree in biophysics from the University of Rochester at the age of 19, studying under the famous German physicist Victor Weisskopf. He earned his M.D. degree at Harvard in 1946, then headed west to work at Caltech, drawn by Linus Pauling, but ended up working with Max Delbrück's group on microbial genetics. His mentor at Harvard, John Enders, found a place for him at Walter Reed Army Medical Center in Washington, DC, where he worked with Joe Smadel on infectious hepatitis and hemorrhagic fevers. Gajdusek then moved back to Harvard to work on viruses with his mentor Enders where he learned to grow viruses in Petri dishes, which would play a prominent role in his study of kuru. Smadel encouraged Gajdusek to travel throughout the world, studying local diseases, and send back reports on outbreaks and blood samples from infected communities. Gajdusek was sure that he had now found his vocation. He wrote to his mother that he was "planning a wild, moving and restless next few years, which will bring me much further from 'settling down' than I ever have been."[10] By the time Gajdusek was 32, he had traveled throughout the world, studying rabies and plague in the Middle East, viruses in North and South American Indians, epidemic hemorrhagic fever in Korea during the Korean War, and encephalitis in the Soviet Union.[11] He studied rabies, plague, arbovirus infections, and scurvy at the Pasteur Institute in Tehran. He scoured the Hindu Kush and the jungles of South America, always in search of that rare disease he could call his own. His

---

[9] Goudsmit, J. (2009). Daniel Carleton Gajdusek (1923–2008). *Nature*, 457(7228), 394. https://doi.org/10.1038/457394a.

[10] Anderson (2008), p. 43.

[11] Rhodes (1998), p. 32.

final traineeship was with Macfarlane Burnet in Australia, where he worked on infectious hepatitis. While there he introduced the Australians to viral tissue culture. Speaking about Gajdusek, Burnet later claimed that Joe Smadel had warned him that "the only way to handle him was to kick him in the tail, hard." Smadel called Gajdusek "one of the unique individuals in medicine who combines the intelligence of a near genius with the adventurous spirit of a privateer."[12] Burnet would later describe Gajdusek as having "an intelligence quotient in the 180s and the emotional maturity of a fifteen-year-old…He is completely self-centered, thick-skinned, and inconsiderate…He apparently has no interest in women but an almost obsessional interest in children…" This latter peculiarity would have tragic consequences for Gajdusek.

Burnet recommended that Gajdusek spend some time in Papua New Guinea to study the native peoples and their diseases. Gajdusek arrived in Port Moresby, capital of Papua New Guinea, in March 1956, and "discovered the work that would occupy him for the next forty years."[13] He reported to the public health director Roy Skaggs. Skaggs informed Gajdusek that Vincent Zigas, a public health officer in New Guinea, had appealed for help in studying kuru, the deadly disease that was claiming an increasing number of mostly women and children in the North and South Fore districts. Gajdusek immediately flew to the outpost Kainantu to team up with Zigas. Vin Zigas had arrived in Papua New Guinea in the early 1950s as a young physician from West Germany,[14] but soon found that Australia did not recognize his medical credentials. The one unit that accepted him as a physician was the newly formed Public Health Service of Papua New Guinea. Gajdusek and Zigas met for the first time in March 1957 at the primitive hospital in Kainantu that Zigas was constructing. Upon meeting Gajdusek for the first time, Zigas described him as "just plain shabby…Even standing still, he seemed to be on the move…with [a] posture of someone who never had time enough to get where he had to be…".[15]

A few weeks after their arrival in the Okapa district, Gajdusek and Zigas patrolled the south Fore region. Traveling through this primitive country, they cut trails through the rain forests with machetes and waded through deep mud. They discovered that every Fore village and hamlet had a history of kuru deaths. Their initial examinations found 200 kuru patients: 50 children and the rest primarily adult women. They calculated that annually, kuru affected

---

[12] Anderson (2008), p. 42.

[13] Rhodes (1998), p. 33.

[14] Gajdusek, C. (1985). Vincent Zigas: 1920–1983. *Neurology*, 33, 1200.

[15] Rhodes (1998), p. 28.

1% of the total Fore population of 30,000 (300 cases/year). The first two patients that Gajdusek saw were typical of the many patients that he would eventually observe; they could no longer walk, and they shivered uncontrollably. They exhibited the classical symptoms of kuru. The course of the disease was unsteady gait in the first month; tremors, involuntary writhing motions (athetosis), and blurred speech by the second month; and almost complete incapacitation in the third and final month. Death was horrible to watch: the victims could not swallow and died of thirst and starvation. Gajdusek and his colleagues were able to draw blood from patients for further analysis (Figs. 6.3 and 6.4).

Zigas and Gajdusek, at least at first, were convinced that kuru was an infectious disease, perhaps a new form of encephalitis. But, as they examined their patients, they found absolutely no evidence of infection. There was no inflammation anywhere in the body, a hallmark of all diseases caused by an infectious agent. There was no fever and no increase in white cells in any of the patients. Zigas had to return to his duties in Kainantu, leaving Gajdusek to continue the hunt for a cause. Gajdusek threw every possible treatment at the disease: "antibiotics, phenobarbital, benadryl or pyribenzamine, cortisone acetate, ACTH, aspirin, or vitamin preparations."[16] There was no

**Fig. 6.3** Gajdusek drawing blood from a kuru patient. (Provided by https://circulating-now.nlm.nih.gov/2015/04/07/d-carleton-gajdusek-and-kuru-in-new-guinea/;    Ryan Cohen, MLIS, Reference and Customer Service Librarian)

---

[16] Rhodes (1998), p. 39.

**Fig. 6.4** Gajdusek and colleagues analyzing data. (D. Carleton Gajdusek Papers. 1918–2009. Located in: Archives and Modern Manuscripts Collection, History of Medicine Division, National Library of Medicine, Bethesda, MD; MS C 565. Ryan Cohen, MLIS, Reference and Customer Service Librarian)

improvement in the patients. Gajdusek thought that the disease might be hereditary. If so, Gajdusek wrote, "we have the highest incidence concentrated 'epidemic' [of hereditary disease] ever seen."[17] He decided that he would have to get tissues—especially the brain—to send to neuropathologists for examination. But this was no simple task. The relatives of the dying patients would have to be convinced to allow him to take away the valuable "edible" tissues.

Gajdusek finally got his opportunity to conduct a complete autopsy on a patient who had died of kuru. As he wrote to Smadel: "[I did] a complete autopsy …at 2 a.m., during a howling storm, in a native hut…The brain…is off to Melbourne for neuropathology, along with pieces of all organs I sampled…"[18] Unfortunately he was unsuccessful in preserving the brain tissue in this first try at an autopsy. He had cut the brain into sections, mashing most of it badly. Before the next autopsy, Smadel sent careful brain-handling instructions. Gajdusek performed his second kuru autopsy in a hospital. This time he carefully followed the instructions on preserving the tissues and sent the brain on to Smadel. Five more brains followed shortly after the first. Smadel assigned the brain analyses to neuropathologist D. Igor Klatzo at the

---

[17] Rhodes (1998), p. 37.
[18] Rhodes (1998), p. 38.

NIH. In his preliminary report, Dr. Klatzo found extensive damage in the cerebellum, but concluded that there was no evidence of an infectious agent or inflammatory process. He concluded that perhaps some toxic substance in the food or environment was causing the condition.

As Gajdusek and his group continued their studies of kuru brains, two incredible findings expanded the developing story of these neurological diseases. First, on examining the kuru brain tissue, Klatzo "saw a type of damage that reminded him of an existing human disease,"[19] a disease two German doctors had reported in the early 1900s in Breslau, Germany. A young woman suffered a breakdown and was taken to a clinic run by famed neurologist Alois Alzheimer. Her symptoms were an unsteady gait, tremors, and uncontrollable writhing, indicating to Alzheimer that she was suffering from some kind of neurological damage. Hans Sr. Gerhard Creutzfeldt examined her brain during the autopsy and found extensive brain damage, but no inflammation. He found large areas where repair cells (glia) had replaced the dead cells. Under the microscope this proliferation looked like "brown stars crowding a dead gray sky."[20] Then, in 1921, Dr. Alfons Jakob reported that a patient complained of aching legs and dizziness and progressed like the young woman in Breslau to the final stage of dementia and stupor. This patient's brain again showed extensive cell death together with proliferation of the star-like glia cells, but again there was no indication of inflammation. In 1922, another physician—Walter Spielmeyer—first coined the term Creutzfeld-Jakob disease (CJD).[21] Together, these first early reports described CJD as a new, rare, neurodegenerative human disease with extensive brain damage but no sign of inflammation. But, both Creutzfeld and Jakob had left out one crucial observation—holes in the brain. Further studies would describe the distinctive mark of the family of diseases that were discovered over the next several decades—the brain was full of holes, similar to a sponge. The condition would eventually carry the descriptive name of spongiform encephalopathy.

In 1959, William Hadlow, a veterinarian trained at Ohio State University, discovered a second serendipitous connection with kuru. In 1958, he was working on loan at the British Agricultural Research Council field station at Compton, southeast of London. His subject of interest was the old but still mysterious disease of sheep called scrapie. The name arose from the primary symptom in sheep where their itch is so intense that they scrape the wool off

---

[19] Rhodes (1998), p. 43.
[20] Rhodes (1998), p. 49.
[21] Henry, R., & Murphy, F. A. (2017). Etymologia: Creutzfeldt-Jakob Disease. *Emerging Infectious Diseases, 23*(6), 956. https://doi.org/10.3201/eid2306.et2306.

their sides. They then develop tremors, begin to stagger, go blind, fall down, and ultimately die. In a bizarre happenstance in France in 1931, French veterinarians demonstrated that scrapie was infectious, and the most common route of infection was the spread from an infected ewe to her offspring and other sheep or goats who came in contact with the birthing fluids. Since humans commonly eat mutton but no disease developed, it was assumed that sheep to human infections did not occur. In a rather bizarre corollary to the French experiments, a veterinarian, Dr. William Gordon at the Compton field station, accidentally conducted a large scale, unintended scrapie transmission experiment. Dr. Gordon had developed a vaccine using tissue from sheep that had come down with a tick-born disease called louping ill. The sheep with louping ill were inoculated with the vaccine and to his horror every one of the sheep came down with scrapie, supporting the French finding that scrapie was transmissible from sheep to sheep. Curiously—and importantly—the formaldehyde in the louping ill vaccine had not destroyed the scrapie agent.[22]

Hadlow's job in Compton was to learn all he could about scrapie and take it back to his laboratory in Montana. Studying the brain sections that he had obtained from scrapie sheep, Hadlow found the expected cerebellar holes. He also reported proliferation of star-like glial cells, but—again—no obvious inflammation. At this point, he thought he had identified a new disease. As sometimes happens in science, an off-hand comment or event can lead to a serendipitous finding that sparks further research. In this case, Hadlow viewed an exhibit by Gajdusek in the London Medical Museum of a "strange brain disease of primitive people in New Guinea."[23] Hadlow was struck by the pathological similarity of the scrapie and kuru brains: both diseased brains were full of holes. He drafted a letter to the scientific journal *Lancet*, comparing the pathologies of kuru and scrapie. As the veterinarian wrote:

Each disease is endemic in certain confined populations, whether this be flock or tribe, in which the incidence is low…Yet…the exact mechanism underlying the 'spread' of scrapie or kuru is obscure…Loss of coordination, which becomes progressively more severe, tremors, and changes in behavior are features of both diseases.[24]

---

[22] Rhodes (1998), p. 59.
[23] Rhodes (1998), p. 62.
[24] Hadlow, W. (1959). Scrapie and kuru. *The Lancet*, 74(7097), 289–290. https://doi.org/10.1016/s0140-6736(59)92081-1.

By the late 1950s, three transmissible spongiform encephalopathies (TSEs) had been identified—kuru and CJD in humans and scrapie in sheep—but researchers still had no understanding of the route of transmission or the causative agent. Several modes of transmission as well as a possible infectious agent were considered by medical researchers and by anthropologists. At first, Western observers thought kuru was a hysteria "directly associated with the threat and fear of what was believed to be a particularly malignant form of sorcery."[25] The first patient sent to the Kainantu hospital elicited a diagnosis of acute hysteria in an otherwise healthy woman. Some members of the Fore community continued to believe this theory. The belief in sorcery was so imbedded in some of the older Fore that there were counter-sorcerers to deal with kuru. As described by Klitzman:

> A counter-sorcerer, for instance, told me he had cured dozens of cases. His treatment consisted of first uttering an incantation, and then dispensing herbal medicines, and prescribing several behavioural changes: for one week, patients could not drink water, eat salt or touch members of the opposite sex. If they still contracted kuru, they simply hadn't followed the advice.[26]

In the first case that Gajdusek and Zigas saw in 1957, Gajdusek wrote to Robert Berndt: "We cannot claim any clues to its pathogenesis…fatal kuru cannot by any stretch of the imagination be identified with hysteria, psychoses or any known…psychologically-induced illnesses."[27]

Some researchers suggested that the high incidence of kuru in certain families and hamlets, together with the fact that it was most prevalent in women and children, might suggest a genetic basis for transmission of the disease. In 1959, Bennett and colleagues postulated that kuru was determined genetically as a dominant trait in females and a recessive trait in males.[28] In contrast, studies principally done by Robert Glasse and Shirely Lindenbaum[29] suggested this model was not possible since many of the victims were, in fact, not genetically related.

In 1961, anthropologists Glasse and Lindenbaum started a comprehensive study of the Fore people to provide additional cultural and genealogical data,

---

[25] Lindenbaum (2013), p. 14.

[26] Klitzman, RL. (2008). Kuru fieldwork of 1981…and beyond. *Philosophical Transactions of the Royal Society B.* 363(1510), 3646–3647. https://doi.org/10.1098/rstb.2008.4016.

[27] Lindenbaum (2013), p. 15.

[28] Bennett, J. V., Rhodes, F.A., & Robson, H. N. (1959). A possible genetic basis for kuru. *American Journal of Human Genetics*, 11(2), 169–187.

[29] Liberski, P. P., Gajos, A., Sikorska, B., & Lindenbaum, S. (2019). Kuru, the first human prion disease. *Viruses*, 11(3), 232. https://doi.org/10.3390/v11030232.

hoping to determine how kuru was transmitted.[30] As suggested previously, they argued that genetics could not be the cause since the disease appeared to be a recent occurrence, started perhaps by one person who experienced a sporadic infection, and many of the victims were unrelated. The anthropologists thus began a search for another explanation. After living with the Fore and traveling amongst several different tribes, the anthropologists were moving closer to answering the question that a neurologist had proposed: "What do the adult women and children of both sexes do that the adult men don't do? They eat their dead."[31] As early as the mid nineteenth century, explorers who dared to enter New Guinea knew the natives practiced cannibalism, but the Fore custom differed from vengeance or hate. Women did most of the eating of the dead as a tribute to their dead relatives, as a feast of love. Dying Fore asked that they be eaten and assigned their bodies to favorites. I eat you' was a common Fore greeting. By the early twentieth century the Fore freely admitted their cannibalism to the Australians and other non-Fore people who entered Papua New Guinea after World War II. The patrol agents knew that cannibalism was a common practice among the Fore tribes in the Okapa region, but the agents lacked an understanding of how this could cause disease. In fact, no one in those early years believed that cannibalism could cause disease. That the appearance of kuru was coincident with the adoption of cannibalism by the Fore people appeared to escape notice. That understanding would eventually require the work of both anthropologists and scientists who entered the area in the 1950s.

The story of the Fore woman Tomisi is worth telling as a documented example of the cannibalism custom.[32]

Fifty-six people—friends and relatives—gathered in Tomisi's garden. Her daughters cut through all the joints and gave the hands and feet to her relatives. They then slit through the skin and removed all the organs. The brain tissue was a particular delicacy. The head was carefully removed, and the skull broken open. Everything, including bones, organs and even feces, would be mixed with edible ferns and cooked in banana leaves. Of the fifty-six participants in Tomisi's garden that day, fifty-three would eventually die of the kuru disease that had taken hold among the Fore. One died of an unrelated disease, and two survived, perhaps having been the most removed from the feast.

---

[30] Lindenbaum, S. (2015). An annotated history of kuru. *Medicine Anthropology Theory*, *2*(1). https://doi.org/10.17157/mat.2.1.217.

[31] Lindenbaum (2015), p. 106.

[32] Rhodes (1998), p. 22.

By the mid-1960s, most researchers were now convinced that cannibalism was the mode of transmission, and the patrol agents were instructed to tell the Okapa communities that they would arrest anyone who continued this custom. Even when cannibalism eventually ended and kuru cases disappeared, some of the Fore continued to believe that sorcery was still a danger. In the 1960s, patrol officers had been instructed to jail members of the Okapa community if cannibalism continued. At first, they continued cannibalism in secret, but then it slowly ended. But they clung to their belief that stopping cannibalism would not stop kuru: it was the sorcerers who caused kuru. The last known case of kuru occurred in 2005, and public health officials announced that the disease was eliminated.[33]

Although cannibalism was widely accepted there was still some resistance to this theory. The most surprising resistance came from Gajdusek, who suggested that mourners at the "feasts" could have contracted the disease through contact with infectious tissues and entry of the kuru agent into the body via cuts and scrapes. Gajdusek's reluctance to accept cannibalism may have been caused by his personal and social connection to the Fore people; he didn't want to view them as primitive people. Alternatively, Gajdusek may have found it difficult to accept these findings from cultural and social studies rather than biological research. Anderson has suggested that Gajdusek finally agreed that cannibalism must have been the route of transmission, and that 1963 was a pivotal year for Gajdusek when he shifted his efforts from attempting to identify an explanation for the disease to focusing on two primary research areas: discovering the identity of the infectious agent, and proving that the disease was not only transmissible among the Fore people, but could be transmitted to other species. This latter effort would demonstrate that this was a new infectious disease.

Scientists knew very little about the nature of the proposed infectious agent. There was, however, information from scrapie studies that provided some clues to what this agent might be. The pathology of scrapie brains was strikingly similar to kuru: there were vacuoles or tiny spaces within the cells, giving the brain a sponge-like appearance. Scrapie could be transferred to healthy sheep and goats, although the incubation time was lengthy, with disease developing only after many months or years. Scrapie had become known as a transmissible spongiform encephalopathy (TSE) caused by a mysterious 'slow virus.' Gajdusek became more convinced that the kuru infectious agent must also be a slow virus when he read about the studies that were first

---

[33] Alpers, M.P. (2008). The epidemiology of kuru: Monitoring the epidemic from its peak to its end. *Philosophical Transactions of the Royal Academy*, p. 3712.

published by Bjorn Sigurdsson, who was studying diseases in sheep in Iceland. In his 1967 paper "Slow-Virus Infections of the Nervous System" Gajdusek noted that Sigurdsson defined slow infection.

> as an infection of the host in which the agent continued to multiply, slowly producing progressive abnormality over the course of months and years, usually with a very long incubation period, and often with a localization of the infectious process to a single organ.[34]

Gajdusek was convinced: the scrapie agent must be a slow virus, and he would set up a laboratory at the National Institutes of Health (NIH) that would focus on inoculation of primates and other animals to create an animal model to study the slow virus theory. Gajdusek had trained with the best and could grow viruses in culture, which would become a large part of his work in the 1970s. He was the right person for this challenge.

His first step was to identify a place near the NIH that would be secluded and off limits to the public, but close enough to the NIH to still have access to equipment and researchers. He found a facility operated by the Fish and Wildlife Service—40,000 acres with the remains of two barns, enough to build a lab. William Hadlow, the veterinarian who studied scrapie, turned down Gajdusek's offer to head up the fledgling lab, but Gajdusek recruited an accomplished virologist—Joe Gibbs—to join the team and head up the virus studies.

Gajdusek and his colleagues realized that to begin laboratory studies to identify and characterize the agent, they would need an animal model. Although their ultimate goal was to transmit kuru to primates, they started modestly by infecting all kinds of small animals such as mice and rabbits. In these animals, the time between inoculation and appearance of symptoms was in the range of months, while in primates Gajdusek assumed the incubation period would be closer to years as they had found with kuru in human infections. Gajdusek knew that it could take a big swath of his career to determine if kuru or scrapie could cause disease in primates. But primates would be the closest he could get to humans and would offer Gajdusek his best opportunity to isolate and characterize the unknown virus.

Joe Gibbs conducted most of the inoculation experiments in the U.S. laboratory. As his work with small animal infections was proceeding, Gibbs started the first primate infection using a brain sample from a kuru-infected boy from

---

[34] Gajdusek, C. (1967). Slow-virus Infections of the nervous system. *The New England Journal of Medicine*, 276(7), 392–400. https://doi.org/10.1056/NEJM196702162760708.

Papua New Guinea. The tissue was frozen and sent on to Gibbs in the USA. The tissue was homogenized with sterile saline in a blender. A small hole was drilled into the skull of the chimp (named Georgette), and the solution injected directly into the cerebrum. Gibbs was having a difficult time experimenting on primates—he became too close to the lively little chimps. As Gibbs wrote:

> I made a point of [doing] animal observations…The chimps were still small enough that you could handle them. Play with them. I would get them out, take one by the hands and throw it up into the air and it would be a couple of flips in the air and I'd catch it.[35]

On June 28, 1964—nine months after Georgette's initial inoculation—the researchers noticed slight changes in the chimp: she was not as active and stayed by herself in a corner. She had a vacant look in her eyes and would occasionally shiver. On July 15, 1965, Gibbs recorded: "Tremors continue. Fell off top of cage today." Georgette had kuru. Once Gibbs diagnosed kuru, Georgette deteriorated quickly. On October 7, Gibbs recorded that Georgette was getting weaker and would never recover. He stated that she must be sacrificed soon. On October 28, Georgette's behavior was filmed for the last time and a final examination conducted. The chimp was anesthetized and painlessly killed by draining her blood. The autopsy was then conducted, with the following result: "pathology of Georgette indistinguishable from human kuru."[36]

Even though Gibbs had demonstrated transmissibility to a primate, the researchers were still no closer to identifying the causative agent. Gajdusek's focus remained on the isolation of the slow virus that was the presumptive agent of scrapie and kuru. In 1961, Gajdusek presented a lecture at the Xth Pacific Science Congress in Honolulu entitled "Kuru: An Appraisal of Five Years of Investigation." With a discussion of the "still undiscernible possibility of an infectious agent," he went on to say: "In spite of all the genetic evidence, both the pathological picture and the epidemiological peculiarities of the disease persistently suggest that some yet-overlooked, chronic, slowly progressive, microbial infection may be involved in kuru pathogenesis".[37] In 1970, Gajdusek became chief of the National Institutes of Health (NIH) Laboratory for Study of Slow, Latent, and Temperate Virus Infections, and throughout

---

[35] Rhodes (1998), p. 89.

[36] Rhodes (1998), p. 98.

[37] Liberski, P., & Gajdusek, D. (1997). Kuru: Forty years later, a historical note. *Brain Pathology*, *7*(1), 555–560. https://doi.org/10.1111/j.1750-3639.1997.tb01073.x.

the 1970s, the laboratory had few competitors. Gajdusek had established himself as the leader in the search for human slow viruses. He stated that "if any one of the wide range of subacute and chronic affections of the central nervous system of unknown etiology were to prove the result of a slow virus infection, we shall win a significant advance in our understanding of diseases of the human brain."[38]

Gajdusek hated spending time in his NIH laboratory and left it to the myriad of young and veteran scientists that he attracted to guide the scientific work. He used his time to focus on a search for additional wild places that might hold other diseases yet to be conquered. His laboratory group was, in most instances, grateful for the time without his presence. In the laboratory he grew irritable and gruff, lashing out if someone was not working rapidly enough or had taken a wrong turn in the research. Gajdusek warned his colleagues that no one should stay around him for more than a few years. There were episodes of shouting and insistent demands, together with occasional tantrums. The laboratory had revealed a side of Gajdusek that no one in his group had anticipated and certainly did not enjoy. When he was once again in Papua New Guinea he wrote: "Personally, I'm having fun, productive fun—and that reduces the need for Wagnerian display."[39]

On October 14, 1976, Gajdusek wrote in his journal:

I just learned that I have won the Nobel Prize. Major reaction: worry! Can I manage to complete the work I am trying to do…and retain humility and creativity? I must figure out some way to escape this notoriety and regain equanimity in solitude.[40]

The Nobel Committee notified Gajdusek that he was sharing the prize with Baruch Blumberg, someone he barely knew. They were being recognized "for their discoveries concerning new mechanisms for the origin and dissemination of infectious diseases (Fig. 6.5)." Gajdusek was dismayed to learn that none of his colleagues shared the prize with him. He was especially saddened that Vin Zigas or Joe Smadel had not been recognized. None of his colleagues ever publicly expressed resentment for being left out of the Prize.

Back in Papua New Guinea, the means of spreading kuru had been identified as cannibalism, and once this practice declined, kuru began to disappear. By 1982, kuru deaths had declined to almost zero (Fig. 6.6). Medical research

---

[38] Anderson (2008), pp. 140–141.

[39] Anderson (2008), p. 144.

[40] Anderson (2008), p. 185.

**Fig. 6.5** Gajdusek receives his Nobel Prize. (Provided by Ryan Cohen, MLIS, Reference and Customer Service Librarian, National Library of Medicine. D. Carleton Gajdusek Papers. 1918–2009. Located in: Archives and Modern Manuscripts Collection, History of Medicine Division, NLM, Bethesda, MD; MS C 565)

activities in the country shifted to the study of other health problems under the direction of Michal Alpers, director of Gajdusek's Institute of Medical Research. The studies included respiratory diseases, malaria, HIV, and women's health, and this work continued for another 20 years. Meanwhile, Gajdusek and his group still had not been able to definitively show that scrapie or kuru was caused by a slow virus. There had been some exciting work beginning in the 1960s that the principal agent might be protein in nature. Gajdusek's interest began to focus on what the nature of this protein might be. Gajdusek now wanted everyone in his laboratory to work on the molecular biology properties of scrapie. But the unconventional or slow virus theory refused to go away, and additional groups appeared in the USA and Britain in competition with Gajdusek's group. Unfortunately, while Gajdusek was one of the world's foremost experts in exotic viruses, he was not an expert in molecular and protein biology. This kept him on the perimeter of the exciting prion work that would take over the TSE story. All the while, Gajdusek focused on strange diseases in strange places. As noted by Anderson about Gajdusek: "Increasingly, his anthropological enthusiasms, clinical commitments, and biology sensitivity appears eccentric and recondite…".[41]

---

[41] Anderson (2008), p. 198.

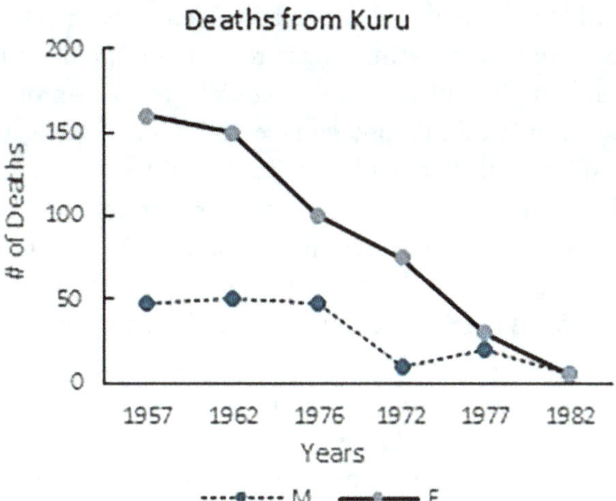


**Fig. 6.6** The decline in kuru deaths

Several laboratories by the 1960s and 1970s had begun investigating the physical characteristics of the scrapie agent. As the research intensified, what these scientists found was that the infectious agent did not behave like any virus that had been described to this point. The scrapie agent survived boiling for several hours; treatment with chloroform, phenol, and acetyl ethyl imine; extraction with ether; numerous cycles of freezing and thawing; and treatment with DNAse and RNAse that should inactivate nucleic acids. As stated by Ian Pattison, "it is likely that if the transmissible agent of scrapie is a living virus, it is a virus of a kind as yet unrecognized."[42] As early as 1966, Tikvah Alper and her colleagues in London reported that the scrapie agent had a low molecular weight more like a protein than a virus, and stunningly suggested that the agent could be a "protein without nucleic acid."[43] Also in 1966, another researcher, this time a mathematician, followed up on the Alper study. The mathematician J.S. Griffith stated: "the idea that the scrapie agent is a protein therefore presents difficulties since it is not generally thought" that proteins can self-generate.[44] Griffith went on to present three ways that this could occur and suggested "that the existence of a protein agent would cause the whole theoretical structure of molecular biology to come tumbling down." Griffith's second way was remarkably close to what actually happens (as we

---

[42] Pattison, I. (1965). Resistance of the scrapie agent to formalin. *Journal of Comparative Pathology*, *75*(2), p. 164. https://doi.org/10.1016/0021-9975(65)90006-x.

[43] Alper, T., Cramp, W., Haig, D. M., & Clarke, M. (1967). Does the agent of scrapie replicate without nucleic acid? *Nature*, 214(5090), 764–766. https://doi.org/10.1038/214764a0.

[44] Griffith, J. S. (1967). Nature of the scrapie agent: Self-replication and scrapie. *Nature*, 215(5105), 1043–1044. https://doi.org/10.1038/2151043a0.

shall see later). He predicted that the scrapie agent was an aberrant form of a normal host protein that sometimes gets produced spontaneously. This altered form could serve as a template and induce the production of more aberrant forms which accumulate and cause damage to the brain tissue. This was in fact just speculation by Griffith, a mathematician, but TSE researchers took note.

To have a better understanding of the ongoing protein versus virus controversy, it's worth taking a brief foray into what was known at the time about protein synthesis and what later was called Crick's Central Dogma. On February 28, 1953, Francis Crick and James Watson published a one-page article in Nature entitled "Molecular Structure of Nucleic Acids."[45] This theory suggested that deoxyribonucleic acid (DNA) in the now famous double helix contains a sequence of billions of bases (smaller molecules) that contain all the information necessary to construct an entire living thing. Since the two strands of DNA are complementary, each contains all the same information, and each, if separated from the other, can form a new DNA molecule identical to the original. As Crick and Watson mentioned in the last sentence of their 1953 paper, "It has not escaped our notice that the specific pairing we have postulated immediately suggests a possible copying mechanism for the genetic material." There it was: the basic mechanism of reproduction, or life. The process from DNA to protein goes something like this: from the DNA, other molecules in the cell read sections of the DNA (called genes) and produce a single-stranded molecule called messenger ribonucleic acid (mRNA). This mRNA is read like a ticker tape by other RNA forms, decoding the sequence in a process called translation. Every three bases code for a specific amino acid, the building block of proteins. As each section is read, a new amino acid is brought in and attached to the growing protein chain. At the same time as the chain elongates, it begins to fold upon itself into conformations including corkscrew-like structures called $\alpha$-helices and flat $\beta$-sheets. The function of each protein depends on both its primary sequence of amino acids and its secondary folded conformation. Proteins typically fold into predestined conformations specific to that protein which define its function in the body. In 1957, after this process was understood, Crick proposed his "central dogma of molecular biology" which states that all life starts with DNA from which comes RNA and on to proteins. The theory stated that the process never goes backward, so every protein within a living thing must come from DNA.[46]

Gajdusek and the scrapie researchers were baffled. They had postulated in their kuru work that the causative agent was a virus. Once inside a cell, the

---

[45] Watson, J. D., & Crick, F. H. C. (1953) A structure for deoxyribose nucleic acid. *Nature*, 171, 737–738.

[46] Crick, F. (1970). Central dogma of molecular biology. *Nature*, 227(5258), 561–563. https://doi.org/10.1038/227561a0.

virus hijacks the cellular DNA or RNA and uses the now accepted central dogma to produce additional copies of itself. It all fit. But there was now new evidence that the causative agent was a protein. As Griffith suggested, after synthesis from DNA to RNA, this protein sporadically switches its conformation and acts as a catalyst or stimulus to cause all proteins like itself to switch to the same disease-causing conformation. It's these rogue proteins that potentially cause the degenerative diseases such as scrapie, CJD, and kuru. Even Crick in his 1970 paper had expressed particular interest in the nature of the scrapie agent, posing the question "what exactly was the chemical nature of the agent of the disease scrapie?"[47]

The idea of a self-replicating protein was a totally new concept in the infectious disease world, and the idea fascinated Gajdusek. He suggested that "the kuru and scrapie viruses may well prove to be the first infectious 'microorganisms' without nucleic acids for their genetic information."[48] He began to reflect on what type of mechanism might be involved if the scrapie agent was a protein and focused on the process of crystallization where molecules stack together in layers. These aggregates then precipitate out of solution. A component such as an aberrant protein could serve as the nucleating agent, allowing the formation of new crystals from fragments of the ones previously formed. The curious aspect of this story is that Gajdusek was thinking back to the plot of Kurt Vonnegut's *Cat's Cradle*, published in 1963. In this story, a brilliant scientist invents a new form of ice—Ice-9—that causes all the water on earth to freeze solid, basically ending the earth as we know it. Gajdusek envisioned that a piece of the scrapie agent could twist together, forming fibril-like structures. Lansbury and Caughey continued the "ice-9" metaphor in a 1995 paper entitled "The Chemistry of Scrapie Infection: Implications of the 'Ice-9' Metaphor."[49] The authors stated, "A rare accidental misfolding would account for sporadic CJD; seeding—transmission of the infectious protein—would speed the process up." Another well-known popular scientist and science writer, Lewis Thomas (President of Sloan-Kettering Cancer Center), also expressed curiosity about the nature of the scrapie agent, and suggested that this could be one of life's "biological riddles":

> To be able to catch hold of, and inspect from all sides, a living, self-replicating form of life that no one has so far been able to see or detect by chemical methods, and one that may turn out to have its own private mechanisms for producing

---

[47] Crick, F. (1970).

[48] Anderson (2008), p. 189.

[49] Lansbury, P. T., & Caughey, B. (1995). The chemistry of scrapie infection: Implications of the "ice 9" metaphor. *Chemistry & Biology*, 2(1), 1–5. https://doi.org/10.1016/1074-5521(95)90074-8.

progeny, novel to earth's life, should be the chance of a lifetime for any investigator. To have such biological riddle, sitting there unsolved and neglected, is an embarrassment for biological science.[50]

## Another Human-to-Human Transmission: Iatrogenesis

One of the main goals of Gajdusek's work in the 1960s and 1970s had been the study of the transmission of kuru, CJD, and scrapie to non-human primates and other species. By the early 1970s, Gajdusek and Gibbs reported the successful transmission of kuru to chimpanzees, CJD to chimpanzees, and scrapie to monkeys. Gajdusek's laboratory at the NIH was now considered the world center of TSE work. Gajdusek's primate work convinced him that other transmissions of TSEs to humans would inevitably appear. His prediction would regrettably come true as iatrogenic, or physician-caused CJD, appeared in the 1950s and 1960s.

The term iatrogenesis means "brought forth by a healer." The term comes from the Greek *iatros* (healer) and *genesis* (origin). Iatrogenic transmission of disease has been known at least since the time of Hippocrates. Some modern forms of iatrogenesis relevant to the discussion of TSEs are the transfer of infectious material from a cadaver to a patient to treat an adverse condition or use of non-sterilized surgical equipment.

The first known iatrogenic transmission of a TSE occurred as the result of an eye surgery in the USA in 1971. The patient was a 51-year-old woman with corneal damage. A donor became available: a middle-aged male with a two-month history of tremors and memory loss. The doctor punched out the cornea from the donor eyeball and transplanted the cornea into the diseased woman's eyeball. Unknown at that time, the donor had died from CJD; unfortunately for the patient, the optic nerve is a direct connection from the eye to the brain, providing a channel for infection. Eighteen months after the transplant, the woman had difficulty swallowing. As the disease progressed, she stumbled and jerked, and eventually, two years after the transplant, she died. On autopsy, her brain looked like the brain from the donor—both were full of holes as in the spongiform diseases. As far as everyone knew, this was the first time that CJD had been transmitted from a CJD patient to a healthy human. It was the first warning that transfer of human tissue from CJD patients could result in iatrogenic or iCJD. Although this first case was from

---

[50] Anderson (2008). p. 189.

an eye surgeon, the two most significant means of transmission would soon appear.

In the 1950s, human growth hormone (GH) was isolated from cadaver pituitary glands, small glands attached to the base of the brain. In 1958, a physician demonstrated the remarkable result that injection of this GH could change a child suffering from dwarfism to normal height. By 1963, the NIH set up a national program to collect the hormone, purify it, and distribute it to pediatricians. By 1976, TSE researchers realized that the CJD agent might be present in the cadaver-derived hormone, but synthetic GH was still several years in the future. Officials grappled with the difficult question of what to do until the synthetic hormone was available; they ultimately decided that the benefit outweighed the danger for the few years until synthetic GH was available. One of the first recipients of cadaver GH was a young boy who started therapy when he was two years old. By the age of 20, a miracle had happened: the boy had grown to a height of five feet four inches, a result unheard of before GH was made available. Then the unthinkable happened. In his 20th year, the young man developed dizziness, and within a short period his speech and coordination were significantly impaired. He spiraled downhill at a rapid rate and died within six months. On autopsy, he had the classic signs of CJD. The pediatrician immediately sounded the alarm with the NIH. In 1984, the USA halted all treatments with cadaver-derived GH. Fortunately for the patients, synthetic GH was available within months of the death. But other deaths began to be identified in the UK, France, and other countries. Gajdusek and his colleagues warned the medical community of the potential for a CJD epidemic in a *New England Journal of Medicine* article.[51] Unfortunately, given the long incubation period of TSEs, it was still too soon to tell what would happen. By 1996, the death toll stood at about 80 cases of previously healthy young people worldwide. As Gajdusek stated at a 1992 conference: "It was only luck that saved thousands of people when we contaminated, inadvertently, the human growth hormone….Only fortuitously was it not more like two hundred thousand…deaths!" He called this "high-tech neocannibalism." He continued: "We are once again…reminded that human tissues are a source of infectious disease, and that any therapeutic transfer of tissue from one person to another carries an unavoidable risk of

---

[51] Brown, P., Gajdusek, D. C., Gibbs, C. J., & Asher, D. (1985). Potential epidemic of Creutzfeldt–Jakob disease from human growth hormone therapy. *The New England Journal of Medicine*, 313(12), 728–731. https://doi.org/10.1056/nejm198509193131205.

transferring the infection." By 2012, there were 226 cases due to the use of GH isolated from cadavers, with the highest numbers in France and England.[52]

A second means of human–human transmission was the transfer of dura mater tissue from cadaver brains. Dura mater is the outermost covering in the brain and spinal cord. It's not uncommon for this layer to be damaged during spinal or brain surgery. If the dural damage is too great to simply sew up, a graft or transplant of cadaver tissue was used. The first case of iCJD resulting from a dura mater transplantation was reported in the USA in 1986. Since that time, more than 250 cases have been reported, with over half in Japan.[53] The main culprit was a product called Lyodura which was freeze-dried cadaver tissue that was simply stored on hospital shelves and made ready to use by soaking in water. What was not realized was that this product, prepared from human cadavers, with no accounting of the cause of death of the donor, was contaminated with CJD.[54] When the connection between the graft product and the transmission of CJD to human patients was discovered in the mid-1980s, the production process was changed to reduce the risk. Hospitals were mandated to strictly screen donors, keep diligent records, and use validated, stringent sterilization procedures. In addition, since 2018, new entirely synthetic materials in place of cadaver-derived dura mater have been used to avoid CJD contamination.[55]

## The Search for the Infectious Agent

Following Tikvah Alper's finding that the agent had a size similar to a protein, and the mathematician Griffith who suggested the rogue protein model, researchers from all over the world joined the search for the infectious agent that caused kuru, scrapie and CJD. It was a real scientific adventure story with fragments and hints, overlaid with rivalries, competitions, and maybe a Nobel Prize for the discovery. The primary questions on everyone's mind: "Did the

---

[52] Brown, P. P., Brandel, J., Sato, T., Nakamura, Y., Mackenzie, J., Will, R., Ladogana, A., Pocchiari, M., Leschek, E., & Schonberger, L. B. (2012). Iatrogenic Creutzfeldt-Jakob Disease, final assessment. *Emerging Infectious Diseases*, 18(6), 901–907. https://doi.org/10.3201/eid1806.120116.

[53] www.cdc.gov/mmwr/volumes/67/wr/mm6709a3.htm#:~:text=The%20most%20common%20medical%20conditions,hemorrhage%20(25%3B%2016%25).

[54] Brooke, F.J., Boyd, A., Klug, G.M., Masters, C.L., & Collins, S.J. (2004). Lyodura use and the risk of iatrogenic Creutzfeldt-Jakob disease in Australia. *Medical Journal of Australia*, 180, pp. 177–181.

[55] Schmalz, P. G., Griessenauer, C. J., Ogilvy, C. S., & Thomas, A. J. (2018). Use of an absorbable synthetic polymer dural substitute for repair of dural defects: A technical note. *Cureus*. https://doi.org/10.7759/cureus.2127.

scrapie agent contain nucleic acid or not? If not, what on earth was it made of and how did it preserve and transmit information?"

Patricia Merz entered the TSE field in the 1970s to examine tissue sections using electron microscopy (EM), predicting that it might be possible to actually see the scrapie agent. Merz had not yet finished her Ph.D. in Philadelphia when she was uprooted by her husband-to-be and moved to the NY State Institute for Basic Research on Staten Island. The Institute had just opened in 1969; it was established to study brain diseases. The TSEs were just becoming a hot topic, and Merz jumped in. She chose to study scrapie, which had been extensively studied in mice and hamsters, and there was at least some pathological information from autopsy analysis of scrapie samples. The EM focuses a beam of electrons on the sample, providing magnification many times higher than a light microscope. Visualization of the sample takes place on a fluorescent screen. The image is usually dim, forcing one to sit in a dark room—sometimes for hours—to view the sample and take multiple pictures with the camera associated with the EM instrument. Before the sample can be visualized, Merz had to go through a cumbersome series of steps to prepare the sample for examination. She homogenized the scrapie brain tissue, coated a gold grid with the sample, then stained it with various dyes to examine the microstructure of the sample. The grid was perhaps 1 mm in diameter and would have 400 or more segments in a grid pattern. Once Merz had placed the sample in the scope holder, she would examine segment by segment to view the tiny world of cellular products from the scrapie sample (Fig. 6.7). As Merz stated:

> You can study photographs in books…But if you deal with…particulate material and stain them, you have absolutely no guidelines, you have to spend a long time building up frameworks of what things are, what they could be, what they might be, is it real, is it debris…Just you and the microscope. There's nothing else that guides you.[56]

After several months of learning how to use the EM and identifying the appropriate stain, Merz was ready to begin studies of the scrapie-infected tissue. She chose the Hitachi 8, a very small EM, then moved to a more powerful EM to get the magnification up as high as 68,400 fold. By this time, she was living in the EM room. Her first photos showed "sticks" only in the scrapie samples, not in non-infected tissue. The sticks looked like "short, broken lengths of twisted thread." Her first reaction was that the sticks were like

---

[56] Rhodes (1998), p. 154.

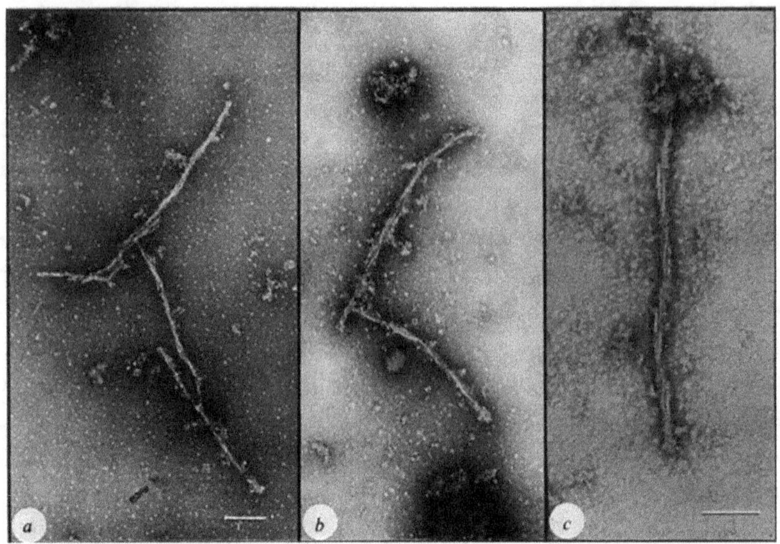

**Fig. 6.7** Merz's EM photos of scrapie-infected tissue. (Courtesy of L. Manuelidis., R.A. Sommerville and colleagues, Nature, 306, 1983)

the protein clumps that bind together to form amyloid plaques, first identified in kuru samples by Klatzo. Amyloid plaques are also found in Alzheimer's disease (AD). As Merz began to show her photos to colleagues, none of the experts thought that her sticks were amyloid. So, were these "sticks" a new cellular structure specific to TSEs? In her first publication with her colleague Robert Somerville, she named the sticks scrapie-associated fibrils—SAF. But, going against the opinion of other colleagues, the authors continued to suggest that there was a connection between the SAF and amyloid. Merz continued studies with both scrapie and CJD tissue. Her pictures continued to show the abnormal fibrils. All of her studies showed definitively that her fibrils were not just cellular debris but were something significant with respect to the TSEs. Finally, in 1980, she had an idea of what she had found.

> In all the long history of the spongiform encephalopathies, no one had ever seen the disease agent. I was doing the dishes when it dawned on me that with SAF I might actually be looking at the agent itself. My stomach clenched and I just threw up my dinner.[57]

While Merz was examining the pathology of scrapie and CJD tissues, another researcher, determined to be the first to isolate the scrapie agent,

---

[57] Rhodes (1998), p. 159.

entered the field. Stanley Prusiner, a biochemist and neurologist, had lost a patient to CJD in 1972 and became intrigued by the disease. He realized that this was a wonderful problem for him to tackle—one that could possibly lead to national recognition. He became obsessed with winning a Nobel Prize and unabashedly set off running to be the first to identify the scrapie agent.

Prusiner was born in Des Moines, Iowa in 1942, and attended a college preparatory high school in Cincinnati, Ohio when his family moved there. By his own admission, he was an average student, not particularly interested in his high school studies: Prusiner found high school dreary, and his teachers seemed to have little interest in capturing the attention of their students. He went on to the University of Pennsylvania where he earned his B.S. degree in chemistry. While an undergraduate at Penn, he studied brain swelling in rats under the mentorship of Sidney Wolfson, who encouraged Prusiner to stay on at Penn for his medical training. After two years with Wolfson, he switched his studies to brown fat cells, working with Britton Chance, probably the best-known scientist at Penn. To complete his medical studies, he worked with Olle Lindberg at the Wener-Gren Institute in Stockholm, Sweden. He spent one year at the University of California in San Francisco (UCSF) and three years at the NIH, before returning to UCSF to begin his career. It was at UCSF that he saw that first CJD patient, and he became intrigued by the unusual structures in the brain tissue. At that point, he decided to learn all he could about these degenerative diseases, and he joined the growing number of researchers searching for the "slow virus" that was the presumptive cause. To identify the infectious agent, he developed "a biochemical research program devoted to isolating and characterizing the scrapie agent."[58] And as early as the mid-1970s, he decided that "purification of the scrapie agent would be my singular focus."[59]

Prusiner had been intrigued by the publications from Alper, Griffith, and others that suggested that the scrapie agent contained no nucleic acid and might, in fact, be proteinaceous. In 1982, Prusiner submitted a paper to the prestigious journal *Science* entitled "Novel Proteinaceous Particles Cause Scrapie"[60] describing his latest findings on the nature of the infectious agent. Together with studies from several other laboratories, he declared that the agent was resistant to enzymes that would destroy nucleic acids but was susceptible to at least two enzymes that disrupted protein structure. At this point,

---

[58] Prusiner, S.B. (2014). *Madness and memory.* Yale University Press, p. 21.

[59] Prusiner (2014), p. 22.

[60] Prusiner, S.B. (1982). Novel proteinaceous infectious particles cause scrapie. *Science,* 216(4542), 136–144. https://doi.org/10.1126/science.6801762.

Prusiner didn't completely rule out the presence of nucleic acid, but he focused on protein as the primary component. Even though there was no conclusive evidence to define the agent, he went so far as to provide a name for the component: prion (pronounced "pree-on") signifying a proteinaceous infectious particle and thereby staking claim for all time to his "discovery" of the scrapie agent (Fig. 6.8).

The journal accepted the paper, but with numerous concerns before they would publish it. Three of the four reviews of the manuscript were modest at best, and the fourth—a three-and-a-half-page critique—was scathing. This last review ended with "The article has a rather unpleasant, almost sarcastic tone frequently belittling the work of others."[61] Despite these reviews, *Science* accepted the article after Prusiner's revisions. But months went by and there was still no word on a publication date. Prusiner then found out that "someone" (not identified in Prusiner's autobiography) had notified the *San Francisco Chronicle* about his new "life form," and on February 19, 1982, a story headlined "Tiny Life Form Found,"[62] was published in the *Chronicle*. This was before publication in a peer-reviewed journal (*Science*), which is unacceptable for today's scientists. Prusiner boasted that the newspaper published the article on page one, with his (Prusiner's) picture on the upper left and President

**Fig. 6.8** Stanley Prusiner in his lab. (Courtesy of the UCSF Archives and Special Collections)

---

[61] Prusiner (2014). p. 90.

[62] Prusiner, S. B., Groth, D., Bolton, D. A. E., Kent, S. J., & Hood, L. (1984). Purification and structural studies of a major scrapie prion protein. *Cell*, 38(1), 127–134. https://doi.org/10.1016/0092-8674(84)90533-6.

Reagan on the right. Regarding the delayed publication in *Science*, Prusiner declared that the journal had hesitated in publishing his article to avoid any embarrassment if another researcher found nucleic acid in the scrapie agent. No doubt, the poor reviews were also a factor. However, the journal felt that they now had no choice—they published the controversial article in April 1982.

Prusiner knew he had jumped the gun in his *Science* publication, and if his theory of a protein as the causative agent was wrong, it would be embarrassing to say the least. So, in his article, he hedged with this statement: "current knowledge does not allow exclusion of a small nucleic acid within the interior of the particle." He further proposed two models: "a small nucleic acid surrounded by a tightly packed protein coat," or "a protein devoid of nucleic acid, that is, an infectious protein." As Gajudsek would state: "He hedged much more than I did in those early 1980s."[63]

Outcry from other researchers in the field was intense. In Edinburgh, Alan Dickinson and his colleagues at the Neuropathogenesis Unit were outraged and felt that Prusiner was attempting to capture the limelight and move research away from unconventional viruses and toward a protein-only theory. Dickinson was one of the leaders of the group that continued to argue that the infectious agent contained nucleic acid that was "hidden inside a thick protein coat." Hoping perhaps that he could establish priority for himself by naming the infectious particle, he called it a "virino."[64] Another researcher in the virus group also supported the coated nucleic acid, and they continued to pit their genetic interests against Prusiner's biochemical proposals. In a surprising move, one of Prusiner's colleagues—Frank Masiarz—left the Prusiner laboratory, stating that "there's no point in creating a name for something we don't even know exists yet…[Stan] tends to jump to conclusions and give no credence to the facts that would discount his…interpretation of the results."[65]

In naming the infectious agent as the prion and replacing the terms "unconventional virus" or "unusual slow virus-like agent," with the term "prion," he knew that "naming something before you discover it is a risky business…You have to hope that as the data come in they fit your definition, or at least that your definition can be made to fit the data."[66] As Prusiner points out in his autobiography, the words thrown at him by other scientists included impulsive, presumptuous, reckless, ambitious, aggressive, callous, manipulative,

---

[63] Rhodes (1998). p. 163.
[64] Anderson (2008), pp. 194–195.
[65] Anderson (2008), p. 194.
[66] Rhodes (1998), p. 144.

and egotistical.[67] These characterizations would stick to him throughout his career, creating enemies within the field who resented their virus research being trampled on. For years he would be booed, interrupted, and criticized at national and international conferences, and in some cases simply not invited. One of the most startling accusations came from Gary Taubes, a science writer from *Discover* magazine. Taubes had started out pro-Prusiner, but he soon uncovered information that would move him to the side of Prusiner's opponents. Taubes spent hours interviewing Prusiner, and in the end became his loudest critic. In his article "The Name of the Game is Fame, but Is It Science",[68] Taubes highlights Prusiner's efforts to cover up the fact that no one had definitively eliminated a requirement for nucleic acid in the disease process. But worse, Taubes uncovered manipulations by Prusiner to suppress work by others. In one particularly unethical event, Prusiner had gone to the editor of a prestigious journal, and the editor admitted that he had held up a publication by a Prusiner competitor to publish Prusiner's article on essentially the same topic. Prusiner's article was eventually rejected and the article by the competitor published. Taubes continued: "That was a very dark moment for Stan, because there it was, it was true, and there was nothing he could do about it. He was caught with his hand in the cookie jar."[69] Obviously the Taubes article hit a major nerve: Prusiner devoted an entire chapter in his autobiography, attempting to solidify his prion theory and discredit Taubes.[70] Unfortunately, his jousting with the media would continue throughout his career, and he attempted whenever possible to avoid speaking to the press. As Richard Rhodes noted, Prusiner refused to comment or be interviewed for Rhodes' book *Deadly Feasts*. The 1982 article by Prusiner contributed to keep him in the controversial limelight, but also stimulated research on identifying the nature of the scrapie agent. Prusiner knew he was gambling, and if he was wrong, "wrong guesses don't win Nobels."[71]

In an elegant series of experiments by Prusiner and his colleagues and the growing number of competing research groups, a picture began to emerge that a major component of the scrapie and kuru diseases was a misfolded protein. In December 1982, Prusiner purified a protein from scrapie-infected brains that appeared to track with infectivity. In the following year, they reported that PrP (the acronym for the prion protein) looked like rod-shaped

---

[67] Prusiner (2014), p. 93.

[68] Taubes, G. (1980). The name of the game is fame, but is it science? *Discover*, 7(12), 28–31.

[69] Rhodes (1998), p. 204.

[70] Prusiner (2014), pp. 143–152.

[71] Rhodes (1993), p. 162.

particles under a microscope. Prusiner labeled his particles "prion rods." Although Prusiner had little evidence, he didn't pass up the opportunity to state unequivocally that his "rods" were a part of the infectious prion, but clearly, his prion rods and the fibrils or rods that Merz first described in 1978 were not the same. Prusiner would never admit that Merz' SAF and his PrP rods were the same, and he refused to even reference Merz's research in his publications. Further research from other laboratories eventually confirmed that they were identical.

Prusiner's group continued the prion studies and identified one major macromolecule from scrapie-infected brain. The molecule was a protein of approximate molecule weight of 27–30 kDa,[72] which they designated PrP 27–30. Importantly, the authors were able to actually show that this molecule was a protein by determining the sequence of amino acids (the building blocks of proteins) at the N-terminus or first part of this molecule. They emphasized that the ability to sequence a portion of the protein suggests that the prion is a major component of the scrapie agent. But still as the authors point out: "the mechanism of prion replication remains enigmatic." In the following year, Prusiner and Weissmann were able to identify the DNA sequence for PrP.[73] To their surprise, not only was the PrP gene in infected hamster brains, but it was also present in normal hamsters and normal humans as well. This was an incredibly important piece of the puzzle, since it showed unequivocally that the normal PrP and the disease form had identical amino acid sequences. In other words, the normal and disease PrP were products of the same gene! In every test they performed both proteins behaved similarly, except for one very important test. The normal protein was readily digested by proteases, whereas the diseased form was resistant to digestion! In addition, since folding occurs after the DNA has sent on its code, then perhaps…maybe…the normal PrP changed to a diseased form after it was a complete protein. But what caused this change from normal to diseased?

Prusiner's group continued to examine how the normal PrPc could be transformed into the diseased PrPsc. At this point, they assumed that the change occurred after protein synthesis and after the protein had folded into its usual state. But they had no idea yet about how the transition from normal to disease state occurred. In yet another incredibly important experiment,

---

[72] Prusiner et al. (1984), 127–134.

[73] Basler, K., Oesch, B., Scott, M., Westaway, D., Wälchli, M., Groth, D., McKinley, M., Prusiner, S., & Weissmann, C. (1986). Scrapie and cellular PrP isoforms are encoded by the same chromosomal gene. *Cell*, 46(3), 417–428. https://doi.org/10.1016/0092-8674(86)90662-8.

using a technique called Fourier transform infrared spectroscopy, they demonstrated that PrPc had a high (42%) α-helix content (coil-like formation of the protein's amino acids) and very low β-sheet (flat sheet-like arrangement). In contrast, PrPsc had a whopping 54% β-sheet (Fig. 6.9). Prusiner proposed that "this conformational transition is a fundamental event in the propagation of prions."[74]

Prusiner next moved on to some elegant genetic experiments. First, in a 1989 paper, Prusiner and colleagues reported the development of a line of mice in which the normal mouse PrP gene had been replaced by a gene encoding the hamster PrP protein (these are called transgenic mice). It was known that normal mice were resistant to hamster scrapie, but when the transgenic mice expressing the hamster gene were challenged with brain tissue from infected hamsters, the mice all developed hamster scrapie. As a control, the transgenic mice lacking the hamster PrP were resistant to the hamster scrapie. In 1992 and 1993, Prusiner and Weissmann eliminated the normal mouse prion gene completely (known as "knockout" mice) and demonstrated that these mice lacking any PrP remained happy and healthy when injected with mouse scrapie. In other words, these mice could not get scrapie without the

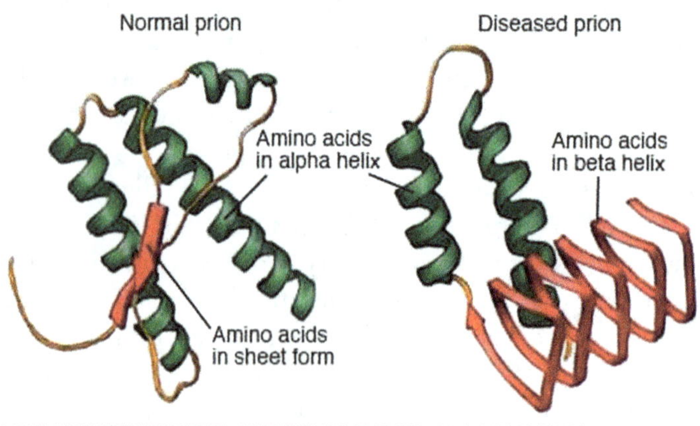



**Fig. 6.9** Structures of a normal versus diseased prion. (Used with permission: William E. Hoffman, copyright agent, Mayo Foundation for Medical Education and Research, all rights reserved)

---

[74]Pan, K., Baldwin, M. A., Nguyen, J., Gasset, M., Serban, A., Groth, D., Mehlhorn, I., Huang, Z., Fletterick, R., Cohen, F. E., & Prusiner, S. (1993). Conversion of alpha-helices into beta-sheets features in the formation of the scrapie prion proteins. *Proceedings of the National Academy of Sciences*, 90(23), 10962–10966. https://doi.org/10.1073/pnas.90.23.10962.

presence of the normal protein. There was no PrPc for the scrapie agent to cause to misfold.

Additional results from the Prusiner and Weissman groups demonstrated that the scrapie PrPsc and the normal PrPc were products of the same gene, that is, they had the same sequence of amino acids. This finding reinforced Prusiner's prediction in 1993 that PrPc becomes PrPsc through some modification that occurs after the protein is synthesized. As Prusiner stated:

> Our results suggest that there is only a single PrP gene, and its sequence and organization make it unlikely that the different properties of the PrP isoforms can be explained by alterations in the amino acid sequence; thus, it seems more probable that the isoforms arise from post-translational modifications or variations in protein conformation.[75]

Throughout the 1980s and early 1990s, Prusiner's "turf battles," as he called them, continued with his colleagues. His former collaborator—Weissmann—and his laboratory group had completed the experiments that resulted in the first knockout mouse, the mouse lacking any PrP. As Prusiner stated in his autobiography, he was shocked when he found out that he was not listed as a coauthor on the first of the two important publications describing the creation of mice lacking the PrP gene.[76] Prusiner was also left off the patent application by Weissmann. Prusiner challenged both actions; Weissmann simply answered that his group had done the experiments, and he felt that it was time for him (Weissman) to begin to develop his own career. It was at this point that Prusiner's approach to research began to emerge: his management style was focused on contractual arrangements and limitation of access to new information by others, completely opposite to the free exchange of research materials and ideas that Gajdusek and others believed was so essential to furthering a research program. He hired a public-relations consultant to market him and his work. He set up a company to commercialize any products that came from his laboratory, and he filed more than 100 patents. His company, InPro Biotechnology Inc. in San Francisco, was designed to "optimize and commercialize prion diagnostic and disinfectant products…with a motto of

---

[75] Basler et al. (1986). 417–428.

[76] Büeler, H., Aguzzi, A., Sailer, A., Greiner, R., Autenried, P., Aguet, M., & Weissmann, C. (1993). Mice devoid of PrP are resistant to scrapie. *Cell*, 73(7), 1339–1347. https://doi. org/10.1016/0092-8674(93)90360-3.

saving lives through proteomics."[77] His eyes were clearly fixed on winning that Nobel Prize for himself.

## Mad Cows and Hamburgers

By the 1970s and into the 1980s people knew that humans suffered from a spontaneous TSE with the strange name Creutzfeld-Jacob disease (CJD). Human-to-human transmission was also possible as in kuru, but kuru had been controlled once it had been determined that cannibalism was the cause, and the second human-to-human transmission (iatrogenic caused by human tissue transplants) was reduced to almost zero once the source of the transmission (human cadaver tissue) had been identified. Importantly, no animal to human transmission had been detected…yet. That was about to change, with terrible consequences.

On April 25, 1985, a dairy farmer in Sussex, England called his veterinarian to let him know that one of his cows was behaving "oddly." The cow had become aggressive and dangerous to handle. Over the next several weeks, the cow began stumbling and staggering and was eventually slaughtered. Over the next several months, seven more cows sickened, and by 1986, three other herds were infected. In late 1986, researchers reported that the brains of the dead cows had spongiform damage and other structures similar to scrapie and other TSEs. A new disease was officially defined: bovine spongiform encephalopathy or BSE. The popular media would give it another name that described the aggressive behavior of the cows: "mad cow disease." By the end of 1986, the disease had infected herds in England and Wales, with about 420 confirmed cases. Officials began to analyze the case distribution in an attempt to determine a cause. After ruling out every other possible cause, from genetics to vaccinations, there was one possibility remaining—food contamination. They knew that all cases were in dairy herds, not beef cattle, and there was one common food fed to dairy cattle called meat-and-bone meal. This protein supplement was made from the ground-up, cooked, and dried remains of dead cattle, including those called downer cattle—animals that had died of an undiagnosed disease such as BSE. So, if this were true, then BSE from that first cow in 1985 was ground up and fed to additional cows who got sick and died, and an epidemic had begun. Much like the eating of remains of dead relatives transmitted kuru, cannibalism in the cow herds was propagating BSE.

---

[77] Anderson (2007), pp. 198–199.

In 1980, Gajdusek and colleagues had reported that kuru and CJD could be transmitted to spider monkeys by feeding them infected brain tissue.[78] But meat-and-bone meal had been fed to cattle in Britain for decades without any BSE. What had changed? Possibly, perhaps likely, significant alterations had occurred in processing the cattle parts in order to increase profits. This led to a ban imposed in 1988 on the feeding of ruminant-derived protein to cows and sheep. Sheep were included in these official decrees since it was assumed that BSE might actually have come from scrapie-infected sheep, passed from sheep to cattle through the meat-and-bone feeding. And, since scrapie had not spread to humans even though mutton was commonly on the menu, it was assumed—tragically—that BSE could not be transmitted to humans by eating infected beef.

The government response at this point was limited. For example, the farmers were only getting half of the market value for their sick animals, which prompted many farmers to ship off their slaughtered cows to the renderers at the first sign of illness. In addition, a government committee report in 1989 suggested that once the feeding of cows was taken care of, cattle would become a "dead-end" host, and BSE would die out. But by the time this report appeared, BSE cases were already up to 500 per month. By 1989 there were 900 per month, and the total confirmed cases by 1995 had risen to over 140,000, far higher than the officials had predicted. Moreover, the committee recklessly concluded that "the risk of transmission of BSE to humans appears remote."[79] Alan Dickinson, a key player in TSE research, and his colleague Hugh Fraser were furious. They had warned for years that "we should not assume, at this stage of our knowledge, that scrapie agents are never transmissible to man from infected meat." Dickinson and Fraser were warned not to speak to the media.

In 1989, the British government imposed a ban on the use of entrails and internal organs (offal) in processing beef. But there was no guidance as to how the internal organs would be removed. One veterinarian described her trip to a slaughterhouse after this ban:

I saw one hundred fifty, maybe two hundred [cow] heads being thrown into a truck in a heap, where you could see the brain material seeping out through the holes when the heads were upside down, running down onto the heads below and contaminating whatever was beneath. When the heads arrived at the plant,

---

[78] Gibbs, C. J., Amyx, H. L., Bacote, A., Masters, C., & Gajdusek, D. C. (1980). Oral transmission of kuru, Creutzfeldt-Jakob disease, and scrapie to nonhuman primates. *The Journal of Infectious Diseases*, 142(2), 205–208. https://doi.org/10.1093/infdis/142.2.205.

[79] Rhodes, R. (1997, March 9). Mad cows and Americans. *Washington Post*, https://www.washingtonpost.com/archive/lifestyle/magazine/1997/03/09/mad-cows-and-americans/5aeb7ac5-8d44-4804-91d7-10d269239d3b/

> the meat from the cheeks…was trimmed off for use in hamburger, for human consumption.[80]

Whether by negligence or intent, the government was misleading the people into believing that beef was safe to eat.

In 1988, the government raised compensation to farmers to 100% of healthy market value. But the European Union (EU) was not convinced that the British had contained the outbreak and, in response, they restricted export of cattle from Britain to only calves. With the number of diseased cattle continuing to increase, Dr. Richard Lacey, a leading food scientist, called for all cattle (six million) to be slaughtered at enormous cost. Before it was over 4.4 million cows were shipped to the slaughter house.[81] At the same time, attempting to calm the British people, the Minister of Agriculture—John Gummer—ensured a place in history when he offered his daughter a hamburger—which she refused to eat—on live TV.[82] At the same time, the government came out with a statement that today would simply be unbelievable. The report suggested that the best way to reassure the public that BSE had no effect on humans would be to monitor the number of CJD cases over the next 20 years to demonstrate no increase due to BSE. Twenty years! They were proposing a human experiment to see how many people would die! If there was a significant increase, well, their hypothesis would be wrong. Again, Lacey was outraged: "In order to find out how big the problem is, we are going to see how many people die."[83]

It didn't take 20 years. In 1993, the first case of what would be called variant CJD (vCJD) was reported. In May 15-year-old Victoria Rimmer began to exhibit signs of CJD. Two more cases appeared in 1994. The three young victims died in 1995 and 1996. This was not supposed to be happening. Gajdusek exclaimed in 1996 that this was "cause for concern. Ten adolescents [with CJD] would be an epidemic."[84] By the beginning of 1996, there were ten young people with CJD. But the British government remained firm. As the Conservative Prime Minister John Major said, "I should make it clear that humans do not get mad cow disease."[85] Then, on Wednesday, March 20,

---

[80] Rhodes (1998), p. 182.

[81] British Broadcasting Corporation, October 18, 2018.

[82] Frankel, G. (1990, May 19). Hysteria over "mad cow disease" sweeping Britain. Washington Post. https://www.washingtonpost.com/archive/politics/1990/05/19/hysteria-over-mad-cow-disease-sweeping-britain/71ad572c-72db-4f0d-8024-3895d86ade60/.

[83] Rhodes (1998), p. 187.

[84] Rhodes (1998), p. 190.

[85] Rhodes (1998), p. 211.

1996, British Health Secretary Steven Dorell, who had steadfastly denied that BSE posed any danger to humans, stood ashen-faced before the House of Commons and announced that BSE is "the most likely explanation at present for the ten cases of CJD that have been identified in people aged under 42."[86] The shock and dismay were palpable. By this time, the British populace was terrified that a new plague had arrived. The British newspaper *The Guardian* stated in an editorial that "The Roast Beef of Old England…had suddenly been revealed as a Trojan horse for our destruction."[87] A story appeared in the *London Observer* in 1996, a few days after the human transmission was announced, imagining such a plague:

> It is 20 March 2016…Now, Britain's National Euthanasia Clinics churn on overtime, struggling to help 500 people a week to a dignified death before brain disease robs them of reason and self-control…The Channel Tunnel is blocked with five miles of French concrete. The health service is crippled; blood transfusions are impossible because undetectable prions…infect most donors…The fabric of the nation is being torn apart.[88]

The public fear of BSE even made it into popular culture. Less than one month after the announcement by Dorrell that BSE had jumped to humans, Oprah Winfrey hosted a program titled Dangerous Food, discussing the fear of humans contracting vCJD. Her guest argued that the risk in the USA of a BSE epidemic, and a consequent outbreak of vCJD, "was significant, owing to the widespread practice of adding 'rendered' animal parts—consisting of the ground-up tissues and bones of cattle, sheep, goats, pigs, birds, and other animals—into cattle feed as a cheap source of protein." Winfrey was alarmed and stated that "It has just stopped me cold from eating another burger."[89] Her statement led to a lawsuit by the beef industry underscoring the fear on the part of the public as well as the beef industry. (Winfrey was subsequently found not guilty.) Additionally, in an episode of the USA show West Wing, the President is informed that a presumptive positive BSE case has been identified in Nebraska. The President must decide whether to inform the public or say nothing and wait the 72 h for tests to be completed. He understands the dilemma and the consequences for the economy. The

---

[86] Rhodes (1998), p. 212.

[87] Fort, N, (1996, March 23). Household fetish turns into Trojan horse. *The Guardian.* https://theguardian.newspapers.com/image-view/260892052/?match=1&terms=roast%20beef%20of%20old%20England

[88] Rhodes (1998), pp. 229–230.

[89] Duignan, B. (n.d.). How Oprah got sued for dissing a burger. Encyclopedia Britannica. https://www.britannica.com/story/a-brief-history-of-food-libel-laws#:~:text=%E2%80%9CDangerous%20Food%2C%E2%80%9D%20which%20was,one%20month%20before%20the%20broadcast%2C.

President's response: "The second we say positive, beef futures collapse, and we lose $3.6 billion in beef exports. Fast food is deserted, supermarkets pull beef, it's panic…in the meantime, we wait."[90]

Contributing to the public panic were the reports coming from multiple sources that predicted large numbers of potential human cases. How many hamburgers had been consumed; how many steaks were still in people's freezers; how much beef had been sent to other countries? In 1998 *The Independent* published an article titled "Scientists Predict CJD Deaths for 30 Years," and projected that deaths from vCJD would reach somewhere between 1000 and 10,000.[91] The *New York Times* suggested that the number of deaths could hit as high as 100,000 over the next several decades.[92]

Gajdusek, who had done so much work with kuru and the infection of children, simply could not contain his fury. He warned that any species fed on the meat-and-bone meal could be carrying and spreading the disease. "They [the government] need to assess the risk and deal with it realistically."[93] He expressed a frightening prediction that all the pigs in England fed on this diet could be carrying the disease. They simply hadn't shown up yet because pigs are not kept alive beyond two or three years. "Probably all pigs in England are infected."[94]

The emergence of BSE in Britain in the 1980s was one of the most dreadful failures of disease control, owing in part to the limited knowledge of either the cause or the containment. Despite warnings from the scientific community, the government, the food industry, and consumers were all caught off guard by the discovery that the infectious agent crossed from cattle into humans through the ingestion of beef. Eventually, in 1996, the British government employed extreme measures to control the slaughter of cattle and implemented bans on feeding of mammalian parts to cattle to control the potential epidemic. The ban continued for ten years before it was eventually lifted on May 1, 2006. In addition, the EU banned the export of all British cattle born before 1988, and in 1996 prohibited not only the export of live animals, but any parts of those animals that might enter the food chain. During the height of the crisis and as cases began to decline, the UK government came under extreme scrutiny and criticism for its delayed response, specifically how slow it was to acknowledge the problem and to inform the public. In December

---

[90] *West Wing* (2001), season 3, episode 8. https://www.dailymotion.com/video/x6tlcg4.

[91] Arthur, C. (1997, January 16). Scientists predict CJD deaths for 30 years. *The Independent*. https://www.independent.co.uk/news/scientists-predict-cjd-deaths-for-30-years-5585075.html.

[92] Estimates of future human death toll from mad cow disease vary widely. (2001, October 30). *The New York Times*. https://www.nytimes.com/2001/10/30/health/estimates-of-future-human-death-toll-from-mad-cow-disease-vary-widely.html.

[93] Rhodes (1998), p. 220.

[94] Rhodes (1998), p. 220.

1997, two months after the Nobel Prize was awarded for the discovery of prions, it was announced there would be an official investigation into the history of the response by the British government and the steps taken to deal with the outbreak.[95] Every governmental agency involved in the outbreak was blasted for its lame response to what turned out to be a tragic loss of life in the UK and allowing the disease to spread to other countries due to their inability to take drastic steps to contain the nightmare.[96] It most certainly led to the near bankruptcy of the British cattle industry, and was a key factor in the defeat of the Conservative government in 1997.

The panic among the public was fueled by predictions from experts in the field as well as the media. In 2001, an article in the *New York Times* reported that epidemiologists in London predicted a rate of human deaths at 100 per year, with no more than a few thousand over the next several decades. Other experts, however, predicted as many as 100,000 deaths before the epidemic is over.[97] According to an article in the *New Scientist* in 2003, original estimates of deaths from vCJD from a UK research group were as high as ten million. That number was readjusted to 50,000, then reduced even further to 7000 in 2003. The scientists involved in this study suggested that deaths could be even fewer, but there was still a lot of uncertainty in the models used.[98] Fortunately, BSE in cattle peaked at over 37,000 cases in 1992 (dotted blue line in Fig. 6.10),[99] and vCJD in humans (solid red line) peaked at 28 cases in 2000. It is worth including one last word from Gajdusek who had been warning the experts for years that TSEs could crop up anywhere; the possibilities were everywhere. This final comment from Gajdusek underscores the dangers of TSE transmission (Fig. 6.11):

You know the bone meal that people use on their roses? It's made from downer cattle. Ground extremely fine. The instructions on the bag warn you not to open it in a closed room. Gets up your nose. [Gajdusek] looked at me meaningfully. 'Do you use bone meal on your roses?' I told him I did…[he responded] I wouldn't if I were you.[100]

---

[95] Wikipedia contributors. (2023). United Kingdom BSE outbreak. *Wikipedia*. https://en.wikipedia.org/wiki/United_Kingdom_BSE_outbreak.

[96] Wikipedia contributors. (2023). United Kingdom BSE outbreak. *Wikipedia*. https://en.wikipedia.org/wiki/United_Kingdom_BSE_outbreak.

[97] Blakeslee, S. (2001, October 30). Estimates of future human death toll from mad cow disease vary widely. *The New York Times*. https://www.nytimes.com/2001/10/30/health/estimates-of-future-human-death-toll-from-mad-cow-disease-vary-widely.html.

[98] Bhattacharya, S. (2003, February 26). Predicted deaths from vCJD slashed. *New Scientist*. https://www.newscientist.com/article/dn3440-predicted-deaths-from-vcjd-slashed/.

[99] *Data tables: Creutzfeldt-Jakob Disease International Surveillance Network*. (n.d.). https://www.eurocjd.ed.ac.uk/data_tables.

[100] Rhodes (1998), p. 242.

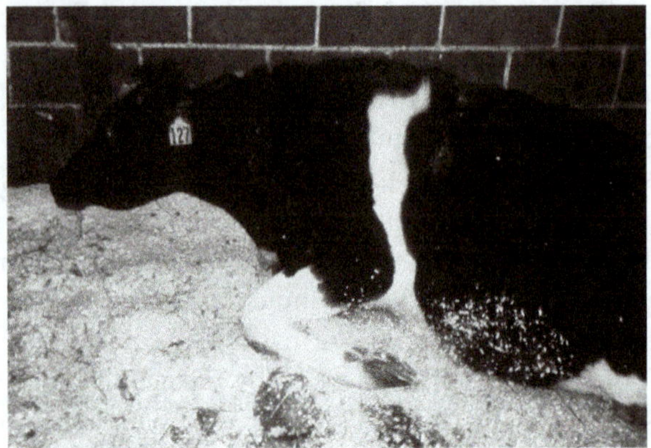

**Fig. 6.10** Cow suffering from BSE. (This image or file is a work of a United States Department of Agriculture and is in the public domain)

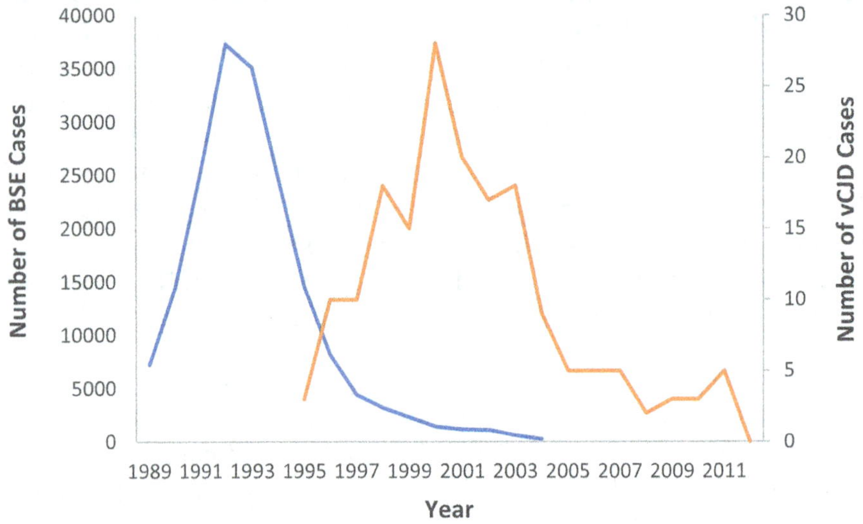


**Fig. 6.11** Decline of BSE (dotted blue) and vCJD (solid red line) deaths

# The 1997 Nobel Prize in Physiology or Medicine

Since that first landmark publication in 1982 when Prusiner first used the term prion, he had garnered a number of prestigious awards, including election to the National Academy of Sciences in 1992 and the Albert Lasker award in 1994. The Lasker award was not only one of the most prestigious awards in the sciences, but often presaged the winning of a Nobel Prize. However, the Prize presented yet another battle with the media for Prusiner, and in October 1994 he was forced to break his "don't talk to the media stance" and sit for an interview with a *New York Times* reporter. The *Times* well-known science writer Gina Kolata conducted the interview which from the very beginning became confrontational and lasted only 20 min.[101] When the article appeared two days later, it was as bad as Prusiner had imagined. The headline read "Viruses or Prions: An Old Medical Debate Still Rages."[102] Below that was the line: "Lasker Jury Takes One Side of the Controversy." The writer went on to state: "For more than a de cade, Dr. Stanley B. Prusiner…and his collaborators have zealously promoted [an] iconoclastic theory in the face of skepticism and hostility from many scientists."

The chair of the Lasker Foundation simply responded that the award should be seen as "reflecting the courage of the Foundation."

Then, as predicted by the Lasker award, the Medical Nobel Assembly at the Karolinska Institute in Stockholm announced on October 7, 1997 that Stanley Prusiner was the recipient of the 1997 Prize in Physiology or Medicine for "his discovery of prions—a new biological principle of infection."[103] Since his first prion article in 1982, Prusiner had published over 180 articles and had garnered more than $40 million in federal and private grants. He had contributed to or directed most of the important studies that supported prions as the scrapie agent. Certainly, by this time, many in the scientific world were convinced that prions were at least a part of the infectious particle, believing that misfolding of a normal prion protein resulted in holes in the brains of infected victims. As discussed by Rhodes in the *New Yorker* (1997), the controversial theory was "that the infectious agent was not a virus or a bacterium but a protein devoid of the nucleic acids DNA and RNA" (Fig. 6.12).[104]

When Prusiner's Nobel Prize was announced, it was evident that the controversy had not gone away. A plethora of national and international headlines decried the Nobel award as too controversial, with insufficient evidence to support the prion as the TSE infectious agent over the involvement of a virus. Many felt that in siding with the prion explanation, the Nobel Committee was inadvertently skewing the funding that might go to research on other theories. Support for the virus theory was still strong and vocal. As recently as 1995, Laura Manuelides had suggested that scientific findings still pointed to the involvement of an "unconventional virus," and that prions might perform a gatekeeper function at some point in the infection. She went

---

[101] Prusiner (2014), p. 185.

[102] Kolata, G. (1997, October 12). Eye on the Nobel: They should give a prize for ambition. *The New York Times.* https://www.nytimes.com/1997/10/12/weekinreview/eye-on-the-nobel-they-should-give-a-prize-for-ambition.html

[103] https://www.nobelprize.org/prizes/medicine/1997/summary/

[104] Rhodes, R. (1997, November 24) Pathological science, *The New Yorker.*

**Fig. 6.12** Prusiner accepting his Nobel Prize. (Courtesy of the Nobel Foundation, photo by Lars Aström)

on to say that "almost everything about the TSEs points to a slow virus…"[105] At a conference in Paris in 1995, Manuelides expanded on the evidence for a virus, repeating the idea that PrP could act as the gatekeeper during infection. Some of the key facts that supported a virus theory were (1) BSE and other TSE forms of the disease come in various strains, much like viruses such as influenza; (2) the scrapie agent persists during a long dormant period when proteins are normally cleared quickly; and (3) TSEs are transmitted across species barriers, what Manuelides called "a hallmark of an infectious agent." Finally, BSE is almost certainly transmitted orally, ending up in the gut where the enzymatic juices would normally digest the prion protein; viruses commonly survive this bath of strong acid and enzymes.

The Nobel committee, in an uncharacteristic move, felt a responsibility to respond to the public about their choice of Prusiner. Karolinska neurologist Lars Edstrom told the *New York Times* that "there are still people who don't believe that a protein can cause these diseases, but we believe it. From our point of view, there is no doubt." The vice chair of the Nobel Assembly, Ralf

---

[105] Rhodes (1998), pp. 253–254.

Pettersson even implied that the doubt expressed by Prusiner's naysayers might have contributed to the spread of BSE to humans.[106] Prusiner responded to his critics that "skepticism [is] important in science. It forces researchers to provide convincing evidence."[107] He went on to say that he didn't necessarily feel vindicated by the award, since "concepts are vindicated by the constant accrual of data and independent verification of data." Manuelidis was incredulous. She remarked "How could there be no doubt? Is this science? She continued: "There must be a special strain of the TSE agent that causes egomania."[108] By Manuelidis' count, "only four of the 14 TSE laboratories in 1997 wholeheartedly support the prion theory; nine others consider it unlikely, and one is undecided."[109] One clear outcome of the Nobel Committee's decision was that it would influence funding for future TSE research. As Bruce Chesebro at the NIH noted: "it would be tragic to stifle the future research needed to identify the precise nature of the transmissible agent responsible for this disease or to impede the search for drugs…to prevent or cure them."[110] But of course, it would be easier for government agencies to justify funding a laboratory or a program that had just received a Nobel Prize.

It was not just Manuelidis' colleagues who remained incredulous; the media was unconvinced, at best. Headlines from all over the world used words such as controversial work, unproven points, and skepticism. In perhaps the most scathing attack, Prusiner's old nemesis Taubes wrote that the Nobel was given for Prusiner's persistence. Taubes goes on to state:

> Good science, not just Nobel Prize-caliber science, depends on hypothesis and test, and then the rigorous demonstration that the preferred interpretation of the data was the only interpretation. In other words, remarkable results demand remarkable evidence. In the case of Prusiner's prize, the Nobel Committee has settled for enthusiasm and single-mindedness.[111]

Rhodes concluded that the award to Prusiner will provide a soapbox from which a Nobelist could encourage "a full range of studies—including continuing to search for a virus…"[112] Prusiner never supported the existence of a virus as a part of the infectious agent, and the Prize gave him an opportunity to reduce funds allocated for that area of research.

---

[106] Rhodes (1997), p. 54.

[107] Rhodes (1998), p. 250.

[108] Rhodes (1998), p. 251.

[109] Rhodes (1997), p. 55.

[110] Rhodes (1997), p. 59.

[111] Taubes, G. (1997, October 11). Nobel gas. *Slate Magazine.* https://slate.com/news-and-politics/1997/10/nobel-gas.html.

[112] Rhodes (1997), p. 59.

# Prion Studies Continued

Field work on kuru in Papua New Guinea ended in 2012,[113] the last BSE case in cows in Britain was recorded in September of 2021,[114] and the last case in the USA as recently as May 2023.[115] The last vCJD in humans occurred in 2019, bringing the total deaths worldwide to 233.[116] All of these TSE diseases are now considered very rare, with sporadic CJD occurring as infrequently as one in a million per year. Thus, the urgency of research into what is now an extremely rare group of diseases is greatly diminished. Yet research at various laboratories continues, funded primarily through government grants. For example, the MRC Prion Unit and Institute of Prion Diseases in London and the National Prion Disease Pathology Surveillance Center in the USA continue to monitor outbreaks of TSEs as well as conduct basic research.

Given that the TSEs are rare, the question then arises: why would scientists continue their research on prion diseases? Those in the field answer that question in several ways. First, detailed structural studies on purified prions could lead to new therapies and diagnostic tests. Several elegant studies have produced models of the prion that demonstrate the sites required for aggregation as well as protein misfolding, including work from Kurt Wüthrich who won a Nobel Prize in Chemistry in 2002 for structural studies of several proteins including prions.[117] Most recently, Byron Caughey and colleagues have provided evidence that the α-helices of the normal PrP unravel during infection, resulting in increased β-sheet structure first described by Prusiner. The unraveling exposes new sites within the protein that appear to provide the structures necessary for seeding the misfolding of normal PrP and to promote the aggregation of the misfolded proteins. The authors stated that "the pathogenic rod-like PrP fibrils have sticky ends that capture normal PrP and unravel

---

[113] UCL Faculty of Brain Scientists. (2022, March 21). *Kuru*. MRC Prion Unit and Institute of Prion Diseases. https://www.ucl.ac.uk/prion/kuru.

[114] Ritchie, D. L., Peden, A. H., & Barria, M. A. (2021). Variant CJD: Reflections a quarter of a century on. *Pathogens*, 10(11), 1413. https://doi.org/10.3390/pathogens10111413.

[115] Baker, H. (2023, May 24). Extremely rare case of mad cow disease detected in the US. *Livescience.com*. https://www.livescience.com/animals/land-mammals/extremely-rare-case-of-mad-cow-disease-detected-in-the-us.

[116] About Variant Creutzfeldt-Jakob Disease (VCJD). (2024, May 13). *Variant Creutzfeldt-Jakob Disease (vCJD)*. https://www.cdc.gov/variant-creutzfeldt-jakob/about/index.html.

[117] Riek, R., Wider, G., Billeter, M., Hornemann, S., Glockshuber, R., & Wüthrich, K. (1998). Prion protein NMR structure and familial human spongiform encephalopathies. *Proceedings of the National Academy of Sciences*, 95(20), 11667–11672. https://doi.org/10.1073/pnas.95.20.11667.

them as if they were binding to a narrow…strip of Velcro."[118] Identification of the exact amino acid sequences involved in this aggregation could assist researchers in developing drugs that could bind to the newly exposed sites and interfere with aggregation, and potentially halt the progression of the disease.

In a second area of research, scientists are actively looking for evidence that a cofactor such as a virus might be required for infection. Many researchers have argued that purification of PrP and demonstration that this synthetic prion alone can cause disease would finally end any talk of cofactors. Rosie Mestel stated in a review article in *Science* in 1996:

> To dispel lingering doubts that a virus might somehow be responsible for these diseases after all, researchers have to synthesize normal PrP in conditions guaranteed to be free of viral nucleic acid, and then show that the protein can fold into its rogue state and infect animals.[119]

Claudio Soto reported in 2010 that researchers are close to purification in a "clean" system. He reported:

> it's been 43 years since Prusiner first proposed the prion hypothesis, and after decades of controversy, it appears that the scientific community has accepted that misfolded prions are the active component of the infectious agent. With the recent demonstration from two laboratories[120,121] that infection can be produced by prions produced in vitro with defined components, it would appear that researchers have found the "Holy Grail."

However, Soto added a statement that indicated that some researchers were still not convinced that PrP acted alone: "it [still] remains unclear whether another molecule besides the misfolded prion protein might be an essential element of the infectious agent." In fact, several laboratories have recently reported that small molecules (RNA and phospholipids) can act as cofactors

[118] Kraus, A., Hoyt, F. H., Schwartz, C. L., Hansen, B., Artikis, E., Hughson, A. G., Raymond, G. J., Race, B., Baron, G. S., & Caughey, B. (2021). High-resolution structure and strain comparison of infectious mammalian prions. *Molecular Cell*, *81*(21), 4540–4551.e6. https://doi.org/10.1016/j.molcel.2021.08.011.

[119] Mestel, R. (1996). Special news report: Putting prions to the test. *Science*, 273(5272), 184–189. https://doi.org/10.1126/science.273.5272.184.

[120] Makarava, N., Kovacs, G. G., Bocharova, O. V., Savtchenko, R., Alexeeva, I., Budka, H., Rohwer, R. G., & Baskakov, I. V. (2010). Recombinant prion protein induces a new transmissible prion disease in wild-type animals. *Acta Neuropathologica*, 119(2), 177–187. https://doi.org/10.1007/s00401-009-0633-x.

[121] Wang, F., Wang, X., Yuan, C., & Ma, J. (2010). Generating a prion with bacterially expressed recombinant prion protein. *Science*, 327(5969), 1132–1135. https://doi.org/10.1126/science.1183748.

to facilitate the formation of prions with high levels of infectivity.[122,123] Several possible roles have been proposed for these accessory molecules: the cofactor might facilitate interaction with other cellular components, the cofactor might act as a catalyst for prion replication, or the cofactor might stabilize the conformation of the misfolded protein.

Ever since Gajdusek first described kuru and suggested that the disease was caused by a slow virus, there are still groups that are convinced that a virus— perhaps a small virion or viroid—could be hidden by a protein shell (perhaps a PrP shell) which would protect the virus from degradation by nucleases. In a 1996 article by Bruce Chesebro, the author suggests that "so far no virus has been found, and the exact nature of the transmissible TSE agent remains the central unsolved mystery of this field."[124] Despite the demonstrations that the infectious agent can be created in vitro without nucleic acid, Manuelidis' laboratory has continued its claim that viruses are the infectious agent—not prions. As recently as 2016, the group presented evidence that the CJD and scrapie infectious agent is viral in nature. The authors stated: "These findings have fundamental ramifications, and open new avenues for the…prevention and treatment" of TSEs.[125] Most recently, Hideyuki Hara and coworkers have continued the debate and demonstrated that influenza virus could induce conversion of PrPc to PrPsc in cultured cells. These studies suggest that infection with influenza A viruses or other related viruses might be a cause of or are associated with the pathogenesis of sporadic CJD.[126] All of these studies continue to support the potential role of viruses in protein misfolding and aggregation and suggest that research should continue to definitively determine if viruses are a component of the infectious TSE agent.

Researchers have also been focusing on potential treatments for CJD that might halt the progression. In exciting recent work, scientists at the National Institute for Health Research UCLH Biomedical Research Centre in London have developed an antibody that has shown promising results in potentially

---

[122] Supattapone, S. (2020). Cofactor molecules: Essential partners for infectious prions. In *Progress in Molecular Biology and Translational Science* (pp. 53–75). Academic Press. https://doi.org/10.1016/bs.pmbts.2020.07.009.

[123] Miller, M. I., Wang, D., Wang, F., Noble, G. P., Ma, J., Woods, V. L., Li, S., & Supattapone, S. (2013). Cofactor molecules induce structural transformation during infectious prion formation. *Structure*, 21(11), 2061–2068. https://doi.org/10.1016/j.str.2013.08.025.

[124] Chesebro, B. (1997). Human TSE disease — viral or protein only? *Nature Medicine*, 3(5), 491–492. https://doi.org/10.1038/nm0597-491.

[125] Botsios, S., & Manuelidis, L. (2016). CJD and scrapie require agent-associated nucleic acids for Infection. *Journal of Cellular Biochemistry*, 117(8), 1947–1958. https://doi.org/10.1002/jcb.25495.

[126] Hara, H., Chida, J., Uchiyama, K., Pasiana, A. D., Takahashi, E., Kido, H., & Sakaguchi, S. (2021). Neurotropic influenza A virus infection causes prion protein misfolding into infectious prions in neuroblastoma cells. *Scientific Reports*, 11(1). https://doi.org/10.1038/s41598-021-89586-6.

halting the progression of CJD.[127] In a second study, researchers at the Laboratory of Persistent Viral Diseases, Rocky Mountain Laboratories in Montana, used a genetic approach to block normal PrP expression in mice and demonstrated that these mice had extended survival times following infection with scrapie tissue.[128]

Finally, perhaps the most important and exciting area of research that keeps prion research alive and well is the finding that other neurodegenerative disorders such as Alzheimer's disease (AD) have "prion-like" misfolded proteins and formation of amyloid plaques as part of their diseases as discussed below.

## A Prion Revolution?

One of the most feared outcomes of growing old is the development of dementia with the progressive loss of a range of mental abilities. There are more than 50 recognized dementia disorders, but the most common is Alzheimer's disease (AD), with a worldwide prevalence of over 35 million people as of 2010. Dementia can be traced back at least to 2000 B.C. and the ancient Egyptians. Plato, a philosopher in Greece from 428–347 B.C., stated that the principal cause of dementia is old age itself, and the mental stability in humans is destined to degrade. In more recent times, Dr. Alois Alzheimer, one of the first physicians to describe the pathological features of dementia, conducted the first autopsy on an Alzheimer patient, Auguste Deter, in 1906. At the age of 51, Auguste began to exhibit uncontrollable behavior. Her symptoms rapidly deteriorated, and she died of pneumonia at the age of 55. On autopsy, Auguste's brain contained plaques in the neurons, and tangles in nerve fibers. Plaques had been found previously in the brains of older adults, but the tangles were new. In someone this young, both findings were remarkable.

By the early 1980s, researchers in the prion field knew that kuru, CJD, and scrapie all have abnormal PrP proteins that form insoluble scrapie-associated

---

[127] Mead, S., Khalili-Shirazi, A., Potter, C., Mok, T., Nihat, A., Hyare, H., Canning, S., Schmidt, C., Campbell, T., Darwent, L., Muirhead, N., Ebsworth, N. M., Hextall, P., Wakeling, M., Linehan, J. M., Libri, V., Williams, B., Jaunmuktane, Z., Brandner, S., . . . Collinge, J. (2022). Prion protein monoclonal antibody (PRN100) therapy for Creutzfeldt–Jakob disease: Evaluation of a first-in-human treatment programme. *Lancet Neurology*, 21(4), 342–354. https://doi.org/10.1016/s1474-4422(22)00082-5.

[128] Raymond, G. J., Zhao, H., Race, B., Raymond, L. D., Williams, K., Swayze, E. E., Graffam, S., Le, J., Caron, T., Stathopoulos, J., O'Keefe, R., Lubke, L., Reidenbach, A. G., Kraus, A., Schreiber, S. L., Mazur, C., Cabin, D. E., Carroll, J. B., Minikel, E. V., . . . Vallabh, S. M. (2019). Antisense oligonucleotides extend survival of prion-infected mice. *JCI Insight*, 4(16). https://doi.org/10.1172/jci.insight.131175.

fibrils (Merz's SAF). In addition, several groups had proposed that an abnormal form of PrP (PrPsc) caused normal forms to misfold, damaging cells until they died. The dead cells were cleared out, leaving the characteristic holes of spongiform degeneration. Gajdusek couldn't help but notice that the TSEs and other neurodegenerative disorders such as Alzheimer's share some pathological features: both involve abnormal protein folding followed by formation of amyloid plaques. The major brain destruction—the holes—is similar in both CJD and AD. Gajdusek and his colleagues realized that they were looking at two similar but different kinds of brain disorders: infectious (TSE) and noninfectious (Alzheimer's). Connecting a rare group of diseases (TSEs) with the most common and incurable dementias that affect humans brought many new researchers and new funding into the field. In their first major study to examine the similarity between CJD and Alzheimer's, Gajdusek and colleagues studied the pathology and transmissibility of the two disorders. While both exhibit the classic signs of amyloid plaques, the primary difference between these two diseases was that CJD and other TSEs could be readily transmitted to a variety of non-human species, while Alzheimer's was not transmissible.[129] Gajdusek still believed that there was a connection between the disorders. As he stated: "a new era has come to neurology as we have realized that both Alzheimer's disease and the normal aging brain are examples of nontransmissible brain amyloidosis."[130]

Building on what was known about the pathology of neurodegenerative disorders, scientists began to search for the causative agents in these diseases. They knew that tangles and amyloid plaques were well-described characteristics of AD,[131] but what was the underlying cause? In 1975, Kirshner and colleagues had purified a protein that they called tau (tubule-associated protein). From other studies they knew that tau is found primarily in neurons and assists in maintaining the proper shape of cells. In the AD brain, tau is changed by the addition of phosphate, causing tau to misfold. The free modified tau molecules aggregate to form the tangles leading to cell death. These and many other studies have now shown that several degenerative diseases such as AD share a pathology in common with prion diseases: each has specific proteins that misfold and aggregate, leading to progressive loss of neurons in specific

---

[129] Brown, P., Salazar, A. M., Gibbs, C. J., & Gajdusek, D. C. (1982). Alzheimer's disease and transmissible virus dementia (Creutzfeldt-Jakob disease). *Annals of the New York Academy of Sciences*, 396(1), 131–143. https://doi.org/10.1111/j.1749-6632.1982.tb26849.x.

[130] Rhodes (1998), p. 198.

[131] DeTure, M., & Dickson, D. W. (2019). The neuropathological diagnosis of Alzheimer's disease. *Molecular Neurodegeneration*, 14(1). https://doi.org/10.1186/s13024-019-0333-5.

areas of the brain.[132] These striking similarities between AD and prion diseases have led researchers to focus on altering protein misfolding and/or aggregation in an effort to halt disease progression.

It didn't take long for Prusiner to jump on the neurodegenerative disorder bandwagon. Once again, he was taking control of the field by calling all of these diseases "prion disorders." As he stated in his autobiography:

"the prion revolution...now encompasses Alzheimer's (AD), Parkinson's, the frontotemporal diseases, ALS, and Huntington's disease." Furthermore, "The new vistas that have been opened by investigations demonstrating that prions cause these neurodegenerative disorders promise to enhance our understanding and facilitate research. Once so mysterious, degenerative brain diseases are beginning to reveal the mechanisms by which they compromise nervous system function."[133]

In his 1997 Nobel lecture Prusiner alluded to this "revolution": "understanding of prion formation could open new approaches to developing effective therapies for the more common neurodegenerative diseases, including Alzheimer's disease, Parkinson's disease, and amyotrophic lateral sclerosis (ALS)."[134]

In a 2020 review describing both differences and similarities between prion diseases and other neurodegenerative diseases such as AD, the authors point out that we still don't understand how these diseases start.[135] There are indeed remarkable similarities between prion diseases and AD, including misfolded proteins that act as catalysts to propagate further misfolding leading to cell death. However, there are several significant differences between the prion diseases and other degenerative diseases. For example, after AD patients begin to exhibit dementia symptoms, there is a relatively long period of time until death (i.e. years). In TSE, after the first symptom appears, patients die relatively quickly (months). And perhaps the most controversial difference: whereas TSEs are readily transmissible from human to human, the neurodegenerative diseases do not appear to be infectious. One thing is certain: the

---

[132] Gomez-Gutierrez, R., & Morales, R. (2020). The prion-like phenomenon in Alzheimer's disease: Evidence of pathology transmission in humans. *PLOS Pathogens*, 16(10), e1009004. https://doi.org/10.1371/journal.ppat.1009004.

[133] Prusiner (2014), p. 260.

[134] Kwon, D. (2015). The great brain drain. *Scientific American*, 313(5), 17–18. https://doi.org/10.1038/scientificamerican1115-17.

[135] Jaunmuktane, Z., & Brandner, S. (2020). Invited review: The role of prion-like mechanisms in neurodegenerative diseases. *Neuropathology and Applied Neurobiology*, 46(6), 522–545. https://doi.org/10.1111/nan.12592.

remarkable shared protein misfolding mechanisms first identified in TSEs, followed by numerous studies in neurodegenerative diseases, has opened up an exciting new area of investigation focused on development of strategies to halt the progression of these debilitating and deadly diseases. Even as we discover new and exciting approaches to treatments and cures, questions still remain. As the TSE researcher Paul Brown has suggested, there are two important questions that have not yet been answered: once this misfolding begins, "how is it self-sustaining? But, the more interesting problem is how does it begin in the first place?"[136]

## Where Are They Now?

**Stanley Prusiner.** As of 2024, Prusiner is currently the Director of the Institute for Neurodegenerative Diseases and Professor of Neurology at UCSF. He has authored or co-authored over 500 scientific publications and 300 review articles, edited 11 books, and holds 50 patents. He has delivered over 725 invited lectures. Currently, he is the President-elect of the American Neurological Association. In addition to his Nobel Prize, Prusiner has received various awards, honorary degrees, and admission to honor societies. The notable honors include election to the National Academy of Sciences in 1996, and to its governing board in 2007; membership in the Institute of Medicine; and the National Medal of Science in 2010. Dr. Prusiner continues to conduct research at UCSF, and currently has a research publication in press entitled "Different $\alpha$-Synuclein Prion Strains Cause Dementia with Lewy Bodies and Multiple System Atrophy" (Fig. 6.13).

    **D. Carleton Gajdusek.** Like many of his Nobel laureate colleagues, Gajdusek reveled in every minute of the fame that was associated with winning the prize. Although he wondered at first about retaining his humility, he began working less with his research group at the NIH and spent much of his time traveling and speaking widely. His resume lists awards, invited lectures, and more than 130 honorary degrees. Gajdusek was highly sought after as a visiting scientist at several institutions in Europe and Asia. He continued to receive awards and recognitions long after his Nobel Prize award. At the 200th anniversary of Harvard's Medical School in 1982, Gajdusek was honored with one of the Medical School's first honorary degrees (Fig. 6.14).

    Gajdusek officially retired from his NIH position in 1997. He spent most of his time at the Alfred Fessard Institute of Neurobiology near Paris, the

---

[136] Rhodes (1998), p. 199.

Fig. 6.13  Stanley Prusiner in 2024. (Licensed under the Creative Commons Attribution 4.0 International. Author Cmichael67)

University of Amsterdam, and the University of Tromsø in Norway. He was a visiting scientist in China, and between 1998 and 2002 he visited 11 other countries. Staff remaining in the laboratory continued to conduct TSE research for another eight years, as Joe Gibbs attempted to bring Gajdusek's program to an orderly conclusion. Finally, with the retirement of Paul Brown, who was the last investigator engaged in TSE research on the NIH Bethesda campus, Gajdusek's laboratory ceased to exist. The NIH continues to maintain some laboratory records and a collection of research samples.

Then tragedy struck. For several decades, Gajdusek had been bringing boys from New Guinea back to the USA as his "foster children." Back in 1960, while reflecting on his relationships with the young boys who had been his "workers," he stated: "All have worked wonderfully for me and I cannot adequately repay them. How much I would like to have them in the U.S. and share my home with them!"[137] Throughout the years since then he had welcomed the young boys into his home, and he kept meticulous journals that eventually became part of his required summaries and reports to the NIH. Then FBI and members of Congress became aware of strange, perhaps pedophile, activities that were going on in his Maryland house. In 1996, one

---

[137] Anderson (2008), p. 223.

**Fig. 6.14**  Daniel Carleton Gajdusek. (Courtesy of the National Library of Medicine; photo in public domain)

boy came out publicly with an accusation against Gajdusek for sexual abuse. Five other boys were most likely also abused, although these boys did not come forward. An additional boy—the son of one of his coworkers—eventually came out with graphic and tragic details of sexual abuse by Gajdusek, documented in a film called *The Genius and the Boys*.[138] Gajdusek was eventually charged with child abuse and two counts of violating a Maryland law prohibiting oral sex.

Gajdusek was arrested on April 4, 1996, and ultimately agreed to a plea deal in which he admitted to two counts of child abuse. In the agreement, the federal government would end any further investigations, including possible immigration violations involving the children and his misuse of NIH funds. Despite a number of letters from colleagues who pleaded for leniency, Gajdusek was sentenced to 18 months in prison. He also agreed to spend the rest of his life outside the USA.

Gajdusek was unrepentant about his relationships with the boys, stating that sex with young boys was accepted in many countries, and felt that "American law was prudish."[139] Gajdusek was also a proponent of "intergenerational sex," prompted perhaps by what might have been an abusive

---

[138] The Genius and the Boys documentary (2009). https://www.youtube.com/watch?v=_fgJ85zxd7U

[139] Independent reporter. (1996, August 4). The fall of a family man. *The Independent*. https://www.independent.co.uk/news/the-fall-of-a-family-man-1308277.html.

relationship with his uncle. As reported by *The Independent* in August 1996, Gajdusek claimed:

> I would, at this moment, have every youth sleep with his sister, get seduced by his older brother and male teacher, practice with his male and female cousins, aunts, uncles and teacher and maid—anything!—only to know sex as fun and frivolity, as rhythm and passionate play—from an early age—from the very onset of puberty.[140]

He included the boys in his family through legal adoption, and he spent much of his personal money, including at least a portion of his Nobel award money, on their education.

Gajdusek died in Tromsø, Norway on December 12, 2008, at the age of 85. At the time of his death, he was working with and visiting colleagues in Tromsø. As Goudsmit noted in his obituary: "Gajdusek will be remembered for both his scientific contributions and his overwhelming presence."[141] As Richard Rhodes observed in his book *Deadly Feasts*, Gajdusek was "A compulsive talker who spills ideas nonstop for hours—good talk, often brilliant talk and consummate story-telling, but more than some listeners can bear." Like Lord Jim, "he passes away under a cloud, inscrutable at heart, forgotten…an obscure conqueror of fame."[142]

## Is Amyloid the Final Chapter?

From that first autopsy of the brain from a child who died of kuru, Igor Klatzo identified a structure called amyloid plaques in the child's brain sections. These plaques reminded him of the structures found nearly four decades earlier by Creutzfeld and Jakob. Then in 1974, Patricia Merz was looking through her electron microscope at scrapie brain slices. And in those images cast on the fluorescent screen there they were again—those unusual fibril-looking plaques. Through staining with dyes, Merz concluded that they were indeed amyloid. So, what was this unusual collection of fibrils, and what role did it play in scrapie and other TSEs? Merz' best guess at this point was that the plaques (or fibrils) were an accumulation of some unknown material that

---

[140] Independent. (1996, August 4). The fall of a family man. *The Independent*. https://www.independent.co.uk/news/the-fall-of-a-family-man-1308277.html.

[141] Goudsmit (2009) p. 457.

[142] Anderson (2008), p. 230.

was accumulating outside cells in the brain, ultimately causing cell death. In fact, these plaques have now been reported in over 20 degenerative brain diseases including AD and TSEs.

Further research showed that these plaques contain protein, and through exquisite imaging techniques it is now known that these proteins are the abnormal forms of prions and other prion-like proteins that undergo a shift from normal to abnormal forms. The abnormal proteins have reconfigured their shape to include a high amount of the flattened form (β-sheet) within the protein. These abnormal forms lie on top of one another, building up like Legos to form the dreaded amyloid plaques or fibrils.

Current research now focuses on attacking either the formation of the amyloid plaques or breaking them up once formed. The Food and Drug Administration approved the first drug in the form of an antibody against the plaque protein in 2024. The drug appears to bind to the amyloid and prevents further growth and cell damage. The question now is whether this approach can be used to treat all of these amyloid-containing brain diseases by halting formation of the destructive plaques.

As mentioned earlier, AD is one of the most feared diseases in our aging population, but it's only one of the many brain diseases that have been identified to date. New strategies for diagnosis and treatments are increasing as more researchers enter the field. There are, of course, still questions surrounding these diseases. Are there other structures underlying the disease progression that have yet to be found? What about diagnoses: can we identify disease in the early stages of dementia? Perhaps the most troubling question was raised recently: a group at Queen Mary University in London[143] reported that using imaging techniques, they can predict whether a person is going to develop dementia as early as nine years before symptoms develop. Would you really want to know? Perhaps now, with treatments appearing, you would.

## Bibliography

Anderson, W. (2008). *The Collectors of Lost Souls.* Johns Hopkins University Press eBooks.

---

[143] Queen Mary University of London (2024, June 7). First-of-its-kind test can predict dementia up to nine years before diagnosis. *ScienceDaily.* https://www.sciencedaily.com/releases/2024/06/240606152250.htm#:~:text=Summary%3A,used%20methods%20for%20diagnosing%20dementia.

Klitzman, R. L. (1998). *The Trembling Mountain: A Personal Account of Kuru, Cannibals, and Mad Cow Disease*. Perseus Publishing.

Lindenbaum, S. (2013). *Kuru Sorcery*. Paradigm Publishers.

Prusiner, S. B. (2014). *Madness and Memory*. Yale University Press.

Rhodes, R. (1998). *Deadly Feasts: The "Prion" Controversy and the Public's Health*. Simon and Schuster.

# 7

# Afterword

*It is high time laymen recognized the misleading belief that scientific enquiry is a cold dispassionate enterprise, bleached of imaginative qualities, and that a scientist is a man who turns the handle of discovery: for at every level of endeavour scientific research is a passionate undertaking, and the Promotion of Natural Knowledge depends above all upon a sortie into what can be imagined but is not yet known.[1]*

Peter Medawar

The individuals we've examined in the previous chapters are all unique, but taken together they illustrate that progress in science is more complex than just the science itself. They're all brilliant scientists, acknowledged geniuses, if you like. But they're much more than that; genius isn't the whole story. Their lives are beset by cross-currents of war and nationalism, passion and jealousy, competition and ambition, perseverance and luck. The object of this book has been to pull back the curtain to illuminate the factors that led them to their discoveries, the impact of those discoveries on science and society, and what they did with their discoveries and with their Nobel fame. Alfred Nobel himself is an enigma, a brilliant inventor who left his vast fortune to the world, a lonely hermit who wrote poetry, "the merchant of death" who supported peace. Fritz Haber was an ambitious chemist whose scientific gift now feeds the world, and whose love for Germany gave the world poison gas. Otto Hahn and Lise Meitner were dedicated scientists whose search for knowledge opened Pandora's box with the discovery of nuclear fission. Werner Heisenberg

---

[1] Medawar, P.B. (1963) Times Literary Supplement (London), October 25, p. 850.

wanted to study pure mathematics, but when he was diverted into physics, he completely changed our understanding of nature, after which he used his science to serve the Nazis. John Bardeen and Walter Brattain were the brilliant scientists who discovered the transistor. William Shockley tried to take the credit for their discovery, went on to plant the seeds that became Silicon Valley, then turned to darker thoughts of racism. Carlton Gajdusek discovered a disease spread by cannibalism in a remote tribe in the jungles of New Guinea and later went to prison for pedophilia. Stanley Prusiner convincingly showed that some neurological diseases are caused by protein misfolding, but he crossed ethical boundaries in his pursuit of a Nobel Prize.

All these lives and discoveries shed light on the complexity of science and how it really works and show that despite its faults and missteps, science "is the most successful enterprise human beings have ever engaged upon."[2]

## Progress in Science

As Medawar has said,

> a scientist must indeed be freely imaginative and yet skeptical, creative and yet a critic. There is a sense in which he must be free, but another in which his thought must be very precisely regimented; there is poetry in science, but also a lot of bookkeeping.[3]

Faced with this tension between imagination and precision, how does science actually make progress? There are many ways to answer this question, but fundamentally, science is incremental, constantly building on what's gone before. As Isaac Newton once famously said, "If I have seen further than others, it is by standing upon the shoulders of giants."[4] Or, as Adriano Aguzzi remarked regarding research on prion diseases:

> The term incremental is often used pejoratively to describe futile research, yet the history of prion science shows that many gradual discoveries can cooperate to transform an esoteric hypothesis into a comprehensive, solid body of mainstream science. This narrative is by no means specific to prions. The most

---

[2] Medawar, P.B. (1988) *Preface to the limits of science*. Oxford.

[3] Medawar, P.B. (1996). *The strange case of the spotted mice and other classic essays on science*. Oxford University Press, USA.

[4] Turnbull, H.W. (Ed.) (1959). *The correspondence of Isaac Newton: 1661–1675*, Volume 1, London, UK: Published for the Royal Society at the University Press. p. 416.

impactful paradigm shifts in science may originate from extraordinary intuition, but it is the rigorous and often thankless work of smaller labs that prepares the terrain for such shifts to occur.[5]

Among the scientists we've discussed, there are several good examples of how science progresses. One is offered by Otto Hahn and Lise Meitner and their discovery of fission. They started with the work by Enrico Fermi and others who showed that when lighter atoms are bombarded with neutrons, they absorb the neutrons and transition to heavier atoms located one space farther along the periodic table. Working from Fermi's results, Hahn and Meitner hypothesized that by bombarding uranium with neutrons, they would create what they called "transuranes," which are atoms heavier than uranium that don't exist in nature. But when they did the experiment—repeatedly—it failed. All they got was barium, a much lighter atom about half the size of uranium. Faced with this apparently negative result, Meitner and Robert Otto Frisch generated a new hypothesis: when uranium atoms are bombarded with neutrons, they split into smaller pieces and the fragments have lots of energy. Frish did the experiment and found that the energy of the fragments agreed with their prediction. This supported their hypothesis but wasn't the whole story. Later, during the atomic bomb program in World War II, scientists discovered the transuranes that Hahn and Meitner had been seeking. They are called neptunium and plutonium, and the atomic bomb that fell on Nagasaki was made from plutonium. Science is never finished. Scientists must always keep testing and remain ready for new results.

Another elegant example of how science progresses is the series of experiments that Prusiner did in the discovery of prions. Carlton Gajdusek had shown that spongiform encephalopathies were transmissible and Patricia Merz had found proteinaceous fibrils in infected brains, but other scientists argued that they couldn't be the infectious agent. After all, soon after the discovery of DNA, Francis Crick proposed the "Central Dogma:" genetic information goes from DNA to RNA and then to protein, so an infectious agent must contain DNA or RNA. But attempts to find DNA or RNA in the infectious agent that caused scrapie all failed. Confronting these negative results, Prusiner hypothesized that not only was the scrapie agent a protein (he called it a prion), but it was a misfolded form of proteins already existing in a normal brain. This suggested an experiment: if the normal protein is absent, there should be no disease. Prusiner tried the experiment with a "knockout" mouse

---

[5] Aguzzi, A. (2024). Prion science and its unsung heroes. *Science*, 383(6680). https://doi.org/10.1126/science.adn9424.

having no gene to make the protein, and the mouse was immune to disease, as he had hypothesized. Did that prove that the prion hypothesis is correct? Yes, to a point, but as with Hahn and Meitner's discovery of fission, it's never possible to prove a hypothesis correct. Research is still going on to find a virus, or something else, that plays a role in prion diseases.

Does this imply that one can never find absolute truth? In some sense, yes. Scientists always have to keep testing and be ready for what they find. For example, despite over 200 years of experiments, it took Einstein's theory of relativity to explain an almost unobservable deviation in the motion of the planet Mercury from Newton's predictions…and, by the way, make the GPS (global positioning system) possible. So, can science be trusted? Yes, within its limitations, in part because scientists are genuinely seeking the truth, and in part because science is self-correcting. For most scientists, as Robert Musil writes:

> Thirst for knowledge is like an addiction or a yearning for love or a lust to kill, as it throws a character off balance. It is not true that the scientist goes after the truth. It goes after him. It is something he suffers from.[6]

Or as Jim Ericson put it, "We do not suggest that science invented intellectual honesty, but we do suggest that intellectual honesty invented science."[7] But even when something seems to be "proved," scientists must keep testing and be ready for whatever they find. So often, tiny advances have proved to have major impacts.

Many of the incremental advances that take place between Aguzzi's "impactful paradigm shifts in science" appear in the form of experimental tools that others use. For example, Hahn and Meitner's discovery of fission depended on the Geiger counter that Hans Geiger developed in Ernest Rutherford's lab 30 years before. William Shockley's development of the junction transistor depended on the crystal refining and growing techniques developed by Gordon Teal. Prusiner's experiments depended on the knockout mice that had been developed by his collaborator Charles Weissmann. From time to time, new experimental methods have been so important that they've been awarded Nobel Prizes themselves. For example, in 1930 E.O. Lawrence invented the cyclotron, the world's first large "atom-smashing" machine, for which he got the Nobel Prize in Physics in 1939. In 1983, Kary Mullis

---

[6] Musil, R. (1930). *Der Mann ohne Eigenschaften*. Roman.

[7] Ericson, J. (1975). Rationality of scientific revolutions. Harre, R. (Ed.). Problems of scientific revolutions. Quoted in footnote 42.

invented polymerase chain reaction (PCR), which is now the basis for so much of genetic testing and biomedical research and for which Mullis was awarded the Nobel Prize in Chemistry in 1993. But most of the incremental advances are small and viewed from afar we never notice them. A lot of work goes on between the "aha" moments in science.

Sometimes, scientific progress gets a big boost from luck. The classic example is Alexander Fleming, who discovered penicillin when some petri dishes that he found in the sink showed bacteria dying around spots of mold. The discovery of fission provides another good example of serendipity. Hahn and Strassmann never expected to see nuclei splitting, and only very reluctantly and hesitantly did they realize that was what they were seeing.

Sometimes the progress of science is slowed by the reluctance to accept new ideas. As Francois Jacob, Nobel Laureat in Medicine, expressed it, "In the course of time, as one develops expertise, one becomes a kind of prisoner of what one does and what one knows,"[8] or as Max Planck once said:

A new scientific truth does not triumph by convincing its opponents and making them see the light, but rather because its opponents eventually die, and a new generation grows up that is familiar with it.[9]

This has been paraphrased as: "science progresses one funeral at a time." In Planck's case, it took 21 years, a whole generation, for the Nobel Committee to accept the reality of his quanta, even though Einstein had proved the existence of quanta in the photoelectric effect 15 years earlier. In an extensive study of publication data in *Pub Med*, Pierre Azoulay, et al. showed the decline of research in a particular area of biomedical research when a luminary leaves the field and the increase of research into new ideas related to that area.[10] It's interesting to point out in this connection that all the scientists that we've discussed, save Hahn and Meitner, were under 40 years of age when they made their major discoveries. The members of the Academies that award the Nobel Prizes, on the other hand, are typically more senior, and this may have contributed to their reluctance to accept new ideas. They were certainly slow to accept Einstein's relativity and Planck's quanta, but with the mushroom clouds still fresh in people's minds, the prize-worthiness of fission in 1945 was hard to deny!

---

[8] Jacob, F. (1997). *La souris, la mouche et l'homme*. Jacob.

[9] Planck, M. (1950). *Scientific autobiography and other papers*. New York: Philosophical library.

[10] Azoulay, P., Fons-Rosen, C., & Graff Zivin, J.S. (2019) Does science advance one funeral at a time? *American Economic Review*, 109(9), 2889–2920.

# Competition or Collaboration?

Scientific progress is driven by a combination of competition and collaboration. The importance of collaboration is obvious, since science is incremental and the work of each scientist is built upon what others have done before. But the importance of competition in the lives of scientists cannot be overstated: a discovery can be made only once and credit goes to the one who is first. Scientific institutions as well as scientists must determine how to balance cooperation and competition to provide the best research environment for the progress of science. The traditional forms of competition and collaboration have evolved over the past few decades as the research enterprise has expanded from individuals working alone with occasional letters to friends to the instantaneous global exchange of ideas and research results. This has resulted in the emergence of more intense competition and new forms of collaboration, from laboratory-based teams to university-industry partnerships. Historically, scientific research has been performed by one senior investigator testing his or her hypotheses with the assistance of a team of laboratory-based students and junior scientists, collaborating in the sharing of their ideas, results, and laboratory resources to solve a specific problem. But often there is competition or even jealousy within a team. These teams, in turn, compete with laboratories at other universities or institutes. Today, the world faces a number of complex global problems like climate change that require the collaboration of many teams. Collaboration can bring together innovative thinkers in a field to study a specific topic, while competition can promote creativity and drive individuals to push boundaries, achieve breakthroughs, and continuously improve their work. Both collaboration and competition play crucial roles in advancing knowledge and fostering innovation. The examples in this book provide excellent examples of how collaboration and competition have affected scientists and their research.

Fritz Haber won his Nobel Prize for the discovery of a way to produce ammonia from atmospheric nitrogen. In 1898, Dr. William Crooke declared that the world would run out of natural sources of nitrates to make fertilizer by the early twentieth century. It was up to the chemists to discover the much needed "chemical manure." Haber's training as a chemist had provided him with the knowledge and tools needed to "fix" nitrogen from the air, using chemistry to break the strong bond between nitrogen atoms and combine them with hydrogen to form ammonia to make fertilizer. But there was strong competition from more senior chemists. Wilhelm Ostwald, already a Nobel Laureate, was the first to announce that he had produced ammonia. He filed for a patent and contacted a German industry (BASF) to partner with him to ramp up ammonia production. But when BASF tested his nitrogen-fixing equipment, they found that in fact his "products" were actually contaminants in his apparatus. Although he had correctly determined that the process would require extreme heat and pressure together with a catalyst, he simply dropped the project and never worked on it again. Then Haber and Walther Nernst entered the contest. Both designed equipment they needed for the process, but they argued (publicly) about whose results were correct. In the end, Haber's process produced significant amounts of ammonia, and BASF eventually partnered with Haber to create a large-scale production plant. Nernst, the senior investigator, dropped the project, as had Ostwald, and moved on to other research, but the intense competition led to production of ammonia that is estimated to have saved millions, perhaps billions of people over the course of the next several generations.

Another example of the importance of competition is the German atomic bomb project. Shortly after the discovery of fission, the German military leaders assembled the leading physicists remaining in Germany and created the Uranium Club with a goal of building a German atomic bomb. Kurt Diebner, Director of the Nuclear Research Council of the German Army Research Office, was put in charge. He organized a plan for cooperation using the members of the Uranium Club to cover various aspects of the program according to their expertise. As an experimenter, Diebner ran a reactor experiment at the research station in Gottow and put Heisenberg in charge of theory, a role for which he was eminently qualified. But Heisenberg, by virtue of his Nobel Prize, became a competitive leader in the program and started a reactor experiment of his own at Leipzig, a role for which he was eminently not qualified. Heisenberg and Diebner had no respect for one another and refused to cooperate. Heisenberg's experiments were unnecessary, duplicative, and consumed so many resources that no reactor project could be successful. By the end of the war, the German project consisted of one experimental reactor in a cave in Southern Germany that didn't work. In this case, internal competition and jealousy had a negative influence and collaboration was largely absent. In contrast to this, the Allied program had strong leadership and enjoyed good cooperation between the participants. Moreover, perceived competition with—or fear of—the Germans spurred the Allied program on.

William Shockley had a complex relationship with his colleagues at Bell Labs, including both collaboration and competition. He had been charged by Bell Labs to build solid-state amplifiers and switches to replace the vacuum tubes and mechanical devices in the expanding telephone system, so he formed a team including Walter Brattain and John Bardeen. Assigned by

Shockley to find out why his proposed "field-effect" transistor didn't work, Brattain and Bardeen collaborated on experiments while Shockley was off working on other, unrelated problems. In the course of their experiments, Bardeen and Brattain hit upon the first working transistor, but it wasn't Shockley's "field-effect" transistor. At this point, Shockley believed that as the head of the group, he "owned" the right to call the discovery his, despite being "absent at the creation."[11] Shockley was further infuriated when the patent was assigned to Brattain and Bardeen and he was left off. As the rift got worse, Shockley actually excluded Brattain and Bardeen from further developments of the transistor and went off by himself and furiously filled notebooks with ideas that he could claim as his own. One of these became the junction transistor. Was competition good or bad? Eventually, all three were recognized for the Nobel Prize for the transistor, but by then Bardeen had left for the University of Illinois, where he got a second Nobel Prize for superconductivity. When Bell Labs management finally recognized Shockley's limitations as a manager and stuck him in a middle-management position, Shockley left Bell Labs, moved to California, established the first company of its kind in Silicon Valley but never produced a marketable product. The engineers he attracted to California soon formed their own company and ultimately became billionaires. Shockley got left behind. Competition can be good or bad.

Another example of intense competition is the story of the discovery of prions. The concept of misfolded proteins as the infectious agent that causes the class of neurodegenerative diseases is told in detail by Stanley Prusiner himself in his autobiography entitled *Memory and Madness*. Although Prusiner describes several notable collaborators such as Leroy Hood, Thomas Südhof and Charles Weissmann, for the most part Prusiner planned the experiments that his group performed for his discovery. There is little "we" in this story. Previous contributors such as Patricia Merz got short shrift in Stanley's story. Prusiner was on a mission, with no time for colleagues who were not on his path to the Nobel Prize. Even the important collaboration with Weissmann ended in frustration when Weissmann didn't include Prusiner on the patent for the prion gene and didn't include Prusiner as a co-author on one of the key papers. Prusiner also details a shouting match with Gajdusek, who claimed to have suggested a protein as the infectious agent long before Prusiner; Prusiner worried he might lose his Nobel Prize to Gajdusek.

What's remarkable throughout Prusiner's story was his constant battle with journalists, and after the first row, he was determined to never talk to the press

---

[11] Kessler, R. (1997, April 6). Absent at the creation; How one scientist made off with the biggest invention since the light bulb. *The Washington Post Magazine*, p. 16.

again. He realized soon after his first prion publication in 1982 that he also had to deal with outrage from the transmissible spongiform encephalopathy (TSE) community. Many of the TSE researchers felt that he had jumped too quickly to stake his claim in the field by naming what was as yet an unfounded cause of TSEs. He even lost a student who left his lab over the controversy of naming the prion. These battles continued right on through Prusner's acceptance of the Nobel Prize. In two particularly remarkable instances, Prusiner had issues with journal publications. In his 1982 prion paper, the journal *Science* was reluctant to publish the article since there was one terrible peer review, and the journal wasn't entirely convinced that Prusiner was on the right track with his prion theory. Instead, Prusiner, or someone else, leaked the prion name and scientific results to a San Francisco paper which published the story prior to the peer-reviewed publication (which eventually appeared in *Science*). In a second incident, as a reviewer of a competitor's paper, Prusiner rejected the competitor's publication so his article could be published first. He was caught in the act. The *New England Journal of Medicine* rejected his paper and published his competitor's paper.[12] As his battles with colleagues and journalists continued, Prusiner hired a marketing firm, presumably to build his case for the Nobel Committee. In the end, there are still groups who don't think prions are the whole story, but Prusiner has continued down the prion road, even to the extent of calling all misfolded proteins in other degenerative diseases prions. Competition for the Nobel Prize prompted Prusiner to conceive and name the "prion," perhaps even prematurely, but it also served to isolate him from the rest of the field.

Otto Hahn and Lise Meitner were an amazing example of collaboration. They realized as two researchers from different fields (chemistry and physics) that they could be more productive as a team than they could ever have been separately. They worked collaboratively for over 30 years and produced many groundbreaking results during that time. In 1934, Meitner heard about the work being done by Enrico Fermi in Rome to discover the next elements in the periodic table after uranium—the transuranes. Meitner and Hahn had been working on individual projects for several years, but after seeing Fermi's work Meitner approached Hahn with the idea of entering the transurane field and reforming their partnership. After Fermi closed his laboratory and fled to America, Meitner knew that there was one other major research group (the Joliot-Curies in Paris) that was close—very close—to discovering what both laboratories hoped to find: that next "transuranic element," an element beyond uranium that doesn't exist in nature. Despite the fierce competition, both

---

[12] Rhodes (1998), *The making of the atomic bomb*, Simon and Shuster. p. 205.

laboratories periodically published their results, so they were kept abreast of what each group had discovered. But the difference, in the end, was the internal collaboration of a physicist (Meitner) and chemist (Hahn) in Berlin, that resulted in the dramatic discovery of fission. None of the competing labs had this combination.

## Passion in Science

Considering the lengthy preparation that it takes to become a scientist, and the sometimes tedious incremental nature of scientific progress, why do scientists choose science as a career? As Fred Hoyle put it, "I cannot understand what makes scientists tick. They are always wrong, and they always go on."[13] The reasons, of course, are as varied as the individuals we've discussed.

Lise Meitner had a genuine passion for science. She sought an education when women were excluded. She supported herself by teaching French to students during the day and did science experiments at night until she got her first paid position with Max Planck in Berlin. She met every challenge and barrier that was put in her way and continued her climb to become the head of an institute at the KWI. In contrast, her colleague Otto Hahn had little interest in the sciences, enjoying the humanities in his higher education days, as well as nights off celebrating with beer-drinking buddies. He finally decided that he would enter a chemistry career. He looked forward to a comfortable life as an industrial chemist and only "got the bug" for research when he was sent abroad to learn English in Sir William Ramsay's laboratory. There he discovered his first of many elements (radiothorium). It was at this point that his passion took hold, and he continued on to Rutherford's laboratory before accepting a position in Berlin. It was there that he met Meitner, and perhaps their mutual passion for science at that point propelled them on to making a great discovery.

For Fritz Haber, a bad experience in his father's firm convinced him (and his father!) that academia was a better fit for him than commerce. Haber was ambitious, and once he was in academia, competition with Nernst and an opportunity to save the world from Malthus' predicted famine stirred him to "fix" nitrogen from the air to make fertilizer… and get a Nobel Prize. When World War I came, Haber saw an opportunity to save Germany and win power and prestige by developing poison gas. But as Ishmael warns, "Be sure

---

[13] Hoyle, F. (1957) *The Black Cloud*. Valancourt Books.

of this, O young ambition, all mortal greatness is but disease."[14] Haber's ambition was his undoing: he was considered for trial as a war criminal for his work on poison gases and when Hitler came to power Haber, who was a Jew, had to leave his beloved Germany. By this time, only Chaim Weizmann would accept him, and he died in exile in Switzerland, hoping to get to Palestine to take over as the director of Weizmann's new institute.

Gajdusek was another Nobelist who exhibited an intense passion for science at an early age. He was introduced to scientific observation by his aunt and spent hours reading and learning about science. His favorite book was *Microbe Hunters*,[15] a wonderful book that has steered many youngsters (including one of the authors) into science. Gajdusek even labeled each step up to his home laboratory with the name of one of the scientists in the book. He was committed to becoming the best infectious disease doctor, and sought out mentors, including Nobel scientists, who could best prepare him for his profession. He was driven in his early years to find that next disease that no one had studied, that he could call his own. This led him to kuru and eventually his Nobel Prize.

Nationalism is another passionate driving force that motivates scientists, especially in times of war. Louis Pasteur believed that "*La science n'a pas de patrie, mais le savant doit en avoir une.* [Science has no fatherland, but the scientist ought to have one.]"[16] Haber agreed with him: "In peace-time the scientist belongs to humanity, in war-time to his fatherland."[17] While it's difficult to argue that love of one's country is intrinsically wrong, there are boundaries on what constitutes "acceptable" nationalism and some scientists have definitely crossed this boundary. Should Fritz Haber be condemned for defending Germany in World War I? Arguably, it was justified for German scientists to defend their country by using the Haber-Bosch process to make gunpowder when the British Navy cut off their supplies of nitrate from South America, but the introduction of poisonous gas to defend the Fatherland was wrong both for its horrible effects on its victims and for violating the Hague Conventions of 1899 and 1907. Haber argued that death by asphyxiation was no worse than death by bullets and explosives, and if it shortened the war, it would save lives. Haber's wife Clara disagreed. She committed suicide in part because she felt that Haber's poison gas was a perversion of chemistry.

---

[14] Melville H. (1851), *Moby Dick*, Harper. Chap. 16.

[15] de Kruif, P. (1926). *Microbe Hunters*, Harcourt, Brace and World.

[16] Pasteur, L. Discours d'inauguration de l'Institut Pasteur. Voir sur https://citations.ouest-france.fr/citation-louis-pasteur/science-patrie-16266.html.

[17] Haber, F. 1920. Die chemische industrie und der Krieg. *Die chemische Industrie* 43, 350–352.

It's easy to argue that Heisenberg's passion for his country crossed the boundaries of morality. He was aware of the horrible treatment of the Jews by the Nazis when he supported the Nazi regime by traveling to occupied countries to spread the word of German superiority and tried to make an atomic bomb for Hitler.

## The Nobel Soapbox

Since Alfred Nobel's will established the Nobel Prizes, beginning in 1901, 646 scientists in three disciplines—chemistry, physics, and physiology or medicine—have been recognized. With their Nobel awards, these scientists have been labeled by their peers and by the public as geniuses, and as geniuses, many have felt empowered to speak intelligently not just on their specialty but almost any topic. Just hand them the microphone. There are positive outcomes from this hubris, but there are also those who have succumbed to what is now known as the Nobel syndrome or Nobelitis. Perhaps the best description comes from a statement by Milton Friedman (Nobel Memorial Prize in Economics, 1976):

> I myself have been asked my opinion on everything from a cure for the common cold to the market value of a letter signed by John F. Kennedy. Needless to say, the attention is flattering but also corrupting. Somehow, we badly need an antidote for both the inflated attention granted a Nobel laureate in areas outside his competence and the inflated ego each of us is in danger of acquiring…[A] product that has been so successful is not easy to replace. Hence, I suspect that our inflated egos are safe for a good long time to come.[18]

Otto Hahn was awarded his Nobel Prize for the discovery of fission while he was still interned in England after the war. When he first heard that the Allies had dropped an atomic bomb on Hiroshima "he was completely shattered by the news and said he felt personally responsible for the deaths of hundreds of thousands of people, as it was his original discovery which had made the bomb possible."[19] What he did beyond 1945 is an important story of his final years. After their internment, Hahn, Heisenberg, and Max von Laue were eventually taken to Göttingen. They were uncertain whether they

---

[18] Friedman, M, & Friedman, R. (1998). *Two lucky people: Memoirs* (Paperback edition 1999 ed.). Chicago: The University of Chicago Press. p. 454. ISBN 0-226-26415-7.

[19] Bernstein, J. (2013). *Hitler's uranium club: The secret recordings at Farm Hall.* Springer Science and Business Media. p. 115.

were to establish new institutes or establish individual research programs. They knew that they could not work on anything related to military interests under the Allied Control Law #25. Hahn was constantly bombarded with questions about his role in the German atomic bomb project. At last, in frustration, he publicly stated: "I had no interest whatsoever in a military use. I am a man of science and of peace, and the thought that my work will be able to be used for destruction has often tormented me…" In his Nobel Lecture, he gave his first public warning of the dangers of atomic weapons:

> The energy of nuclear physical reactions has been given into men's hands. Shall it be used for the assistance of free scientific thought, for social improvement and the betterment of the living conditions of mankind? Or will it be misused to destroy what mankind has built up in thousands of years? The answer must be given without hesitation, and undoubtedly the scientists of the world will strive towards the first alternative.[20]

In 1946, Hahn accepted the Presidency of the Max Planck Society, established in the ruins of the old Kaiser Wilhelm Society. Hahn was so unsure of his qualifications for such a position that in a letter to his wife in June 1946 he wrote: "If I see the versatility of…von Weizsäcker, Heisenberg and…von Laue, I am always distressed by how small the extent of my knowledge is." In fact, Hahn stepped into the top scientific and administrative position in postwar Germany with a fervor toward making sure that use of atomic weapons as weapons of war would never happen again.

On a small island in the middle of Lake Constance, Italy, the Nobel Prize winners of the three sciences meet annually. At the fifth meeting in 1955, Hahn established his commitment to peace in the Mainau Declaration. As stated in the Declaration, Hahn firmly believed that the world had only one option:

> All nations must reach the decision to freely give up force as the last means of politics. Were they not ready to do so, they would cease to exist…I have done only my scientific duty. What the powers of this world have made of it is their responsibility.[21]

The Declaration was eventually signed by 52 Laureates.

---

[20] Hahn, O. (1946, December 13). *From the natural transmutations of uranium to its artificial fission*. Nobel Lecture. https://www.nobelprize.org/uploads/2018/06/hahn-lecture.pdf.

[21] https://www.lindau-repository.org/permadocs/MainauDeclaration1955EN.pdf.

By 1957, Otto Hahn had become the most trusted scientist in Germany. As described in one newspaper, he was a slight, somewhat bowed figure with an expression of honesty and nobility. The public continued to line up to hear his speeches. The crowd overflowed for one of his last speeches entitled *"Atomic Energy for Peace or for War?"* Hahn stated that the peace of the world hung upon the possession of atomic weapons only by the great powers, who presumably, could be counted on to ban the use of those weapons during war. He ended with, "A little Hitler somewhere in the world could plunge the world into destruction."[22]

At almost exactly the same time that Hahn was winding down in his public statements, another Nobel Laureate stepped onto the peace stage: Linus Pauling, who had won the Nobel Prize in Chemistry in 1954. After his Nobel award, Pauling became an aggressive activist focused on banning nuclear testing. In his book *No More War!*[23] he maintained that negotiation and diplomacy could avoid war: "He asked scientists to become peacemakers." In parallel with Hahn's efforts, Pauling circulated a petition calling for no more nuclear testing and reduction in the buildup of nuclear weapons. Eventually, over 11,000 scientists signed the petition, which resulted in the signing of a treaty by the then-nuclear powers: the USA, Britain, and the USSR. For his work opposing the proliferation of nuclear weapons, Pauling won the Nobel Peace Prize in 1963, the only Nobel scientist to win a non-science Nobel Prize.

Before World War II, Heisenberg explained to his colleagues in America that he wanted to return to Germany because, after the war, win or lose, he would be needed to rebuild German Science. Following Germany's defeat in World War II, since he was the leading theoretical physicist and a Nobel Laureate, Heisenberg was appointed the director of the newly formed Max Planck Institute for Physics in 1946. He began rebuilding the institute in July 1946. Heisenberg continued his studies of cosmic ray physics that he had started in Leipzig and initiated research programs in superconductivity, turbulence, and the theoretical properties of elementary particles. Aside from research, Heisenberg's primary interest was focused on rehabilitating German science and contributing to his concept that science should be supported by the Federal government. Through his efforts, the German Research Council was set up in 1949 with Heisenberg as the president. Unfortunately, Germany could not relinquish its model whereby research was funded locally, and the Council was dismantled. Heisenberg continued at the institute level to influence the direction of German physics, and by the late 1940s he was

---

[22] https://en.wikipedia.org/wiki/Otto_Hahn.

[23] Pauling, L. (1983). *No More War*. Dodd Meade.

recognized as the leading representative of German science in the international arena. He began traveling to other countries with the goal of getting German scientists re-accepted as members of the international scene and renewing personal relations after the isolation and alienation that had occurred during World War II. Heisenberg never fully achieved his goal, either globally or at home where he lacked support for his federalization plan. Given the number of lectures he had given during the war in Nazi-occupied countries preaching Germany's superiority and his own extreme nationalism, he had lost respect from the international science community.

William Shockley is among the top vote-getters for the most outrageous occupier of the Nobel soapbox, although he has been described as one of the most important scientists of the twentieth century. He led the group at Bell Labs that discovered the transistor which led others to develop the integrated circuit and the computer. He won the Nobel Prize together with his colleagues Bardeen and Brattain. He was elected to the National Academy of Sciences, and even his failed Shockley Semiconductor company started Silicon Valley. Then came his fall from grace. He became notorious for his eugenics theories. "He became…a scientific pariah."[24]

Shockley's foray into eugenics began in the 1960s. In 1969, the Nobel Foundation held the dedication of a new science building at Gustavus Adolphus College to be named Nobel Hall. Twenty-six Nobelists including Shockley, attended the dedication. One of his first speeches on "controlled breeding" occurred during a pre-dedication press conference. At this point, it sounded like he was simply restating Malthus' theory of the dangers of overpopulation.[25] Shockley was invited to be a speaker at the first Nobel Conference held in the Nobel Hall. Eight thousand people were in attendance. Shockley called it "the turning point of his life."[26] It was here that he expanded his belief that the least fit (i.e. least intelligent) should be eliminated. He continued: the unfit should be sterilized and their unborn children aborted. He still had not introduced the race card into this theory. But a few years later he stated that Blacks had IQs that were on average 15 points below Whites. He called this the "Negro problem."[27] By the 1970s, invitations to speak at a number of universities were rescinded, and at least one honorary degree was withdrawn.

---

[24] Shurkin, J.N. (2006). *Broken Genius: The Rise and Fall of William Shockley, Creator of the Electronic Age.* Palgrave Macmillan, p. vix.

[25] Malthus, T. (1798). *An essay on the principle of population as it affects the future improvement of society, with remarks on the speculations of Goodwin, M. Condorcet, M., and other writers.* London: J. Johnson in St Paul's Church-yard.

[26] Shurkin, (2006), p. 193.

[27] Shurkin, (2006), p. 214.

In 1972 Shockley was burned in effigy at Stanford University and in 1973 at Princeton University. In 1975 Shockley was invited to speak at another Nobel Conference. In a bizarre move, even for Shockley, he proposed that each Nobel Laureate attending the Conference take a lie detector test and answer the question: do you really believe there is no racial difference in IQ? None of the 30 Nobelists agreed to take the test. He reached the lowest point in his theory when he proposed his Voluntary Sterilization Bonus Plan, which would pay low-IQ women to undergo sterilization, $1000 for each IQ point below 100.[28]

Shockley never backed away from his eugenic theories, and he probably never fully realized how low he had fallen from being one of the most important scientists of the twentieth century. He had often remarked that if he was mentally alert in the last 5 min of his life, he would think about the positive effects of his eugenics theories. Shockley died on August 12, 1989, a broken and lonely scientist.

On the 75th anniversary of the discovery of the transistor, *Science* magazine published an incredible editorial by the Editor-in-chief, outlining the contribution of Shockley to the discovery, then adding that Shockley had "appallingly" devoted the last years of his life to promoting racist views. The editorial continues: "At the time, *Science* did not condemn Shockley for what he was: a charlatan who used his scientific credentials to advance racist ideology." The magazine made the commitment that every time the name Shockley appears in *Science*, a link to this editorial would be included.[29]

Another Nobelist, James Watson, also used his Nobel status to fall into the racist quagmire. His entry into the field of eugenics is worth mentioning as a footnote to Shockley's story. Watson shared the Nobel Prize in 1962 with Francis Crick and Maurice Wilkins for their discovery of the structure of DNA. Scientifically, Watson has contributed significantly to our understanding of genetics and heredity. However, he used that Nobel soapbox microphone to move outside science and make comments that were offensive to almost everyone, including but not limited to Blacks, Jews, and women. And if those weren't enough to convince us that he took every opportunity to go off the rails, he entered the eugenics discussion by stating: "…all our social policies are based on the fact that their [Blacks'] intelligence is the same as ours—whereas all the testing says not really."[30] At the age of 90, when he had

---

[28] Boyer, E. J. (1989, August 14). Controversial Nobel laureate Shockley dies. *The Los Angelos Times*.

[29] Thorpe, H. (2022), Shockley was a racist and eugenisist. *Science*, 378(6621), p. 683.

[30] https://www.nationalgeographic.com/science/article/151005-nobel-laureates-forget-racist-sexistscience.

the opportunity to finally back away from his offensive statements, he simply restated his position: Black people are intellectually inferior to White people.[31] Because of his remarks, the Cold Spring Harbor Laboratory in New York, where Watson had been director until 1994, finally had enough. They severed all ties with the Nobelist and removed his honorary titles.

## Conclusion

Finally, there remain a couple of questions one might ask about the Nobel Prizes themselves. To begin with, are the Prizes good or bad for science? They encourage competition but may discourage communication. Leaving aside aberrations like the award to a Swede for inventing a lighthouse gas regulator, and a couple of discoveries found later to be not quite correct, what about the award to Gajdusek for his work on an obscure disease in a remote jungle in Papua New Guinea? Did that provide the greatest benefit to mankind in that year, as stipulated in Nobel's will? Could the award have benefited a more useful line of research that was being overlooked, like research on hormones leading to the birth-control pill? And what about political considerations in the award of Prizes? Were the Prizes awarded to German scientists after the World Wars given for the best science in those years by neutral Swedes wearing apolitical glasses? Or were these scientists given special consideration with the rehabilitation of Germany in mind? To what extent was Prusiner's award made in haste to spur political action during the "Mad Cow" crisis? Should the Nobel Prizes ever be withheld or revoked for bad behavior? Even Gajdusek's prize wasn't revoked when he went to prison for his pedophilia. The Prize is awarded for scientists' discoveries, not for their behavior. But should a prize be revoked if scientific fraud is proved?

In sum, it seems fair to say that the Nobel Prizes have on the whole been good for science and for the world at large. Despite a few aberrations and possible mistakes, the Nobel Prizes remain perhaps the most universally acclaimed of all awards. And their announcement every October provides a chance not only for the winners to celebrate their achievements, but for the whole world to celebrate science.

---

[31] https://www.vox.com/2019/1/15/18182530/james-watson-racist.

# Index




© The Editor(s) (if applicable) and The Author(s), under exclusive license to Springer Nature
Switzerland AG 2025
V. Shepherd, C. Brau, *Beyond the Genius*, Copernicus Books,
https://doi.org/10.1007/978-3-032-02735-1

GPSR Compliance
The European Union's (EU) General Product Safety Regulation (GPSR) is a set of rules that requires consumer products to be safe and our obligations to ensure this.

If you have any concerns about our products, you can contact us on

ProductSafety@springernature.com

In case Publisher is established outside the EU, the EU authorized representative is:

Springer Nature Customer Service Center GmbH
Europaplatz 3
69115 Heidelberg, Germany